Security and Privacy for Modern Wireless Communication Systems

Security and Privacy for Modern Wireless Communication Systems

Editors

Tao Huang
Shihao Yan
Guanglin Zhang
Li Sun
Tsz Hon Yuen
YoHan Park
Changhoon Lee

MDPI • Basel • Beijing • Wuhan • Barcelona • Belgrade • Manchester • Tokyo • Cluj • Tianjin

Editors

Tao Huang
James Cook University
Smithfield, Australia

Shihao Yan
Edith Cowan University
Joondalup, Australia

Guanglin Zhang
Donghua University
Shanghai, China

Li Sun
Xi'an Jiaotong University
Xi'an, China

Tsz Hon Yuen
University of Hong Kong
Hong Kong

YoHan Park
Keimyung University
Daegu, Republic of Korea

Changhoon Lee
Seoul National University of Science and Technology
Seoul, Republic of Korea

Editorial Office
MDPI
St. Alban-Anlage 66
4052 Basel, Switzerland

This is a reprint of articles from the Special Issue published online in the open access journal *Electronics* (ISSN 2079-9292) (available at: https://www.mdpi.com/journal/electronics/special_issues/Security_Privacy_Wireless).

For citation purposes, cite each article independently as indicated on the article page online and as indicated below:

LastName, A.A.; LastName, B.B.; LastName, C.C. Article Title. *Journal Name* **Year**, *Volume Number*, Page Range.

ISBN 978-3-0365-8228-3 (Hbk)
ISBN 978-3-0365-8229-0 (PDF)

Contents

About the Editors

Tao Huang

Dr. Huang is a Senior Member of IEEE and holds a Ph.D. in Electrical Engineering from The University of New South Wales, Sydney, Australia. Dr. Huang is an Electronic Systems and IoT Engineering lecturer at James Cook University, Cairns, Australia. He was an Endeavour Australia Cheung Kong Research Fellow, a visiting scholar at The Chinese University of Hong Kong, a research associate at the University of New South Wales, and a postdoctoral research fellow at James Cook University. Dr. Huang has received the Australian Postgraduate Award, the Engineering Research Award at The University of New South Wales, the Best Paper Award from the IEEE WCNC, the IEEE Outstanding Leadership Award, and the Citation for Outstanding Contribution to Student Learning at James Cook University. Dr. Huang is a co-inventor of an international patent for MIMO systems. Dr. Huang is a member of the IEEE Communications Society, IEEE Vehicular Technology Society, IEEE Computational Intelligence Society, and IEEE Industrial Electronics Society. He serves as the MTT-S/COM Chapter Chair and Young Professionals Affinity Group Chair for the IEEE Northern Australia Section. He also serves in various capacities at international conferences as TPC chair/vice chair, program vice chair, symposium chair, local chair, and TPC member. Dr. Huang is an Associate Editor of the IEEE Open Journal of Communications Society, IEEE Access, and IET Communications. He is also a Topical Advisory Panel Member and Guest Editor of MDPI's Electronics. His research interests include IoT security, wireless communications, deep learning, intelligent sensing, computer vision, pattern recognition, and electronic systems.

Shihao Yan

Dr. Yan received a Ph.D. degree in Electrical Engineering from the University of New South Wales (UNSW), Sydney, Australia, in 2015. He received a B.S. in Communication Engineering and an M.S. in Communication and Information Systems from Shandong University, Jinan, China, in 2009 and 2012, respectively. He was a Postdoctoral Research Fellow at the Australian National University, a University Research Fellow at Macquarie University, and a Senior Research Associate at the School of Electrical Engineering and Telecommunications, UNSW, Sydney, Australia. He is currently a Senior Lecturer in the School of Science, Edith Cowan University (ECU), Perth, Australia. He is also the Theme Lead for Emerging Technologies for Cybersecurity in the Security Research Institute (SRI) at ECU. He was a Technical Co-Chair and Panel Member of a number of IEEE conferences and workshops, including the IEEE GlobeCOM 2018 Workshop on Trusted Communications with Physical Layer Security and IEEE VTC 2017 Spring Workshop on Positioning Solutions for Cooperative ITS. He was also awarded the Endeavour Research Fellowship by the Department of Education, Australia. His current research interests are in the areas of signal processing for wireless communication security and privacy, including covert communications, covert sensing, location spoofing detection, physical layer security, IRS-aided wireless communications, and UAV-aided communications.

Guanglin Zhang

Prof. Zhang is currently a professor, doctoral supervisor, and vice dean at the College of Information Science and Technology in Donghua University, Shanghai. He has been recognized with several prestigious honors, including the Excellent Academic Leader in Shanghai in 2023, the Eastern Scholar in 2022, the Shuguang Scholar in Shanghai in 2020, and the Young Top-notch Talent in Shanghai awards in 2019. He completed his B.S. degree in Applied Mathematics at Shandong Normal

University, Jinan, China, in 2003. He further pursued his M.S. degree in Operational Research and Cybernetics at Shanghai University, Shanghai, China, in 2006. In 2012, he obtained his Ph.D. degree in Information and Communication Engineering from Shanghai Jiao Tong University, Shanghai. From 2013 to 2014, he served as a Post-Doctoral Research Associate at the Institute of Network Coding, The Chinese University of Hong Kong. Dr. Zhang's research primarily focuses on intelligent IoT, 5G vehicular networking, energy internet, intelligent edge computing, machine learning, and their applications. He has led many projects funded by the National Natural Science Foundation of China, projects of the Municipal Science Committee, projects of the Municipal Education Committee, and projects in cooperation with enterprises. Over the years, he has published more than 100 papers in prestigious international and domestic academic journals and conferences, including IEEE/ACM TON, ACM TOSN, IEEE JSAC, IEEE TCOM, IEEE TWC, IEEE TCC, IEEE TVT, IEEE TII, IEEE TCST, IEEE TNSE, IOTJ, China Communications, ACM MOBICOM, TURC, IEEE GLOBECOM, and IEEE ICC.

Li Sun

Prof. Sun received B.S. and Ph.D. degrees in Information and Communications Engineering from Xi'an Jiaotong University, China, in 2006 and 2011, respectively. Since 2012, he has been with Xi'an Jiaotong University as an Associate Professor and the Deputy Director of Wireless Communications Institute. His research interests include physical layer security and wireless AI. He has published over 150 papers and has more than 60 granted patents. He received the IEEE Communications Letters Exemplary Reviewers Certificate from IEEE ComSoC in 2013 and 2016, the Best Paper Award of China Communications (2017 and 2020), the Best Paper Award of the IEEE 8th International Conference on Computer and Communication Systems (2023), the Outstanding Scientific Paper Award of Shaanxi Province of China (2016), the Outstanding Master Thesis Supervisor Award of Chinese Institute of Electronics (2021), and the First Price of the Teaching Achievement Award of Shaanxi Province of China (2018). He is serving or has served as the Editor of KSII Transactions on Internet and Information Systems, the Guest Editors of IEEE Network, Electronics, and Wireless Communications and Mobile Computing. He was the TPC Co-chair of the IEEE GLOBECOM'17 Workshop on PHY and Cross-Layer Security Solutions for 5G Networks, and the TPC Co-chair of the IEEE ICC'16 Workshop on Novel Medium Access and Resource Allocation for 5G Networks.

Tsz Hon Yuen

Dr. Tsz Hon Yuen is an assistant professor in the Department of Computer Science at the University of Hong Kong. He received his Ph.D. degree from the University of Wollongong in 2010. His research interests include cryptography (such as public key encryption, digital signatures and identity-based encryption), privacy-preserving protocols (such as anonymous credentials and zero-knowledge proof systems) and blockchain (such as payment channels and confidential transactions). He has published more than 50 technical papers, including top journals and conferences such as Crypto, Eurocrypt, CCS and IEEE S&P. He received the Best Paper Award in ESORICS 2014. He is also active in industrial fields, with more than 10 patents. He has served as the PC members and reviewers for many security and cryptography conferences, workshops and journals.

YoHan Park

Dr. Park received his B.S., M.S., and Ph.D. in electronic engineering from Kyungpook National University, Daegu, South Korea in 2006, 2008, and 2013, respectively. Currently, he is an Assistant Professor in the Department of Computer Engineering at Keimyung University in Daegu. Prior to this, he held the same position at Nazarene University from 2017 to 2019, worked as a BK21 Postdoctoral Researcher at Kyungpook National University from 2016 to 2017, and lectured there from 2014 to 2016. In addition, he held a postdoctoral research position at Singapore National University from 2013 to 2014. Dr. Park's research interests include computer networks, mobile security, blockchain, and the Internet of Things (IoT).

Changhoon Lee

Prof. Changhoon Lee is a prominent researcher and professor in the fields of information security and computer science. He received his Ph.D. degree from the Graduate School of Information Management and Security (GSIMS) at Korea University in Korea. With a career spanning multiple institutions, he has held positions such as a research professor at the Center for Information Security Technologies in Korea University, a professor at School of Computer Engineering in Hanshin University, and Dr. Lee is currently a professor at the Department of Computer Science and Engineering at Seoul National University of Science and Technology (SeoulTech) in Korea. Dr. Lee actively contributes to the academic community, chairing conferences, serving on program committees, and acting as a Guest Editor for international journals. His research interests encompass cyber threats intelligence (CTI), information security, cryptography, digital forensics, IoT security, and blockchain. He is a member of renowned societies including IEEE, IEEE Computer Society, IEEE Communications, IACR, KIISC, KDFS, KIPS, KITCS, KMMS, KONI, and KIIT. Dr. Changhoon Lee's extensive expertise and involvement in the academic community make him a respected figure in the field of information security and computer science in Korea and internationally.

Preface to "Security and Privacy for Modern Wireless Communication Systems"

Wireless communication systems face security and privacy challanges, necessitating the development of novel cryptography designs, transmission strategies, network protocols, and regulations. The transition from 5G to 6G has significantly increased connectivity and information flow within wireless networks. Moreover, emerging applications, such as remote real-time medical services and mixed-type communications, demand stringent data confidentiality to protect sensitive information.

The rapid advancement of the Internet of Things (IoT) has revolutionized automation across various domains. However, the diverse design parameters inherent in IoT systems, including the packet length, transmission patterns, and time delays, present unique security and privacy challenges that require innovative solutions. Furthermore, integrating advanced technologies, such as intelligent reflection surfaces, edge/fog/cloud computing, blockchain, deep learning, and cyber twins, introduces opportunities and challenges for information security and users' privacy in wireless communication systems.

This book presents a collection of 15 articles authored by research experts in the field, focusing on the latest advancements in security and privacy for wireless communications. These articles explore various topics, including novel methods, emerging trends, and practical applications. The contributions encompass various application fields, providing valuable insights into addressing the evolving security landscape of wireless communication networks.

We aim to contribute to the existing body of knowledge and inspire future advancements in wireless communication security and privacy. The articles within this book serve as a valuable resource for researchers, practitioners, and policymakers seeking to deepen their understanding of the challenges and opportunities in securing wireless communication systems.

Tao Huang, Shihao Yan, Guanglin Zhang, Li Sun, Tsz Hon Yuen, YoHan Park, and Changhoon Lee

Editors

electronics

MDPI

Article

Time-Dependent Prediction of Microblog Propagation Trends Based on Group Features

Qin Zhao [1,2,3], Zheyu Zhou [1], Jingjing Li [1], Shilin Jia [1] and Jianguo Pan [1,*]

1 Department of Computer, Shanghai Normal University, Shanghai 201418, China
2 Key Innovation Group of Digital Humanities Resource and Research, Shanghai Municipal Education Commission, Shanghai 200234, China
3 Key Laboratory of Embedded Systems and Service Computing of Ministry of Education, Tongji University, Shanghai 201804, China
* Correspondence: panjg@shnu.edu.cn

Abstract: The conventional machine learning-based method for the prediction of microblogs' reposting number mainly focuses on the extraction and representation of static features of the source microblogs such as user attributes and content attributes, without taking into account the problem that the microblog propagation network is dynamic. Moreover, it neglects dynamic features such as the change of the spatial and temporal background in the process of microblog propagation, leading to the inaccurate description of microblog features, which reduces the performance of prediction. In this paper, we contribute to the study on microblog propagation trends, and propose a new microblog feature presentation and time-dependent prediction method based on group features, using a reposting number which reflects the scale of microblog reposting to quantitatively describe the spreading effect and trends of the microblog. We extract some dynamic features created in the process of microblog propagation and development, and incorporate them with some traditional static features as group features to make a more accurate presentation of microblog features than a traditional machine learning-based research. Subsequently, based on the group features, we construct a time-dependent model with the LSTM network for further learning its hidden features and temporal features, and eventually carry out the prediction of microblog propagation trends. Experimental results show that our approach has better performance than the state-of-the-art methods.

Keywords: propagation trends; social networks; group features; dilated CNN; machine learning

Citation: Zhao, Q.; Zhou, Z.; Li, J.; Jia, S.; Pan, J. Time-Dependent Prediction of Microblog Propagation Trends Based on Group Features. *Electronics* **2022**, *11*, 2585. https://doi.org/10.3390/electronics11162585

Academic Editor: Ahmad Taher Azar

Received: 28 July 2022
Accepted: 16 August 2022
Published: 18 August 2022

1. Introduction

With the rapid development of the Internet in China, a Sina microblog has now become an indispensable way for people to obtain and issue information. On the microblog platform, users can express their own opinions with freedom which will be spread and propagated through other users' browsing and reposting. As the microblogs being continuously reposted by other users, some microblogs will finally lead to their explosive spread and become a hot topic, the priority among people's discussion, while some others will never. Therefore, in order to make a microblog better serve the public in many fields such as public opinion supervision, advertising, information push, and corporate marketing [1], the prediction of potential hot microblogs becomes a key research object among people, namely the prediction of microblog propagation trends.

The conventional machine learning-based methods for the prediction of a microblog reposting number mainly conduct extraction and representation of the static features of user attributes and content attributes of the source microblogs to construct its machine learning prediction model, but neglect the dynamic features generated in the process of microblog propagation.

In this paper, we contribute to the study on microblog propagation trends, and inspired by our previous works [2–4], we propose a new microblog feature description and time-dependent prediction method based on group features. When predicting the scale of reposting, In an innovative way, we extract and take some dynamic features generated in the process of microblog propagation and development into account, together with some traditional static features such as user features and microblog features as group features to solve the problem of inaccurate microblog feature descriptions of a traditional machine-learning based research and also to improve the accuracy of reposting number prediction. We first make a description of microblog group features, which specifically includes features from three aspects, namely, commonly used individual features, reposting comment features, and group influence features , and they are all extracted through their corresponding feature extraction methods. Commonly used individual features are extracted through manually constructed feature engineering. We use feature extraction models based on cluster and Dilated CNN to extract reposting comment features, and PageRank algorithm is used to extract group influence features. Finally, with the extracted microblog group features, we construct a time-dependent prediction model for reposting number and conduct the prediction of microblog propagation trends.

The remainder of the paper is organized as follows: Section 2 briefly introduces the domestic and foreign research progress of microblog propagation trend prediction, and the related technologies mainly involved in this paper. On the basis of traditional static feature representation, Section 3 proposes a microblog feature representation and a time-dependent prediction method for a microblog propagation trend based on group features. In Section 4, we conduct relevant experiments and evaluations on the proposed method with real datasets of Sina microblogs. Finally, in Section 5, we outline our contributions, conclude the paper, and forecast our future work.

2. Related Work

2.1. The Domestic and Foreign Research Progress

The traditional prediction methods for microblog reposting number mainly include three approaches, which are the prediction methods based on topological structure [5], the methods based on machine learning [6], and the methods based on points [7], respectively.

The traditional topological structure-based models originate from the Information Diffusion Theory, which is widely applied, mainly in the fields such as recommendation, and monitoring. For the study on the prediction problem of Sina microblog propagation trends, the most typical models include infectious disease models and information cascade models. These prediction methods fully consider the role of the forwarders in the social network formed by microblog users; however, due to the large number of nodes in the network, the complexity of the methods is higher.

The traditional machine learning-based models mainly use machine learning models to learn the hidden features that affect the microblog reposting number so as to carry out predictions [8]. These methods analyze the relevant factors that affect the microblog reposting number, with which extract the relevant features of the source microblog, and then convert the prediction problem into a classification or regression problem. Through the learning of historical data and extracted features, the machine learning-based models are constructed and trained, and finally make predictions to obtain the corresponding target value.

The point-based prediction methods mainly introduce the idea of time decay. Relevant researchers believe that whether the microblog will be popular is related to time, and the propagation trends of microblog usually change from slow to fast, then from fast to slow, and eventually cease to perish. The basic idea is to regard the event of microblog propagation trends variation as a life cycle event of a microblog, and then predict the probability of the occurrence of a microblog's stopping reposting and death, and finally obtain the propagation scale through the maximum likelihood method.

2.2. Related Techniques

2.2.1. Key Information Extraction Technology Based on TF-IDF

TF-IDF [9] is a simple but effective key information extraction technology used to evaluate the importance of a term to one of the documents in a document set. TF-IDF actually calculates the product of the value of TF and the value of IDF. TF is term frequency, which means the frequency of the appearance of terms in a document, while IDF is inverse document frequency, which measures how many documents in the document set contain the term, and is a measure of the general importance of the term in the document set. The fundamental principle of TF-IDF is that, if a term appears frequently in a document but not frequently in other documents in the document set, then we can consider that, for this document, compared to other terms in it, this term is more important and has better distinguishing ability. The calculation of TF-IDF seen in Equations (1) and (2) is as follows:

$$TF = \frac{t}{s} \tag{1}$$

$$IDF = log(\frac{D}{d} + 0.1) \tag{2}$$

where t represents the number of times the term appears in the document, and s represents the sum of the number of times all terms appear in the document, while h and d respectively represent the total number of documents in the document set and the number of documents which contain the term in the document set.

2.2.2. Dilated Convolutional Neural Network (DCNN)

DCNN is a special convolutional network. Compared to the conventional CNN, DCNN contains a parameter called dilation rate [10], which is mainly used to indicate the size of the dilation. DCNN have the same convolution kernel size as ordinary CNN, so, during DCNN's convolution calculation process, no additional convolution kernel parameters are needed. However, due to the existence of dilation rate, DCNN shares a larger range of parameters and a larger receptive field without reducing the resolution or coverage [11].

As is shown in Figure 1, for DCNN with a convolution kernel size of 3 × 3 and dilation rate of 2, the size of its receptive field is the same as the one of ordinary CNN with a convolution kernel size of 5 × 5. However, during DCNN's calculation process, only nine parameters are used, 36% of the parameter number of ordinary CNN with a convolution kernel size of 5 × 5, which is equivalent to providing a wider receptive field with the same convolution calculation cost.

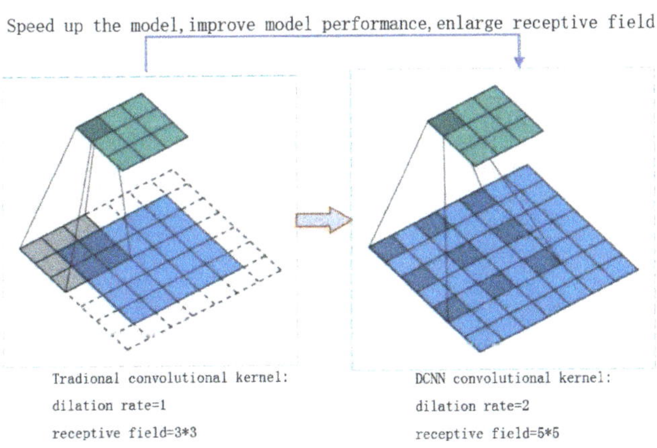

Figure 1. The diagram of the dilated convolution.

DCNN has applications in many fields such as image segmentation, speech synthesis, structural condition inspection, and target detection [12–15]. Due to the fact that high-level convolutional feature maps of the convolution network have larger receptive field and more abstract features, while low-level ones have a smaller receptive field and more detailed features, the usage of the combination of multi-scale feature maps in tasks can contain more information. Therefore, DCNN used in the paper contains multiple sizes of dilated convolutions [16].

2.2.3. Neural Network RNN and LSTM

In the application of machine learning, some tasks require a better ability to process information sequence when a Recurrent Neural Network (RNN) is needed. RNN has a strong ability to process time series data [17], while the process of microblog propagation we studied on in this paper is just a time series process, hence RNN is very suitable as the prediction model in this paper. Although RNN can learn sequential dependency of data, but due to the existence of its gradient vanishing problem, RNN has the defect to store and learn long-term dependence. For this reason, we choose an improved kind of RNN, called Long Short Term Memory Network (LSTM), to construct the prediction model in this paper. LSTM has made up for the shortcomings of the original RNN and is currently one of the most successful and popular RNN architectures, which has been applied to various time series tasks such as natural language processing and sound data processing.

In order to improve the defect that it is difficult for RNN to store and learn long-term dependence, LSTM adds a cell memory controller c to learn long-term features, as is shown in Figure 2. At time t, LSTM has three inputs, which are the current input value x_t, the previous output value h_{t-1}, and the previous cell state c_{t-1}, as well as two outputs, respectively, which are the current output value h_t and the current cell state c_t. Through three gate structures, namely input gate, forget gate, and output gate, LSTM maintains and updates the cell state [18]. In LSTM, temporal information is added or deleted from the cell state by gate structures, which selectively allows information to pass through. Neurons can feed data to the upper layer or the same layer [19].

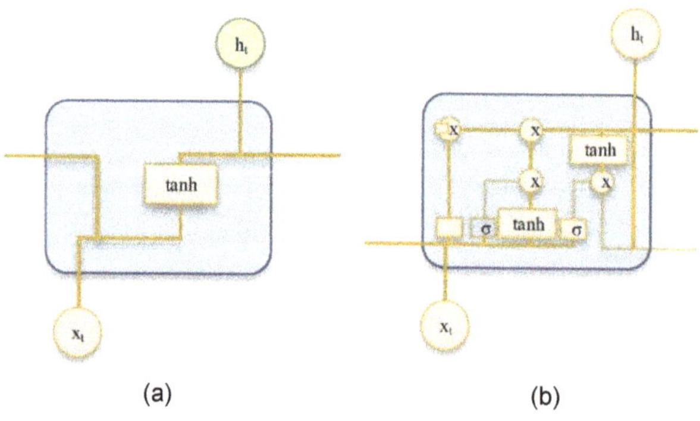

Figure 2. The structure of RNN and LSTM. (**a**) the structure of RNN unit; (**b**) the structure of LSTM unit.

With the current input and the previous cell state, LSTM gradually updates its cell state. Then, the output of the merged layer is trained through the "Relu" layer. Finally, the output layer produces predicted values [20].

3. Methods

In this paper, we use reposting number, which reflects the scale of microblog reposting to quantitatively describe the propagation effect of microblogs on the issue of study on microblog propagation trends. In order to make up for the problem that feature description of traditional machine learning-based prediction methods for reposting number is inaccurate, we mainly use microblog group features, including user and content features, group influence features, and reposting comment features to model [21]. We further learn its hidden features and time-dependent features relying on an LSTM network, and finally predict microblog reposting number.

3.1. Microblog Group Feature Representation

3.1.1. Bloggers and Microblog Content Features

Commonly used microblog features for prediction in current research are mainly divided into two categories [22–24]. The first one is features of users themselves, namely the blogger features, and the second one is features of blogs themselves, namely microblog content features. For Sina microblog, blogger features include the number of blogger's fans, blogger influence, blogger's recent microblog heat [22,25,26], and for microblog content features, current research usually focuses on points including whether the original microblogs contain links, and its hashtags.

The blogger features and microblog content features specifically used in this paper are shown in Table 1, including feature tags, specific meanings, and value ranges.

Table 1. Common microblog features.

Feature	Description	Value
BID	Blogger identity	(0,1)
BIF	Blogger influence	{0–1}
BRH	The blogger's recent microblog heat	[0–1}
BRT	Blogger registration time	[1–10]
ML	Microblog length{0,1}	
EX	Exclamation{0,1}	
NOC	Number of concerns	[0–1]
NOF	Number of fans	[0–1]
QM	Question marks	{0–1}
TT	Topic tags	[0–100]
PTP	Publishing time period	(0, 1, 2, 3, 4, 5)
IU	Includes URL	{0,1}
IH	Includes hashtags	{0,1}
IU	Include username	{0,1}

For some important features, their brief descriptions are as follows:

(1) The blogger's recent microblog heat: There is a certain logical relationship between the heat of one microblog and the heat of its blogger's other microblogs recently issued. Therefore, we use the heat of 10 other microblogs recently issued by the blogger as one basis of the calculation of the blogger's recent microblog heat. The calculation is shown in Equation (3):

$$h = \frac{1}{10}\sum_{m=1}^{10}(r_m + c_m + l_m) \quad (3)$$

where h represents the required feature of the blogger's recent microblog heat. For the m-th other microblog recently issued by the blogger, r_m represents the reposting number of the microblog, c_m represents the number of microblog comments, and l_m represents the praise score of the microblog.

(2) Blogger influence: The microblog propagation trends will be directly affected by the strength of the influence of its blogger, namely, with larger blogger influence, it is easier for the microblog to be spread. Since the following relationship between microblog users is similar to the links between web pages in the Internet, the idea of PageRank algorithm can be used to evaluate the influence of users. The basic idea is that the user's influence is larger followed by more influential users, the user's influence is larger with more fans, and the user's influence is larger with more fans and follow fewer users. According to the research on the topological structure and information propagation of Sina microblog [23], it

is found that it has an obvious small-world experiment, and its degree distribution obeys a power-law distribution. According to the idea that messages can be sent to other people on the network with fewer hops, the calculation of user influence is shown in Equation (4):

$$I(u_i) = (1-d) + d\sum\nolimits_{j=F(u_i)}^{N-1} \frac{I(u_j)}{out(u_i)} \tag{4}$$

where $I(u_i)$ represents the required influence of user i. d is the damping factor, which represents the probability of transferring from one given user to another random user, with its value range between 0 and 1, and the value of d is usually 0.85. $F(u_i)$ represents all user nodes that have an outbound link to the blogger node, namely the user's fan group. N represents the number of all user nodes that have an outbound link to the blogger node, namely the number of user's fans. $Out(u_i)$ represents the out degree of user node u_i.

(3) Microblog length: Microblogs issued by most users are short and fragmented daily life and emotional catharsis, which is hard to result in widespread resonance and reposting. In contrast, those microblogs with more complete expression are more likely to gain the understanding and resonance of other users, and easier to spread. Therefore, we consider the microblog length as one of the microblog content features, and set the classification criteria as whether the length of microblog is more than 15 words.

(4) Whether to include usernames or hashtags: Regarding microblog content features, we consider the problem of whether usernames or hashtags is included. In a microblog, usernames are used to directly quote other users, or to address or talk about a certain user, and hashtags are used to mark specific topics.

(5) Special marks: We consider whether there is an exclamation mark "!" or a question mark "?" at the end of a microblog as part of microblog content features. The exclamation mark is used to mark emotional statements in the text, and the question mark represents a problem in the text. The existence of both marks is more likely to result in the blogger's passing his own emotions to other users or arousing other users responding, which contributes to the spread of the blog.

3.1.2. Key Comment Features Based on Cluster and DCNN

A microblog often expresses different meanings in different temporal and spatial contexts, and sometimes may even contain irony, metaphors, and other information. In this regard, forwarder comments are often needed as supplement to the information of the source microblog to provide temporal and spatial background information which the original microblog lacks. At the same time, users are usually susceptible to comments from other users. Therefore, in this paper, we consider extracting comment features of forwarder group to improve the accuracy of machine learning-based prediction methods.

Since there are too many forwarder comments on a microblog, it is necessary to extract important information of the comments first, and then encode and vectorize them into group comment features of the blog. The process of feature extraction is shown in Figure 3, where there are generally three steps: (1) First, we use the cluster-based key information extraction model to extract key information from the group comments of microblog forwarders; (2) Encoding and vectorizing group comment information into sentence embeddings; and (3) Inputting the sentence embeddings to the DCNN convolution layer for feature extraction and compression. Finally, feature embeddings of the forwarder group comments are extracted, which contains temporal and spatial background information and is a supplement to the source microblog.

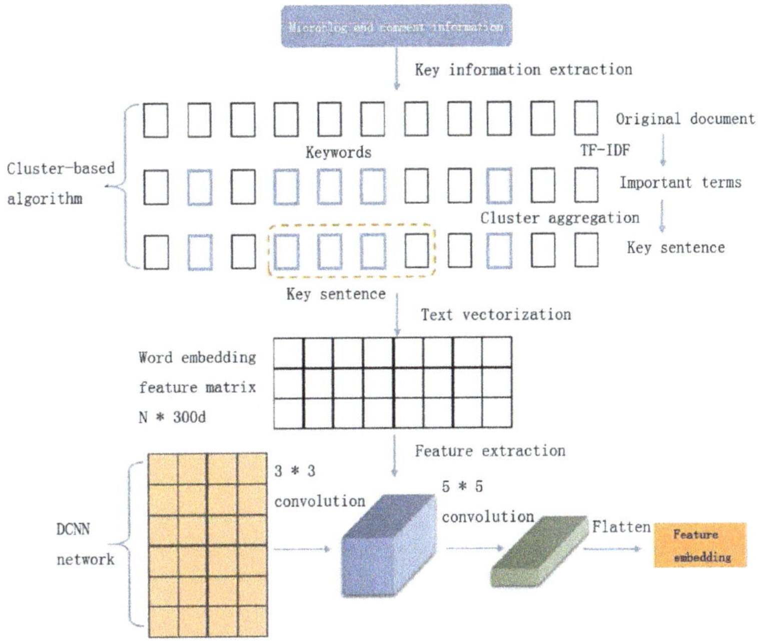

Figure 3. The learning of group comment features.

The specific work of each step is as follows:

Step 1: Key information extraction based on clusters

Forwarder comments are composed of sentences. In the process of microblog propagation, forwarder group comments contain a large number of sentences, so key information of the comments needs to be extracted first. In this paper, we use the cluster-based key information extraction technology to extract the corresponding feature sentences [27]. Our concept of "cluster" in this paper refers to the aggregation of keywords, namely, sentence fragments which contain multiple keywords.

It can be seen in Figure 4 that the framed part in the figure represents a cluster, where the keywords are obtained by calculating the TF-IDF score of terms of the comment sentences. If the distance between two keywords is less than the threshold, then these two keywords are classified into the same cluster. We set the threshold to 4 in this paper. In other words, if there are more than four other terms between two keywords, then these two keywords will be divided into two clusters. Then, we calculate the importance score of the clusters, the calculation of Equation (5) is as follows:

$$C\_IMP = \frac{(NKeys)^2}{len} \tag{5}$$

where C_Imp represents the required importance score of clusters. $NKeys$ represents the number of keywords in the cluster. Len represents the number of terms in the cluster. Taking Figure 4 as an example, in the figure, the cluster in the frame has a total of four terms, two of which are keywords. Therefore, the importance score of this cluster is $(2^2)/4 = 1$. After that, we extract the 10 sentences with the highest cluster scores, and combine them together as the finally extracted comments containing key information which can be further processed later.

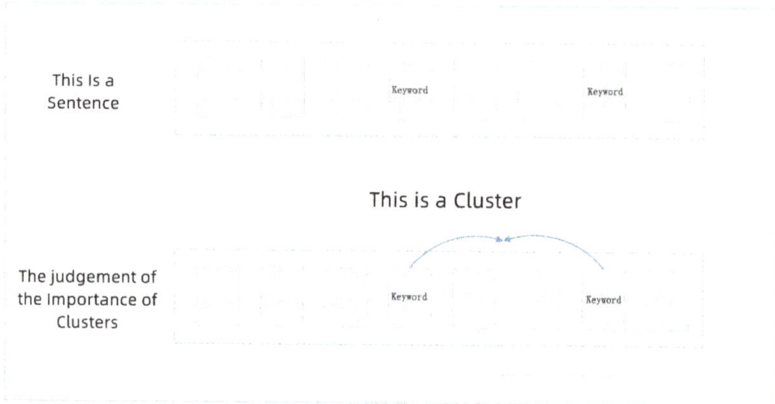

Figure 4. Key sentence extraction based on cluster.

Step 2: Feature encoding and vectorization

Since the computer cannot directly understand the meanings of text information, it is necessary to encode and vectorize the extracted comment features containing key information into a multi-dimensional embedding to facilitate subsequent further processing. The basic idea of word embeddings originates from NNLM [24] (Neural Network Language Model) proposed by Bengio. In this paper, we use open source tool Word2vec of Google in 2013 to solve the word embedding representation problem of microblog comments. Word2vec can quickly and effectively replace text sentences with multi-dimensional embeddings based on a given corpus.

There are two models for Word2vec, which, respectively, are the Continuous Bag-of-Words (CBOW) model and the Skip-Gram (SG) model, whose structures are shown in Figure 5. For a sentence containing L words, where $\ldots, w_{i-1}, w_i \ldots w_L$, respectively, represent the word embedding of each word in the sentence. In the CBOW model, a total of n words before and after the current word w_i (here n = 2) are used to predict the current word w_i. In contrast, the Skip-Gram model uses the word w_i to predict the n words before and after it. Both CBOW and Skip-Gram models include input layer [24], hidden layer, and output layer.

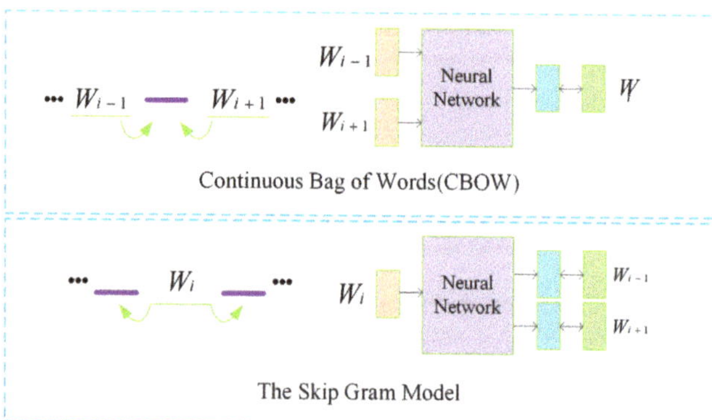

Figure 5. Two models for vectorization.

After preprocessing reposting comments containing key information, Word2vec is used to encode them into multi-dimensional embeddings. We train the Word2vec model, update the weights through the backpropagation algorithm, and use the stochastic gradient descent method to reduce the loss value, and finally obtain the byproduct, word embeddings of the model. Based on the word embeddings trained by the tool word2vec, we convert the words into microblog forwarder comments into word embeddings, and finally convert the key sentences of the comments into sentence embeddings.

Step 3: Feature extraction and compression of DCNN convolutional layer

Finally, we conduct feature extraction and compression on the forwarder comment embeddings. Due to the complexity of microblog language, the effect of usage of ordinary convolutional networks for feature extraction and compression is limited, and there are too many model parameters. Therefore, we choose to use Dilated Convolutional Neural Network (DCNN) and input the sentence embedding representation of reposting comments into the DCNN convolutional layer for feature extraction and feature compression. The three dilated convolutional layers we use are shown in Figure 6.

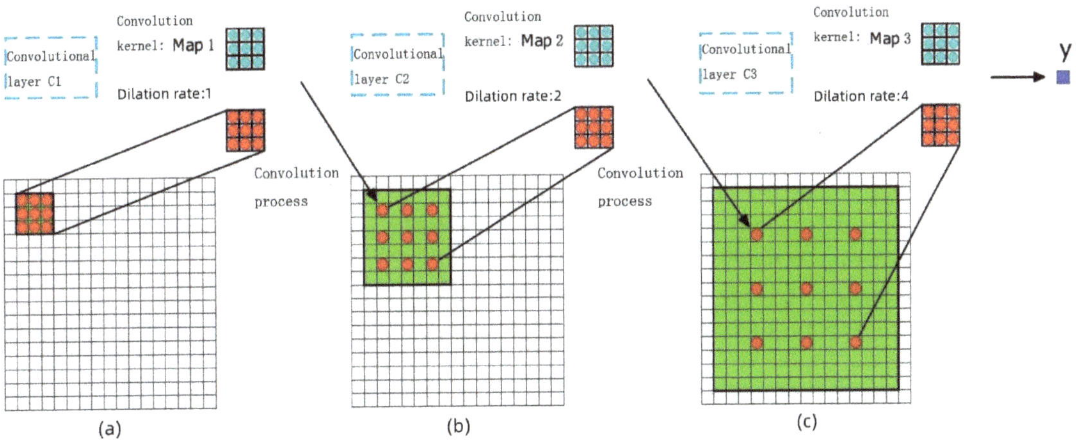

Figure 6. The dilated convolutional layer. (**a**–**c**) are, respectively, the convolution process with dilation rate k = 1, k = 2, and k = 4.

In the figure, for the three dilated convolutional layers C1, C2, and C3, their convolution kernels Map1, Map2, and Map3 are of the same size, which are all 3 × 3 matrices, but the dilation rates of the three convolution kernels are different, with values of 1, 2, and 4. In subgraph (a), a convolution kernel with dilation rate of 1 is used to convolve the input embeddings, and we input the result feature map as the output of C1 to the convolutional layer C2. In subgraph (b), a convolution kernel with dilation rate of 2 is used to convolve the feature map output by the C1 layer, and we input the result feature map as the output of C2 to the convolutional layer C3. In subgraph (c), a convolution kernel with dilation rate of 4 is used to convolve the feature map output by the C2 layer. At this time, the receptive field of the elements in the output y of the convolutional layer C3 has reached 15 × 15, while, with the ordinary convolution operation, the receptive field will only be 7 × 7.

The calculation process of the DCNN convolutional layer is shown in Equation (6).

$$c(t) = f(W^T[X_t^T + X_{t+1}^T + \ldots + X_{t+h-1}^T]^T + b) \quad (6)$$

For the forwarder comment embeddings, the convolution kernel W of dilated convolutional layer is applied to a window of terms of length h, and local features are generated after dilated convolution. In Equation (6), $c(t)$ is the feature value calculated at position t. b is the deviation of the current filter, and $f(*)$ is the nonlinear activation function (ReLU). We use zero padding to ensure that the size of the matrix after convolution meets requirements of the calculation. Then, the pooling operation is performed on each feature map through the maximum pooling layer to perform feature compression on the feature embeddings, and output embedding $p(j)$ with a fixed length. The calculation is shown in Equation (7):

$$p(j) = max_t\{c_j(t)\} \tag{7}$$

As is shown in Figure 6, our model uses multiple filters (with different window sizes) to obtain multiple features, and then outputs a multi-dimensional embedding at the maximum pooling layer network stage. The calculation is as shown in Equation (8):

$$CV = f(W^T[p(j)_1^T, p(j)_2^T \cdots p(j)_{10}^T] + b) \tag{8}$$

where $f(*)$ represents convolution and pooling operations. As a result, the feature embedding representation CV of forwarder key comments is finally obtained, which contains spatial and temporal background information and is a supplement to the source microblog information.

3.1.3. Group Influence Features

User influence refers to the ability of a user's opinions, comments, or behaviors to change the behaviors or opinions of other users. In microblog social networks, user influence has a direct impact on microblog propagation trends. Traditional machine learning-based prediction methods usually consider the personal influence of bloggers, without considering the influence of reposting users group in the process of microblog propagation. For example, if a celebrity user with huge influence reposts a microblog, then the propagation scale of this microblog is likely to be greatly improved [24]. In this paper, we use the PageRank algorithm [28] to calculate group influence to make up for the defect of blogger personal influence in traditional prediction methods.

Some scholars regard the microblog social network as a specific directed graph based on graph theory, each node of which corresponds to each user, and the directed edges in the graph represent the relationship "follow" and "followed" in the microblog network. Since the following relationship between users represented with directed edges is similar to the links between web pages on the Internet, we use the idea of PageRank algorithm to evaluate and calculate user influence. The main idea is that the user's influence is larger followed by more influential users, the user's influence is larger with more fans, and the user's influence is larger with more fans and follow fewer users. The algorithm comprehensively considers the structure of the microblog social network, and the final calculated user influence value can also reflect the user's influence objectively. The calculation Equation (9) of the PageRank value of user influence is as follows:

$$I(u_i) = (1-d) + d\sum\nolimits_{j=F(u_i)}^{N-1} \frac{I(u_j)}{out(u_i)} \tag{9}$$

where $I(u_i)$ represents the required influence of user i. d is the damping factor, which represents the probability of transferring from one given user to another random user, with its value range between 0 and 1, and the value of d is usually 0.85. $F(u_i)$ represents all user nodes that have an outbound link to the blogger node, namely the user's fan group. S represents the number of all user nodes that have an outbound link to the blogger node, namely the number of user's fans. $Out(u_i)$ represents the out degree of user node u_i.

After calculating the personal influence of reposting users through the PageRank algorithm in the microblog propagation process, we accumulate the individual PageRank values of the users in the reposting group, calculate the group influence of reposting users, and serve the combination of group influence features and blogger personal influence features as the final influence features. The calculation of full influence is shown in Equation (10), which calculates the accumulation of the influence of all reposting users before time t_m:

$$FI(u_i) = \sum\nolimits_{t=t_1}^{t_m} \sum\nolimits_{j=F(u_i)}^{N-1} I(u_j) \tag{10}$$

3.2. The Construction of the Prediction Model for Reposting Number

Taking into account the time dependence of the change of microblog propagation trends, in this paper, we choose the extracted microblog group features combined with Long Short Term Memory Network (LSTM) to construct the prediction model. The overall framework of our LSTM prediction model in this paper is shown in Figure 7, which contains four functional models, including input layer, LSTM hidden layer, output layer, and network training, where the input layer is responsible for preprocessing the microblog feature data set to meet requirements of the network input. The LSTM hidden layer is composed of a multi-layer recurrent neural network constructed by LSTM units. The output layer provides the final prediction results of reposting number, and we train the prediction network through the Adam optimization algorithm to update model weights iteratively.

Adam optimization is an effective gradient-based stochastic optimization method, which combines the advantages of AdaGrad and RMSProp optimization algorithms and has excellent performance in network training. Compared to other stochastic optimization methods, Adam is better in terms of speed and calculated amount, occupies fewer computing and storage resources, and the overall performance in practical applications is relatively better.

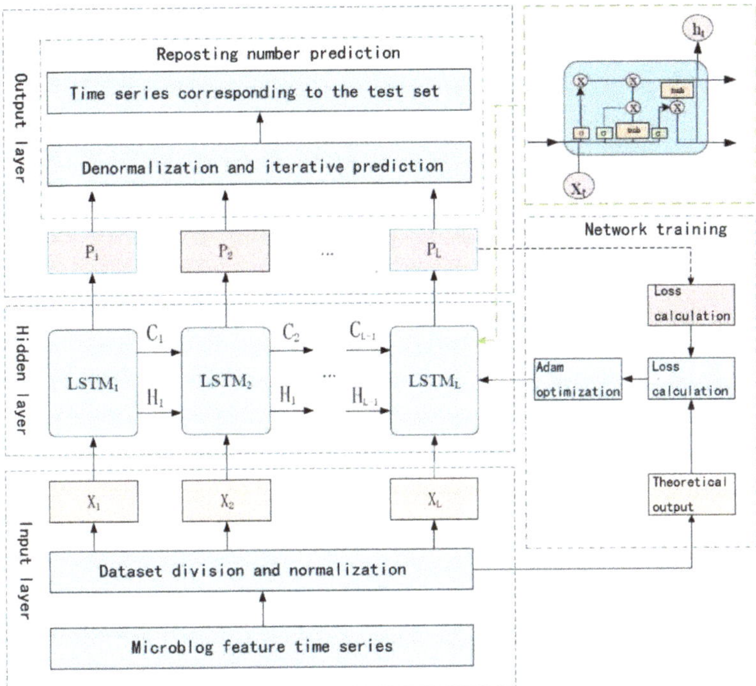

Figure 7. The framework diagram of the LSTM prediction model.

In Figure 7, the upper right corner is the detailed structure of the LSTM unit. LSTM maintains and updates the cell state of the cell memory controller c through three gate structures, including input gate, forget gate, and output gate, and learns long-term features. The internal formulas of LSTM we used are shown in Equations (11)–(15):

$$c_t = f_t c_{t-1} + i_t tanh(W_{xc} x_t + W_{hc} h_{t-1} + b_c) \tag{11}$$

$$O_t = \sigma(W_{xo} x_t + W_{ho} h_{t-1} + b_o) \tag{12}$$

$$f_t = \sigma(W_{xf} x_t + W_{hi} h_{t-1} + b_f) \tag{13}$$

$$h_t = o_t tanh(c_t) \tag{14}$$

$$c_t = f_t c_{t-1} + i_t tanh(W_{xc} x_t + W_{hc} h_{t-1} + b_c) \tag{15}$$

where σ is the activation function, and i_t, o_t, f_t, c_t, and h_t represent, respectively, the input gate, output gate, forget gate, cell state, and the final output of LSTM.

First, we preprocess the multi-dimensional group features of microblog in the input layer. The original microblog feature sequence is defined as $Fo = f_1, f_2, \ldots, f_n$ in the order of timestamps. The multi-dimensional microblog features are preprocessed, time slices are divided, and data set is divided into training set and test set. Supposing that the input length is L, the processed microblog data set is denoted as the sample feature X, the actual reposting number Y, and the corresponding predicted reposting number Y_p output by the output layer, the representations of which respectively correspond to the following Equations (16)–(18):

$$X = \{x_1, x_2, \ldots, x_L\} \tag{16}$$

$$Y = \{y_1, y_2, \ldots, y_k\} \tag{17}$$

$$Y_p = \{y_1, y_2, \ldots, y_k\} \tag{18}$$

where the value of k is 3. The calculation of root mean square error, namely the loss function, is shown in Equation (19):

$$loss = \sqrt{\frac{1}{m}\sum_{i=1}^{m}(Y - Y_p)^2} \tag{19}$$

The calculation and training of the prediction network are mainly done through the back propagation algorithm through time (BPTT) [25], as is shown in Figure 8.

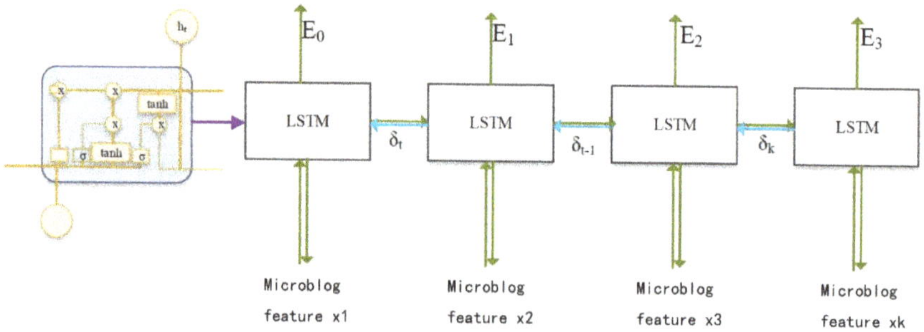

Figure 8. The diagram of BPTT algorithm.

The training flow chart of the model is shown in Figure 9. The training process is generally divided into four steps, the specific process of which is as follows:

(1) First, we calculate the output value of the LSTM unit structure according to the forward propagation.

(2) Secondly, we calculate the error terms of all LSTM unit structures through back-propagation, where the error terms include two propagation directions in terms of time and network structure, respectively.

(3) Then, the network automatically calculates the gradient of corresponding weight according to the calculated error value.

(4) Finally, after setting parameters such as the learning rate, we train the network, and through the gradient-based Adam optimization algorithm, we update the network weights iteratively until the network converges.

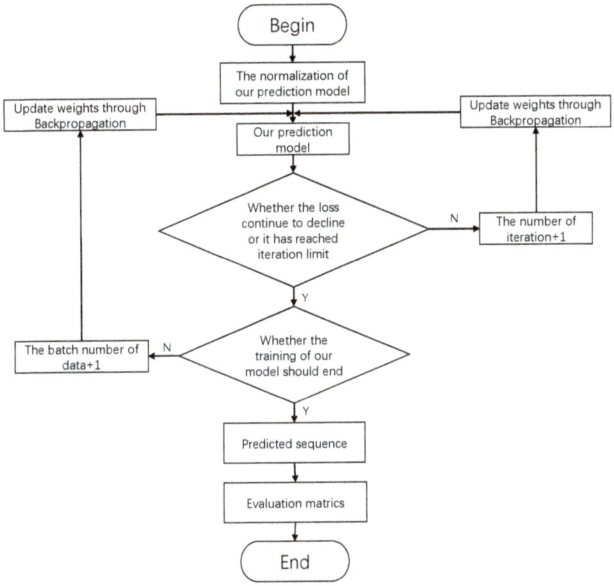

Figure 9. The training flow chart.

To sum up, the overall structure of our prediction method for microblog reposting number based on group features in this paper is shown in Figure 10.

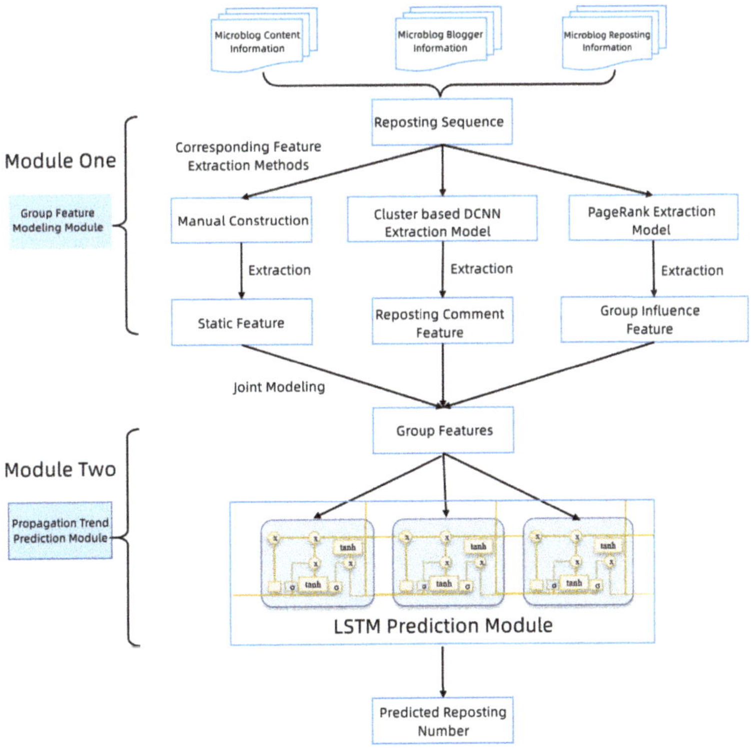

Figure 10. Time-dependent prediction process based on group information.

4. Experiments and Results

4.1. Experiment Preparation and Data Preprocessing

4.1.1. Experimental Environment

The environment and configuration of the hardware and software used in the experiment are as follows:

(1) Hardware Configuration:

1. CPU: inter(R) Core(TM) i5-8265u cpu @160GHz 180GHz RAM: 8GB Memory: 256 Solid+1TB Portable Hard Disk System: windows 10

2. GPU: NVIDA-GeFore GTX1080-Cuda Memory: 700GB Hard Disk System: Ubuntu 15.6

(2) Software Configuration:

Compiler: Python 3.7 Developing tool: Anaconda, Jupyter Notebook, Pycharm Community.

4.1.2. Dataset and Data Preprocessing

In this paper, we use real data from the Sina microblog collected and issued by the team of Tang Jie of Tsinghua University (http://arnetminer.org/Influencelocality, accessed on 15 August 2022). The overview of the original data set is shown in Table 2.

Table 2. Original data set.

Dataset	#Users	#Follow	#Original	#Retweets
Sina microblog	17,776,950	308,489,739	300,000	23,755,810

We select some data from the original data set for our experiments, which contain relevant information such as microblog content and creation time. For the microblog data set, we establish a reposting chain ranked by time according to its reposting time and content. When sampling the data set, in order to ensure the integrity of the reposting chain, the reposting process of each microblog event should already be ended. In Table 3, some relevant attributes obtained through statistical analysis of the data set are shown.

Table 3. Some attributes of the data set.

Attributes of Source Microblog	Number
Number of original microblog	312,310
Reposting time	23,755,812
Number of users	176,695
Average reposting number	78

In Figure 11, the distribution of reposting number of the microblog data set is shown. As can be seen from the figure, the distribution shows a clear power-law distribution trend that is plotted on the scale of logarithm.

To make the original data set meet the requirements of the prediction network input, the data need to be preprocessed accordingly first. The work of preprocessing mainly includes storing microblog data, processing missing values, removing stop words, and word segmentation, where non-numerical data need to be processed as numerical ones first, such as male and female gender replaced with 0 and 1, respectively, and normalization as well as other processing operations are required for numerical data.

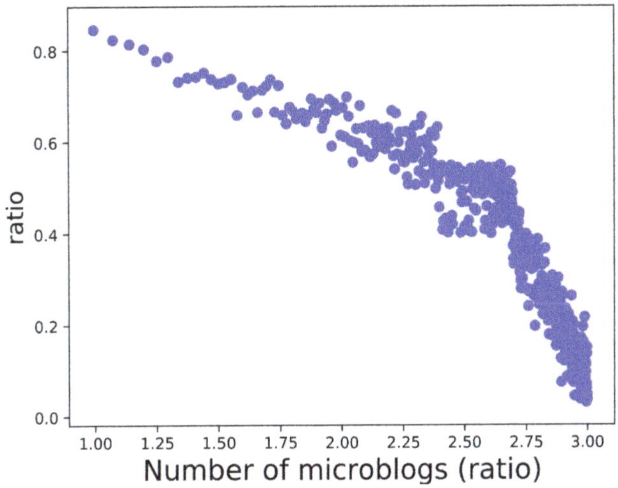

Figure 11. The distribution of a reposting number of the microblog data set.

Since the proposed model is a time-dependent prediction model, and the requirement of the LSTM network input is a 3D format, namely, [samples, timesteps, features], so, after preprocessing of the input data, the format of data needs to be reshaped into a 3D one, and the data be divided into time slices. The specific time slice division process of the data are shown in Figure 12. We select 10, 20, 30, . . . 120 min as a time slice, respectively. Finally, we divide the data set into training set, test set, and validation set.

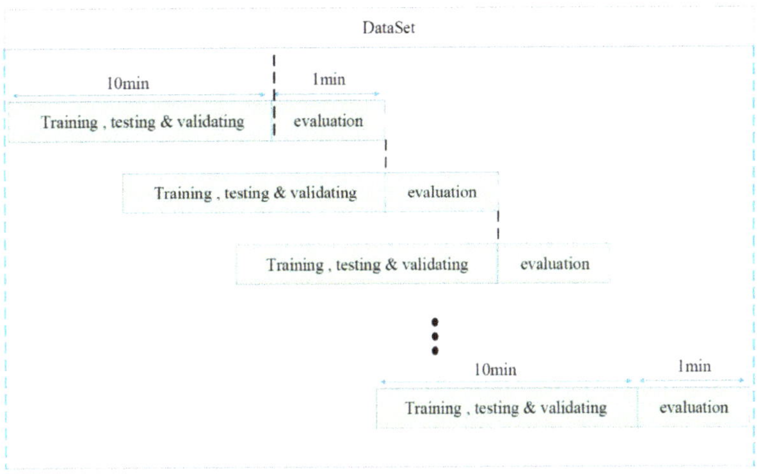

Figure 12. The time slice division of dataset.

4.2. Prediction Model for Reposting Number

4.2.1. The Analysis of Some of Group Features

In Section 3, we have provided a detailed explanation of the group features, where group influence feature is the sum of users' influence values calculated by PageRank in the microblog forwarder group. The users with top 10 personal influence are shown in Table 4, and Table 5 shows the influence data of some users, where the * is used to protect the user privacy.

Table 4. The ranking of user influence.

Overall Ranking	User
1	xin***ji
2	hua***bao
3	hong***k
4	jing***lu
5	xing***yu
6	li***fu
7	xin***kan
8	ai***er
9	qi***zhi
10	wei***xia

Table 5. Some data of microblog influence.

Username	Number of Fans	Number of Following	Number of Microblog	Influence	Rate of Being Reposted
guai***E	137	80	21	0.35	0.165233
fa***a	125	60	301	0.32	0.190243
X***xiao	108	55	173	0.30	0.153745
t***cao	80	55	112	0.22	0.139732
rong***y	73	53	153	0.22	0.122463
dong***er	50	18	25	0.18	0.102345
D***d	36	53	80	0.17	0.112310
tang***y	37	17	29	0.13	0.103345
han***yi	27	85	100	0.13	0.093542
B***zhong	32	80	13	0.14	0.152582

In Figure 13, the relationship between the amount of reposting and user influence is counted, where the horizontal axis and vertical one, respectively, indicate the magnitude of influence and the average amount of microblog reposting. It can be seen from the figure that, as the user influence decreases, the reposting amount of microblog also decreases, indicating a positive correlation between user influence and the reposting amount of microblogs.

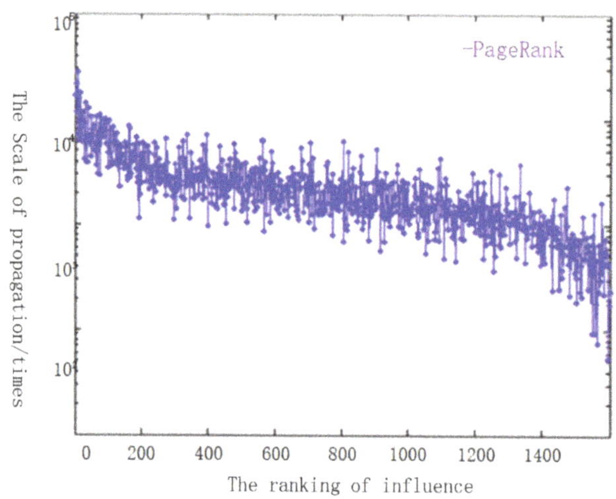

Figure 13. The relationship between the amount of reposting and user influence.

In addition, some other extracted features of bloggers and blogs are also very important. Taking the publishing time period as an example, as is shown in Figure 14, the different publishing times each day also have an impact on the reposting amount of microblog.

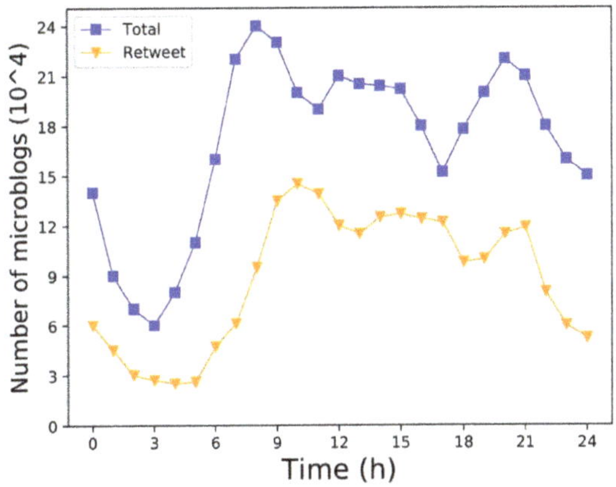

Figure 14. The influence of the publishing time of microblogs.

4.2.2. Training Process and Parameter Selection

In the experiment, we input our extracted features into the model for training and prediction. The hyperparameters of our model, including epoch and learning rate, need to be selected and adjusted through experients, otherwise the performance of our model will decrease. Here, we take hyperparameter epoch and learning rate as an example.

(1) Generally, the generalization ability of the model will increase as the epoch increases. However, an excessively large epoch may lead to the problem of over-fitting, which may decrease the generalization ability of the model on the contrary. Figure 15 shows the performance curve of our model under different epochs. As can be seen from the figure, when the epoch reaches 150, the loss of the model no longer decreases.

(2) The learning rate is another hyperparameter of our model. If the learning rate is too small, the training time of the model will be too long, while, with a learning rate that is too large, it is easy to exceed the threshold, making the model unstable and reducing its performance. Figure 16 shows the relationship curve between learning rate and RMSE. From the figure, it can be found that it is the most appropriate for the learning rate to be 0.1.

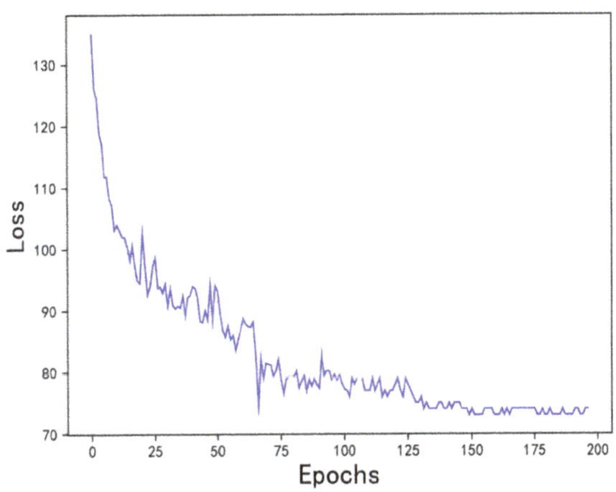

Figure 15. The relationship between Epochs and Loss.

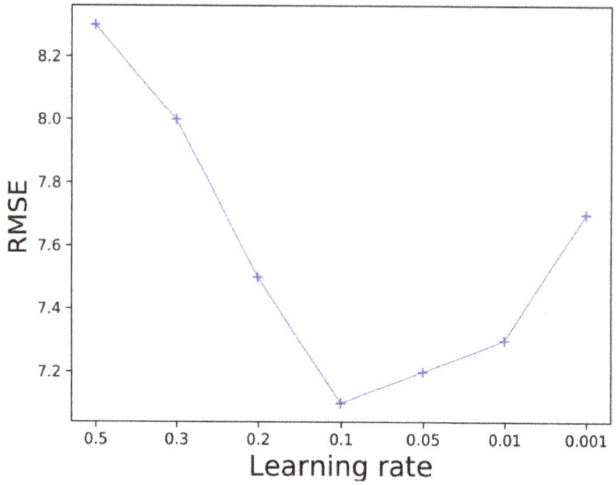

Figure 16. The relationship between Learning rate and RMSE.

Finally, Table 6 shows the final value of all the hyperparameters we use determined through comparative experiments, including epoch and learning rate. Among them, the final learning rate is 0.1. The size of the forwarder comment embedding is 300. The size of the model input is 120. The convolution layer has two layers, the convolution kernel size of which is 3×3 and 5×5, and the number is 128 and 64, respectively. In addition, the number of LSTM prediction units is 30.

Table 6. The parameter setting of the model.

Hyperparameter	Value
epochs	150
learning rate	0.1
Dropout	0.5
Embedding Size	300
BatchSize	120
kennel size	128 (5×5) + 64 (3×3)
LSTM unit	2×30

4.2.3. Results

(1) Evaluation Metrics and Benchmark Methods

We use three evaluation metrics, MAE, $MAPE$, and $RMSE$, to measure the performance of our model, where MAE is used to measure the mean absolute error between the predicted value and the actual value on the data set. For a test set containing n microblog messages, the definition of MAE is in Equation (20):

$$MAE = \frac{1}{n}\sum_{t=1}^{n}|actual(t) - forecast(t)| \tag{20}$$

$MAPE$ is used to measure the mean absolute percentage error between the predicted value and the actual value on the data set. The definition of $MAPE$ is in Equation (21):

$$MAPE = \frac{1}{n}\sum\nolimits_{t=1}^{n}|\frac{actual(t) - forecast(t)}{actual(t)}| \times 100\% \tag{21}$$

RMSE is used to measure the root mean square error between the predicted value and the actual value on the data set. The definition of *RMSE* is in Equation (22):

$$RMSE = \sqrt{\frac{\sum_{t=1}^{n} |actual(t) - forecast(t)|^2}{n}} \tag{22}$$

We compare our proposed method with several benchmark models. The benchmark models are briefly introduced as follows:

RPP is a model based on an enhanced Poisson process, which integrates three aspects of factors, respectively, which are the strength of the message, the time relaxation equation for the message which decays over time, and the enhancement equation for the preferential link phenomenon in message propagation.

The model LR is a simple but efficient classification model in machine learning, which is widely used in practice.

The model S-H is an epidemic prediction model based on logarithm linear regression of a variable proposed by Szabo et al.

The fundamental principle of BP network is to modify the weight and threshold along the direction of rapidly reducing the objective function.

The traditional LSTM network uses static features to predict the reposting number, without considering the dynamic features generated in the process of microblog propagation.

The model MP5 combines the characteristics of decision trees and multiple linear regression, each leaf node of which is a linear regression model. Therefore, the model MP5 can be used for regression problems of continuous value.

The model T-P divides the prediction problem of reposting number into two procedures. In the first procedure, T-P classifies microblog based on the potential reposting number, and, in the second procedure, T-P conduct regression in each subcategory separately.

The model BCI considers the characteristics of two factors, namely historical behavior and content relevance, to predict the problem of reposting number.

(2) Experiment on Real Data Set

The experiment is carried out on the real microblog data set in two parts. In the first part, 80% of the data set is divided into the training set and 20% is divided into the test set. In the second part, 70% of the data set is divided into the training set, and 30% is divided into the test set. Table 7 shows the experiment on the proposed model with benchmark models such as LR, S-H, and RPP, as well as the results of corresponding evaluation metrics RMSE, MAPE, and MAE. It can be seen from the table that, when 70% of the data set is divided into the training set, the RMSE of our proposed model is 7.335, the MAPE is 23.21, and the MAE is 18.77, whose performance is better than the one of any other benchmark models. When 80% of the data set is divided into the training set, the RMSE of our proposed model is 7.233, the MAPE is 22.89, and the MAE is 17.99. Not only does it outperform other benchmark models on every evaluation metric, but compared to results of the case that 70% of the data set is divided into the training set, the results of the situation in which 80% of the data set divided into the training set are also obviously better.

In this paper, we also select some benchmark models at random for additional tests with our proposed method on the data set, and plot the prediction curve of the reposting number, as is shown in Figure 17, where (a), (b), and (c), respectively, are different reposting scales. It can be seen from the figure that, under three different reposting scales, compared to other benchmark models, our proposed method performs better, which explicitly verifies that not only do our extracted microblog group feature representations contain more comprehensive and accurate information, but our proposed time-dependent prediction method based on group features is also more excellent.

Table 7. The results of experiment on propagation trends.

Method	70%			80%		
	RMSE	MAPE	MAE	RMSE	MAPE	MAE
LR	35.84	42.10	36.54	36.03	38.07	35.55
S-H	34.03	50.13	27.03	36.02	49.44	26.51
BP	27.09	28.48	26.05	26.83	28.23	27.32
RPP	17.92	25.41	25.59	17.92	25.41	25.59
BCI	16.82	23.55	25.11	16.32	23.15	24.21
MP5	12.08	35.04	18.02	11.83	32.04	18.17
T-P	10.45	23.66	25.01	10.22	23.23	24.11
LSTM	9.862	24.99	19.38	9.085	23.42	18.34
Proposed model	**7.335**	**23.21**	**18.77**	**7.233**	**22.89**	**17.99**

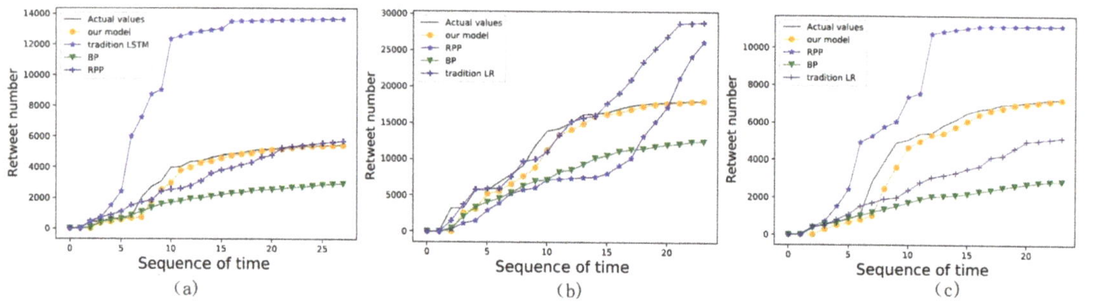

Figure 17. The comparison of model performance. (**a**–**c**), respectively, are different reposting scales.

5. Conclusions

In this paper, we study the propagation trends of microblog events, and, aiming at the problem of inaccurate feature descriptions of traditional machine learning-based predicting methods, in Section 3, a new microblog feature description and time-dependent prediction method of propagation trends based on group features are proposed. The proposed method is evaluated by an experiment on the real dataset of Sina microblog, the results of which prove that not only does the microblog group feature representation extracted in this paper contain more comprehensive and accurate information, but the proposed time-dependent prediction method based on a group feature also has better performance, higher accuracy, faster speed, and better robustness than traditional methods.

The method proposed in this paper also has much room for improvement. In our future work, it is necessary to conduct a further correlation analysis on the main factors and characteristics that affect the trends of microblog propagation, in order to use fewer features as group features in subsequent studies to construct our prediction model, with better performance in experiments at the same time. In addition, we construct our prediction model of microblog propagation trends on the basis of the basic LSTM prediction model, with not enough further improvement on the model itself, which will be a main perspective of our follow-up work. Furthermore, when evaluating our final prediction effects of microblog propagation trends, we use traditional evaluation metrics which lack our consideration on evaluation metrics that characterize other aspects of microblog propagation effects, such as the depth and breadth of propagation. Therefore, we will conduct further research on the establishment of a more comprehensive evaluation metrics system of microblog propagation trends.

Author Contributions: Conceptualization, Q.Z. and J.P.; methodology, Q.Z.; software, Z.Z.; validation, Z.Z. and S.J.; formal analysis, Z.Z.; investigation, S.J.; resources, S.J.; data curation, J.L.; writing—original draft preparation, Z.Z.; writing—review and editing, Q.Z.; visualization, Z.Z.; supervision, Q.Z.; project administration, J.P.; funding acquisition, Q.Z. All authors have read and agreed to the published version of the manuscript.

Funding: This work was supported in part by the National Natural Science Foundation of China under Grant No.61702333, in part by the Opening Topic of the Key Laboratory of Embedded Systems and Service Computing of Ministry of Education under Grant ESSCKF 2019-03, and in part by the Natural Science Foundation of Shanghai under Grant No. 20ZR1455600.

Institutional Review Board Statement: Not applicable.

Data Availability Statement: Not applicable.

Acknowledgments: This work was supported by the Key Innovation Group of Digital Humanities Resource and Research of Shanghai Normal University, and by the Research Base of Online Education for Shanghai Middle and Primary Schools, Shanghai Normal University, both funded by Shanghai Municipal Education Commission, and also by Shanghai Engineering Research Center of Intelligent Education and Bigdata, Shanghai Normal University, funded by Shanghai Municipal Science and Technology Commission.

Conflicts of Interest: The authors declare no conflict of interest.

References

1. Konstas, I.; Stathopoulos, V.; Jose, J.M. On social networks and collaborative recommendation. In Proceedings of the 32nd international ACM SIGIR Conference on Research and Development in Information Retrieval, Boston, MA, USA, 19–23 July 2009; pp. 195–202.
2. Zhao, Q.; Wang, C.; Wang, P.; Zhou, M.; Jiang, C. A novel method on information recommendation via hybrid similarity. *IEEE Trans. Syst. Man, Cybern. Syst.* **2016**, *48*, 448–459. [CrossRef]
3. Zhang, B.; Zhang, L.; Mu, C.; Zhao, Q.; Song, Q.; Hong, X. A most influential node group discovery method for influence maximization in social networks: A trust-based perspective. *Data Knowl. Eng.* **2019**, *121*, 71–87. [CrossRef]
4. Huang, S.; Zhao, Q.; Xu, X.Z.; Zhang, B.; Wang, D. Emojis-based recurrent neural network for Chinese microblogs sentiment analysis. In Proceedings of the 2019 IEEE International Conference on Service Operations and Logistics, and Informatics (SOLI), Zhengzhou, China, 6–8 November 2019; pp. 59–64.
5. Chakrabarti, D.; Wang, Y.; Wang, C.; Leskovec, J.; Faloutsos, C. Epidemic thresholds in real networks. *ACM Trans. Inf. Syst. Secur. (TISSEC)* **2008**, *10*, 13. [CrossRef]
6. Shen, H.; Wang, D.; Song, C.; Barabási, A.L. Modeling and predicting popularity dynamics via reinforced poisson processes. In Proceedings of the AAAI Conference on Artificial Intelligence, Quebec City, QC, Canada, 27–31 July 2014; Volume 28.
7. Li, Y.; Yu, H.; Liu, L. Predictive Algorithm of Micro-blog Retweet Scale Based on SVM. *Appl. Res. Comput.* **2013**, *30*, 2594–2597.
8. Zhang, J.; Tang, J.; Li, J.; Liu, Y.; Xing, C. Who influenced you? predicting retweet via social influence locality. *ACM Trans. Knowl. Discov. Data (TKDD)* **2015**, *9*, 1–26. [CrossRef]
9. Turian, J.; Ratinov, L.; Bengio, Y. Word representations: A simple and general method for semi-supervised learning. In Proceedings of the 48th Annual Meeting of the Association for Computational Linguistics, Uppsala, Sweden, 11–16 July 2010; pp. 384–394.
10. Li, H.; Liang, X.; Song, X.; Cai, Q. Visual Analysis of Spatio-Temporal Distribution and Retweet Relation in Weibo Event. In Proceedings of the 2018 IEEE International Conference on Big Data and Smart Computing (BigComp), Shanghai, China, 15–17 January 2018; pp. 9–16.
11. Amati, G.; Angelini, S.; Gambosi, G.; Rossi, G.; Vocca, P. Influential users in Twitter: Detection and evolution analysis. *Multimed. Tools Appl.* **2019**, *78*, 3395–3407. [CrossRef]
12. Ribeiro, M.H.; Calais, P.H.; Santos, Y.A.; Almeida, V.A.; Meira Jr, W. Characterizing and detecting hateful users on twitter. In Proceedings of the Twelfth International AAAI Conference on Web and Social Media, Palo Alto, CA, USA, 25–28 June 2018.
13. Zhu, H.; Ren, G.; Qin, D.; Wang, W.; Wei, F.; Cao, Y. Predicting user retweet behaviors based on energy optimization. In Proceedings of the 2016 12th International Conference on Computational Intelligence and Security (CIS), Wuxi, China, 16–19 December 2016; pp. 327–330.
14. Yu, Y.; Rashidi, M.; Samali, B.; Mohammadi, M.; Nguyen, T.N.; Zhou, X. Crack detection of concrete structures using deep convolutional neural networks optimized by enhanced chicken swarm algorithm. *Struct. Health Monit.* **2022**, preprint. [CrossRef]
15. Firdaus, S.N.; Ding, C.; Sadeghian, A. Retweet prediction considering user's difference as an author and retweeter. In Proceedings of the 2016 IEEE/ACM International Conference on Advances in Social Networks Analysis and Mining (ASONAM), San Francisco, CA, USA, 18–21 August 2016; pp. 852–859.

16. Pramanik, S.; Wang, Q.; Danisch, M.; Guillaume, J.L.; Mitra, B. Modeling cascade formation in Twitter amidst mentions and retweets. *Soc. Netw. Anal. Min.* **2017**, *7*, 41. [CrossRef]
17. Symeonidis, S.; Effrosynidis, D.; Arampatzis, A. A comparative evaluation of pre-processing techniques and their interactions for twitter sentiment analysis. *Expert Syst. Appl.* **2018**, *110*, 298–310. [CrossRef]
18. Liu, G.; Shi, C.; Chen, Q.; Wu, B.; Qi, J. A two-phase model for retweet number prediction. In Proceedings of the International Conference on Web-Age Information Management, Macau, China, 16–18 June 2014; pp. 781–792.
19. Li, B.; He, M.; Dai, Y.; Cheng, X.; Chen, Y. 3D skeleton based action recognition by video-domain translation-scale invariant mapping and multi-scale dilated CNN. *Multimed. Tools Appl.* **2018**, *77*, 22901–22921. [CrossRef]
20. Lu, X.; Yu, Z.; Guo, B.; Zhou, X. Modeling and predicting the re-post behavior in Sina Weibo. In Proceedings of the 2013 IEEE International Conference on Green Computing and Communications and IEEE Internet of Things and IEEE Cyber, Physical and Social Computing, Beijing, China, 20–23 August 2013; pp. 962–969.
21. Xu, Z.; Ru, L.; Xiang, L.; Yang, Q. Discovering user interest on twitter with a modified author-topic model. In Proceedings of the 2011 IEEE/WIC/ACM International Conferences on Web Intelligence and Intelligent Agent Technology, Lyon, France, 22–27 August 2011; Volume 1, pp. 422–429.
22. Gruhl, D.; Guha, R.; Kumar, R.; Novak, J.; Tomkins, A. The predictive power of online chatter. In Proceedings of the Eleventh ACM SIGKDD International Conference on Knowledge Discovery in Data Mining, Chicago, IL, USA, 21–24 August 2005; pp. 78–87.
23. Peng, H.K.; Zhu, J.; Piao, D.; Yan, R.; Zhang, Y. Retweet modeling using conditional random fields. In Proceedings of the 2011 IEEE 11th International Conference on Data Mining Workshops, Vancouver, BC, Canada, 11 December 2011; pp. 336–343.
24. Bengio, Y.; Ducharme, R.; Vincent, P. A neural probabilistic language model. *Adv. Neural Inf. Process. Syst.* **2000**, *13*, 1137–1155.
25. Stieglitz, S.; Dang-Xuan, L. Emotions and information diffusion in social media—Sentiment of microblogs and sharing behavior. *J. Manag. Inf. Syst.* **2013**, *29*, 217–248. [CrossRef]
26. Lampos, V.; Bie, T.D.; Cristianini, N. Flu detector-tracking epidemics on Twitter. In Proceedings of the Joint European Conference on Machine Learning and Knowledge Discovery in Databases, Barcelona, Spain, 19–23 September 2010; pp. 599–602.
27. Kempe, D.; Kleinberg, J.; Tardos, É. Maximizing the spread of influence through a social network. In Proceedings of the ninth ACM SIGKDD International Conference on Knowledge Discovery and Data Mining, Washington, DC, USA, 24–27 August 2003; pp. 137–146.
28. Langville, A.N.; Meyer, C.D. Deeper inside pagerank. *Internet Math.* **2004**, *1*, 335–380. [CrossRef]

electronics

MDPI

Article

A Study on Secret Key Rate in Wideband Rice Channel

Simone Del Prete *, Franco Fuschini and Marina Barbiroli

Department of Electrical, Electronic and Information Engineering "G. Marconi", University of Bologna, 40126 Bologna, Italy
* Correspondence: simone.delprete4@unibo.it

Abstract: Standard cryptography is expected to poorly fit IoT applications and services, as IoT devices can hardly cope with the computational complexity often required to run encryption algorithms. In this framework, physical layer security is often claimed as an effective solution to enforce secrecy in IoT systems. It relies on wireless channel characteristics to provide a mechanism for secure communications, with or even without cryptography. Among the different possibilities, an interesting solution aims at exploiting the random-like nature of the wireless channel to let the legitimate users agree on a secret key, simultaneously limiting the eavesdropping threat thanks to the spatial decorrelation properties of the wireless channel. The actual reliability of the channel-based key generation process depends on several parameters, as the actual correlation between the channel samples gathered by the users and the noise always affecting the wireless communications. The sensitivity of the key generation process can be expressed by the secrecy key rate, which represents the maximum number of secret bits that can be achieved from each channel observation. In this work, the secrecy key rate value is computed by means of simulations carried out under different working conditions in order to investigate the impact of major channel parameters on the SKR values. In contrast to previous works, the secrecy key rate is computed under a line-of-sight wireless channel and considering different correlation levels between the legitimate users and the eavesdropper.

Keywords: physical layer security; Rice channels; wireless communications; 6G security

Citation: Del Prete, S.; Fuschini, F.; Barbiroli, M. A Study on Secret Key Rate in Wideband Rice Channel. *Electronics* **2022**, *11*, 2772. https://doi.org/10.3390/electronics11172772

Academic Editors: Tao Huang, Shihao Yan, Guanglin Zhang, Li Sun, Tsz Hon Yuen, YoHan Park and Changhoon Lee

Received: 26 July 2022
Accepted: 1 September 2022
Published: 2 September 2022

1. Introduction

Modern cryptography is usually based on mathematical algorithms and can be divided into symmetric and asymmetric encryption systems. Symmetric encryption employs the same key to both encrypt and decrypt messages, while asymmetric cryptography relies on two keys: a public key to turn a plaintext into a ciphertext and a private key to retrieve the plain message. In this framework, the advent of quantum computers might be a threat for modern cryptography systems, that are usually termed to be only computational secure [1]. As an example, RSA, the most popular system for asymmetric cryptography, can easily be broken by Shor's algorithm if run by a quantum computer [2]. Instead, the actual symmetric encryption standard AES, in its version AES-256, is proven to be quantum resistant [3,4]. Moreover, starting from 5G, it is possible to observe a pervasive spread of low-power devices such as IoT devices, which are usually battery powered and have limited computational capacity: the modern RSA system is too lavish to be used on such devices. Therefore, not only is there the need of a quantum resistant set of security techniques, but also methods that can be supported by IoT devices. In this framework, in August 2018, the NIST published a call for an algorithm for lightweight cryptography (https://csrc.nist.gov/projects/lightweight-cryptography, accessed on 12 August 2022), showing the interest from the standardization bodies into the research for new lightweight cryptographic methods.

Physical layer security (PLS) is an umbrella of techniques which is hopefully able to achieve perfect secrecy by exploiting the unpredictable fading characteristics of the wireless channel [1,5]. In addition, PLS has recently been proposed as a key enabler for the security of future 6G communications systems [6,7]. Among the different techniques fostered under the aegis of PLS, physical layer key generation (PLKG) seems to be a mature and promising solution to protect the confidentiality of communications [8,9], in particular when low-power devices are employed in the system (e.g., IoT devices [10]). PLKG allows two users (here referred to as Alice and Bob) to generate a symmetric encryption key, simply by a mutual observation of the wireless channel, which is desirable to be symmetric and random. In the end, there is a nice interest to employ PLS techniques in the future security paradigm.

Several different metrics have been considered to assess the performance of the PLKG protocol, including key randomness, key disagreement and key generation rate. In particular, the secrecy key rate (SKR) may be of special interest, as it represents the maximum number of bits that can be achieved from each channel observation without the possibility of an eavesdropper (referred to as Eve) catching them [11]. Previous studies on the PLKG have mainly focused on the feasibility of the key generation process under the following conditions:

- The wireless channel is usually considered to be affected by Rayleigh fading, which is only suitable for uniform scattering and non-line-of-sight (LoS) communications.
- The possible presence of an eavesdropper is often neglected, although it represents a real limitation to the number of bits that can be reliably extracted from the channel.
- Alice and Bob are assumed to perceive perfectly symmetric channels, whereas this might not be true under real working conditions, as long as they cannot simultaneously sense the channel for whatever reason. In addition, the channel observations collected by Alice and Bob are affected by the noise and/or hardware imperfections.

These assumptions might not be true in a real scenario: higher frequency (e.g., mm-Wave, Tera-hertz) will be used in the future, already starting from the 5G standards [12]. Therefore, the wireless propagation will likely occur mostly in LOS condition, in order to cope with the high attenuation of the high frequency-bands. Moreover, due to the inner broadcast nature of the wireless channel, it is always possible to eavesdrop the communication, and in this specific case, try to steal some bits of the key.

In the literature, the SKR is usually reduced to the mutual information between Alice and Bob [13], i.e., neglecting the presence of Eve in the channel, who nonetheless decreases the number of bits that can be securely extracted. However, in [14], the authors considered the presence of the eavesdropper, but assumed Gaussian channel samples which might not be true in reality. In addition, it is often assumed that the generation occurs in a non-LoS scenario, i.e., under Rayleigh-like fading conditions [15], which is the ideal case for the PLKG thanks to the high entropy of the channel. Few works in the literature have evaluated the PLKG under LoS conditions, e.g., Ref. [16] computed an upper bound on the key generation capability of the two users communicating under LoS conditions. However, they considered the case in which the eavesdropper is capable of estimating the LoS component, and they assumed perfect channel reciprocity.

The goal of this work is to assess the performance of the PLKG through the computation of the SKR. Monte Carlo simulations have been performed under real-case general conditions with the aim of estimating the SKR in a LoS wireless link. In addition, an eavesdropper (Eve) is assumed to be present and sees the Alice–Bob channel with a low, but not zero, correlation: the correlation matrix of the Alice–Bob, Alice–Eve and Bob–Eve channels is an input parameter of the simulation. Moreover, instead of mutual information, the entire SKR, with its upper and lower bound, is computed. Additionally, the channels are generated according to a realistic 3GPP channel model (as it will be described in Section 3.2). Furthermore, the reciprocity is not assumed to be perfect and the impact of non-ideal reciprocity is taken into account by generating highly correlated channels between the legitimate users, but not equals. The simulations are repeated for different channel conditions: different Rice factor, signal-to-noise ratio, and delay spread (DS).

The rest of the paper is organized as follows: the PLKG protocol is shortly introduced in Section 2. Section 3 explains the assessment simulation procedure, whereas Section 4 reports a validation of the simulation procedure under a reference Gaussian case. The results of the assessment are reported in Section 5 and finally some conclusions are drawn in Section 6.

2. Physical-Layer Key Generation Protocol

The aim of the PLKG protocol is to let Alice and Bob autonomously generate a symmetric encryption key, without the possibility for Eve to steal the key. It fundamentally relies on the following general properties of the propagation channel [1]:

- **Reciprocity**: it is known that the wireless channel is almost symmetric between the transmitter and the receiver: this allows Alice and Bob to obtain similar channel samples, by means of mutual and possibly simultaneous observations of the channel.
- **Randomness**: mostly because of the fast fading, the channel fluctuates in time, frequency or space domain, in a random-like fashion, which ensures extracting random-like keys from the channel.
- **Spatial decorrelation**: provided that Eve is placed at a large distance (with respect to the wavelength), she will observe uncorrelated channel samples from Alice/Bob observations, which can strongly limit the actual eavesdropping threat.

PLKG usually consists of four well-known stages [1]:

1. **Channel probing**: Alice and Bob should almost simultaneously sample the channel in order to extract some channel features. The observation pairs must be retrieved within the coherence time of the fading in order to have similar channel samples. Moreover, different features can be extracted: channel observations are often limited to the received signal strength indicator (RSSI), but the whole channel state information (CSI) can also be targeted. Despite being more difficult to obtain, the CSI provides a wider set of features, which leads to an higher number of bits that can be possibly extracted.
2. **Quantization**: since the channel features are analog values, they must be quantized according to some quantization scheme to obtain, at the end of the process, digital encryption keys. A different quantization scheme can be used: uniform or non-uniform, single- or multi-level, differential-based or mean value-based [1,17].
3. **Information Reconciliation**: due to possible imperfection in the hardware, due to noise and non-ideal channel reciprocity, the quantized values might be slightly different between Alice and Bob. These discrepancies can be settled by means of standard error correction techniques through the exchange of public messages between Alice and Bob [18].
4. **Privacy amplification**: at the end of the previous stages, Eve might have acquired some bits of the key agreed between Alice and Bob, depending on the degree of randomness and spatial decorrelation inside the propagation channel. Privacy amplification is therefore enforced, where a new key is distilled by applying a randomly selected Hash function to the key achieved after the reconciliation procedure. Thanks to the properties of the universal hash function, the final key is likely to be fully unknown to Eve [19].

An important metric for the PLKG is the SKR, which was introduced by Maurer in [11]. Suppose that Alice, Bob and Eve, respectively, acquire the channel observations $X^A = [x^A(1), x^A(2), \cdots, x^A(n)]$, $X^B = [x^B(1), x^B(2), \cdots, x^B(n)]$, $X^E = [x^E(1), x^E(2), \cdots, x^E(n)]$, then the SKR has an upper and lower bound expressed by [11]:

$$\mathbb{R}(X^A, X^B \parallel X^E) \geq \max[\mathbb{I}(X^A; X^B) - \mathbb{I}(X^A; X^E), \mathbb{I}(X^A; X^B) - \mathbb{I}(X^B; X^E)], \tag{1}$$

$$\mathbb{R}(X^A, X^B \parallel X^E) \leq \min[\mathbb{I}(X^A; X^B), \mathbb{I}(X^A; X^B \mid X^E)], \tag{2}$$

which is an indication of the maximum number of bit per channel observation that can be extracted without the possibility of Eve guessing the bit [1]. The presence of Eve is taken into consideration in this work, as the information leakage to a possible eavesdropper can actually further limit the SKR. Furthermore, the channel is not assumed to be perfectly symmetric: the channel observations of Alice and Bob are still highly correlated, but not exactly the same. In order to compute the SKR under complete working conditions, a Monte Carlo simulation was therefore carried out, as explained in Section 3.

3. Materials and Methods

The main goal of the work is to assess the value of the SKR under different channel conditions in a system where the encryption keys are generated according to the previously explained PLKG protocol. Figure 1 outlines the presence of the users in the channel, with a particular emphasis on the mutual correlation, whereas a summary of the simulation parameters is reported in Table 1. The target observation is the frequency response of the channel, processed through the filterbank method [20]. Therefore, the vectors of channel observations X^A, X^B, X^E consist of the output of the N_f filters applied to the power spectral density (PSD). Moreover, the filters are supposed to be ideal pass band filters and the PSD is obtained through the square FFT of the channel impulse response (CIR), which is generated according to a wideband tapped delay model [21], where it is possible to tune the delay spread (DS) and the Rice factor K. Furthermore, the PSD observed by Alice Bob, and Eve is generated according to some mutual correlation target. This is accomplished through the Cholesky decomposition, even though it is only theoretically supported in the case of Gaussian samples. The channels are generated in order to achieve a bandwidth of 160 MHz.

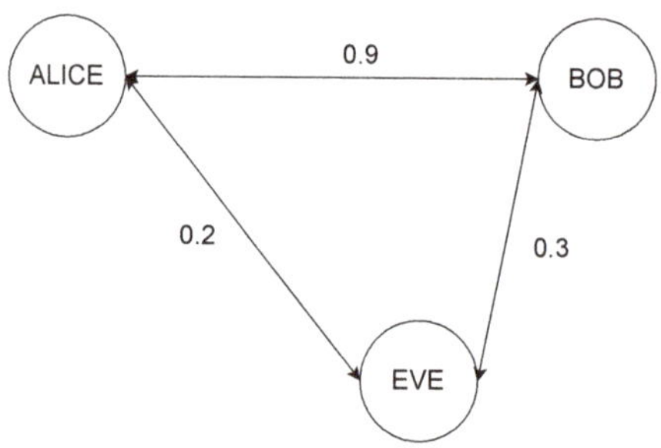

Figure 1. General scheme of Alice, Bob and Eve on the channel: each pair of users sees the channel realizations with a different non-zero correlation.

Table 1. Main simulation parameters.

Parameter	Value
Bandwidth (MHz)	160
Sampling time	2×10^{-9} s
SNR reference value (dB)	10
SNR (dB)	From 0 to 30 with step of 2
Channel realizations	50,000
Nfft	2048
Number of filters	1 or 4
Delay spread reference value (ns)	30
Delay spread (ns)	[10, 30, 100, 300, 600, 1000, 2000, 5000]
K reference value (dB)	10
K array (dB)	from 0 to 30 with step of 2
Alice–Bob correlation	0.99, 0.9, 0.7
Alice–Eve/Bob–Eve correlation	0.1, 0.2, 0.7

The SKR is computed through a Monte Carlo simulation: 5×10^5 channel realizations are generated for the same input values (DS and Rice factor K) and the SKR is computed case by case according to (1) and (2). The different simulation steps are described in the following sections, and a scheme of the procedure is sketched in Figure 2.

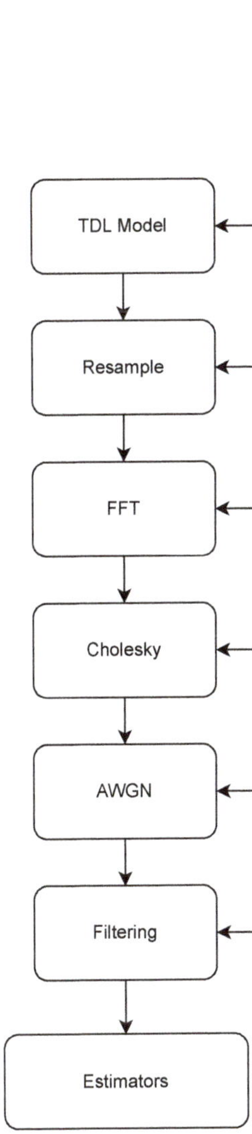

Figure 2. Diagram of the simulation.

3.1. Parameters

The first block in the simulation flow chart outlined in Figure 2 refers to a parameter file listing the parameters required by each simulation snapshot. The main parameters are reported in Table 1.

3.2. Tapped Delay Line Model

The wireless channel is generated according to the Tapped Delay Line "TDL-D" model described in [21]. It is a statistical channel model and consists of a set of paths with a normalized delay and power, which can be tuned to account for different propagation conditions. In particular, the channel model accounts for multipath Rice fading, i.e., the Rice factor and the DS are the tuning parameters of the model.

To generate the channel realizations, the following procedure was applied, as also described in [21]:

1. Modify the power and the delays of the TDL according to the procedure described on page 83 of [21], in order to have a given Rice factor and a DS;
2. As for the first line of the TDL, the component is generated as a Rice random variable with a K-factor equal to the desired one: this represents the LOS component of the channel.
3. For each multipath line, generate a complex Gaussian random variable with a zero-mean and a variance equal to the mean power of each line. As such, it is possible to generate Rayleigh-fading lines with a mean power specified by the average received power of each line of the TDL.

3.3. Resample

The TDL model is then resampled in order to obtain a CIR with a continuous time axis. To this aim, a sample time is selected as the inverse of the channel bandwidth written in Table 1. Each delay of the TDL is transformed into the corresponding time sample, and the complex amplitudes of the taps falling within the same sample are coherently summed up.

3.4. FFT

To obtain the channel frequency response, a simple FFT is performed on the CIR, which is also zero padded to reach "Nfft" samples (see Table 1). For the purpose of this work, the square amplitude of the channel transfer function (CTF), often referred to as Power Spectral Density (PSD) is considered. Therefore, the filtering applies to the PSD.

3.5. Cholesky Decomposition

Cholesky decomposition is a matrix decomposition procedure often employed to generate correlated Gaussian samples. Let $\underline{X} = (x_1, x_2, \dots x_n)$ be a n-dimensional standard Gaussian random vector ($x_i \sim \mathcal{N}(0,1)$) made of uncorrelated samples: its covariance matrix will be the identity matrix. A set of correlated Gaussian random variables can be obtained through the Cholesky decomposition, which decomposes an Hermitian matrix ($\overline{\overline{C}}$) into the product of a triangular lower ($\overline{\overline{L}}$) and a triangular upper matrix ($\overline{\overline{L}}^T$).

$$\overline{\overline{C}} = \overline{\overline{L}} \times \overline{\overline{L}}^T. \tag{3}$$

The vector $\overline{Y} = \overline{\overline{L}} \times \overline{X}$ will then be a Gaussian random vector with a covariance matrix equal to $\overline{\overline{C}}$. The proof of this is simple and follows from the computation of the covariance matrix of $\overline{Y}$:

$$\begin{aligned} E[\overline{Y} \times \overline{Y}^T] &= E[\overline{\overline{L}} \times \overline{X} \times (\overline{\overline{L}} \times \overline{X})^T] = E[\overline{\overline{L}} \times \overline{X} \times \overline{X}^T \times \overline{\overline{L}}^T] = \\ &= \overline{\overline{L}} \times E[\overline{X}\overline{X}^T] \times \overline{\overline{L}}^T = \overline{\overline{L}} \times \overline{\overline{I}}_n \times \overline{\overline{L}}^T = \overline{\overline{L}} \times \overline{\overline{L}}^T = \overline{\overline{C}}. \end{aligned} \tag{4}$$

This method is known to be theoretically grounded for Gaussian variables and according to [22], it is still reliable in case the variables are Gamma distributed. The Rice distribution is approximated by the Nakagami-m distribution and Gamma variables can be obtained as the square of Nakagami-m variables. By means of the `Fitter` (https://pypi.org/project/fitter/, accessed on 12 August 2022) class, the PSD samples were fitted in order to empirically determine the distribution of the samples. By looking at Figure 3 and Table 2, where the Sumsquare error and the parameters (following the`scipy.stats` (https://docs.scipy.org/doc/scipy/tutorial/stats.html, accessed on 12 August 2022) notation) is reported for different distributions, the PSD samples distribution seems to fairly comply with a gamma distribution. Therefore, it is reasonable to suppose that the PSD samples are gamma-distributed and the method of the Cholesky decomposition is still reliable in this case. For instance, by setting the target correlation between Alice and Bob to 0.99 and the correlation between Alice/Bob and Eve to 0.1, the actual correlation levels were then computed from on the channel samples achieved after the Cholesky decomposition, and turned out equal to 0.99 and 0.09.

Table 2. Sumsquare error and parameters of different distributions.

Distribution	Sumsquare Error	Parameters
gamma	0.008232	a = 42.463, loc = −1.146, scale = 0.047
lognorm	0.008412	s = 0.102, loc = −2.142, scale = 3.008
chi2	0.008457	df = 34.237, loc = −0.454, 0.038
norm	0.190884	loc = 0.881, scale = 0.311
rayleigh	1.621860	loc = 0.388, scale = 0.412

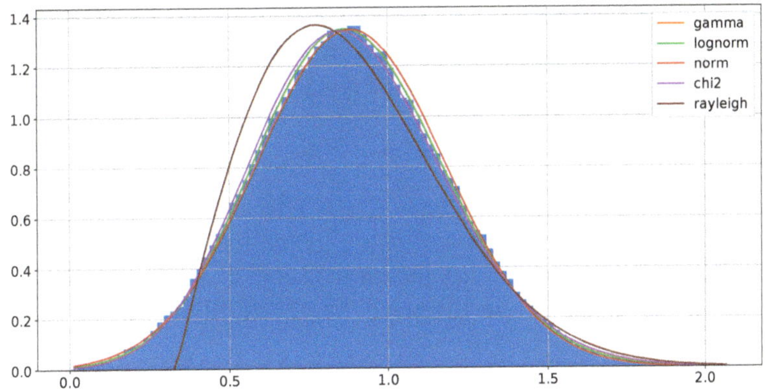

Figure 3. Fitting of different probability density functions to the histogram of the PSD samples.

If the matrix $\overline{\overline{C}} = \overline{\overline{L}} \times \overline{\overline{L}}^T$ is the desired correlation matrix, A_i', B_i', E_i' are respectively Alice's, Bob's and Eve's independent i-th realization of the PSD, the correlated channels (A_i, B_i, E_i) are obtained through s simple matrix multiplication:

$$\begin{bmatrix} a_{i;0} & \dots & a_{i;M} \\ b_{i;0} & \dots & b_{i;M} \\ e_{i;0} & \dots & e_{i;M} \end{bmatrix} = \overline{\overline{\mathbf{L}}} \times \begin{bmatrix} a'_{i;0} & \dots & a'_{i;M} \\ b'_{i;0} & \dots & b'_{i;M} \\ e'_{i;0} & \dots & e'_{i;M} \end{bmatrix} \tag{5}$$

As an example, Figure 4 depicts an example of channel realization, showing that, for high Alice–Bob correlation, the channels in frequency are quite similar, and instead Eve observes an uncorrelated channel.

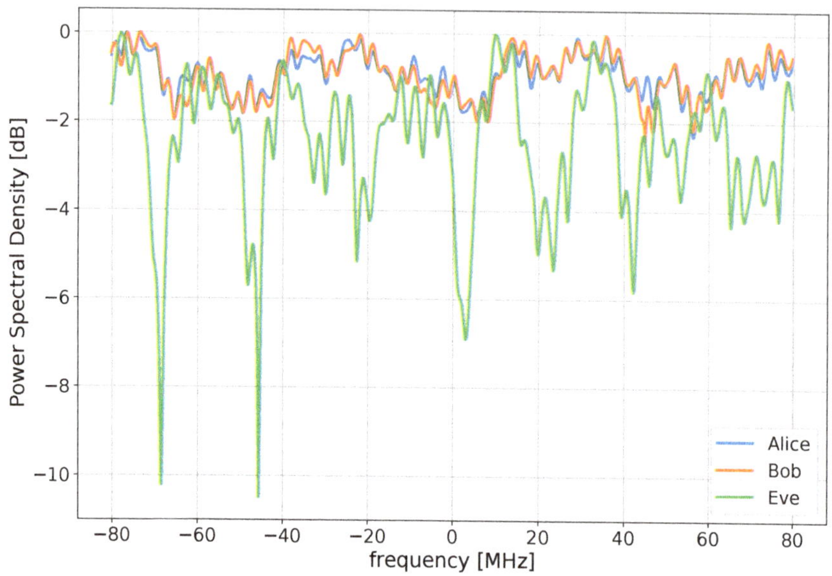

Figure 4. A channel realization obtained with K = 10 dB, delay spread = 30 ns, Alice–Bob correlation of 0.99, Alice–Eve and Bob–Eve correlation of 0.2.

3.6. AWGN

After the correlation of the channel, white noise is added to the PSD according to the signal-to-noise ratio reported in Table 1.

3.7. Filtering

The SKR is computed on the PSD after the filterbank [20] method is applied. For the purpose of this project, the filters are assumed to be ideal pass-band filters and there are either 1 or 4 filters. Each filter acts as a mean operator on the sub band of the PSD (or the entire PSD in case 1 filter is employed), hence the output of a filter is a single number. In practice, if $P(f)$ is the PSD, f_i is the central frequency of the i-th filter and Δf its pass band, then the output of the filter is computed as follows:

$$X_i = \frac{1}{\Delta f} \int_{f_i-\Delta f/2}^{f_i+\Delta f/2} P(f)\, df, \quad i = 1, 2, \ldots, N_f. \tag{6}$$

The filtering is also useful to reduce the dimensionality of the CTF, which is a benefit for the mutual information estimators, as will be clear in the next paragraph. In case 1 filter is employed, the entire 160 MHz is used; instead, when 4 filters are used, each filter has a non overlapping bandwidth of 40 MHz.

3.8. Estimators

Mutual information estimators have been employed to obtain the mutual information required for the computation of the SKR. In particular, the Non-Parametric Entropy Estimator Toolbox (https://github.com/gregversteeg/NPEET, accessed on 15 May 2022) and a python open source estimator of the mutual information based on the channel samples vectors were exploited. Moreover, this allows to estimate the mutual information for a multidimensional sample. However, these kind of estimators requires an exponential number of samples as the dimensionality increases due to the problem known as the curse of dimensionality [23]: therefore, the number of dimensions (number of filters of the filterbank) must be kept low. For the purpose of this work, it was seen that, by using 500,000 channel realizations, the estimators already converge.

4. Gaussian Case and Validation

A preliminary assessment was carried out in the Gaussian case, as the mutual information between Gaussian vectors can be expressed through analytical, closed-form formulas. The goal of this section is to evaluate the effectiveness of the simulator in a case where the mutual information can be expressed by an analytical closed formula. In particular, we derived the expression of the mutual information between correlated Gaussian variables and verified the correctness of the method implemented, particularly of the estimators.

Consider two Gaussian signals affected by AWGN:

$$A = s_a + n_a, \tag{7}$$
$$B = s_b + n_b, \tag{8}$$

where $s_a, s_b \sim \mathcal{N}(0,1)$, $n_a \sim \mathcal{N}(0,\sigma_a)$ and $n_b \sim \mathcal{N}(0,\sigma_b)$ and $\text{corr}(s_a, s_b) = \eta$. Since A and B are the sum of a zero mean Gaussian random variable, they will both be Gaussian with a variance, respectively, σ_A and σ_B. The mutual information between A and B can be therefore expressed as:

$$\mathbb{I}(A;B) = h(A) + h(B) + h(A,B) = \frac{1}{2}\log_2\left(\frac{\sigma_A^2\sigma_B^2}{\sigma_A^2\sigma_B^2 - \eta^2}\right), \tag{9}$$

See Appendix A for the demonstration.

Estimation Procedure

In order to test the estimators, the following procedure is employed. First, independent Gaussian signals are generated, then a correlation is applied according to what has been explained in Section 3.5. After the generation, AWGN is added to the signals:

$$\overline{X_1} \sim \mathcal{N}(0,1), \tag{10}$$
$$\overline{X_2} \sim \mathcal{N}(0,1), \tag{11}$$
$$\overline{n_a} \sim \mathcal{N}(0,\sigma_a), \tag{12}$$
$$\overline{n_b} \sim \mathcal{N}(0,\sigma_b), \tag{13}$$
$$\overline{s_a} = \overline{X_1}, \tag{14}$$
$$\overline{s_b} = \eta\overline{X_1} + \sqrt{1-\eta^2}\,\overline{X_2}, \tag{15}$$
$$\overline{A} = \overline{s_a} + \overline{n_a}, \tag{16}$$
$$\overline{B} = \overline{s_b} + \overline{n_b}. \tag{17}$$

Equation (15) comes from (3) and (5) when two random vectors are considered. The evaluation is repeated for different values of the correlation η: after the generation, the random vectors are given to the estimators to obtain mutual information. Furthermore, $\overline{X_1}$ and $\overline{X_2}$ contain 500,000 samples.

Figure 5 shows the results of the comparison. In particular, the mutual information significantly drops when the correlation is different from 1. Moreover, the estimated curves correspond to the theoretical case, confirming the correct behavior of the estimators. Since the SKR is a combination of mutual information, the same agreement between the theory and the simulation is expected regarding the SKR. This also proves the correctness of the simulation procedure employed.

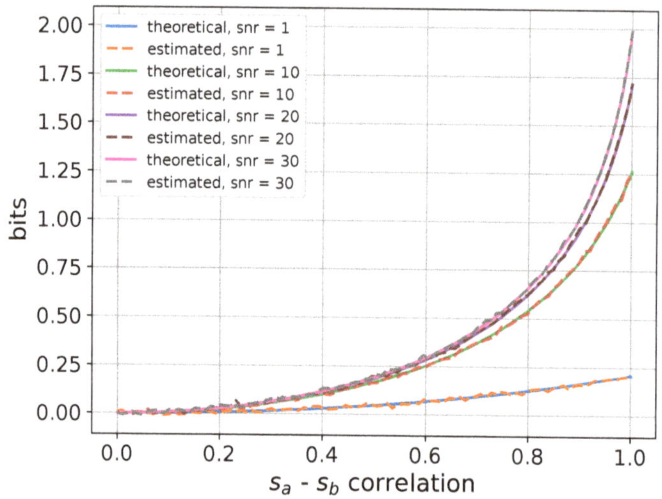

Figure 5. Comparison between the theoretical and estimated mutual information in the correlated Gaussian case, with different SNR conditions.

5. Results and Discussion

Simulations aimed at evaluating the SKR under different channel conditions, i.e., for different values of the Rice factor, of the DS and of the SNR. For the sake of simplicity, the legitimate and the eavesdropped channels are assumed to share the same Rice factor and DS, and Eve is supposed to have the same correlation towards Alice and Bob indifferently.

5.1. SKR and the K Factor

Simulations were run for different values of the Rice factor and correlation between the wireless channels, but always with the same SNR of 10 dB and with a DS of 30 ns. In addition, the estimation was performed both for the one-filter (narrow-band case, Figure 6) and for four-filters (wide-band case, Figure 7) cases. When Alice and Bob share highly correlated channel observations (0.99 in Figures 5 and 6), the SKR lower and upper bound basically coincide: this is not surprising as the lower and upper bound set on the SKR by (1) and (2) come to coincide as soon as Alice and Bob share highly correlated channel observations. Further details can be found in Appendix B. Instead, when the correlation is reduced, the two curves become distinguishable. Moreover, it is possible to highlight a decreasing trend of the SKR with the Rice factor: for a larger K, the channels are more stable and the multipath effects are reduced, thus the channel fluctuations are weaker, the overall randomness inside the channel is lower and hence the SKR is reduced. The reasons for this decreasing evolution of the SKR can be found by looking at Figure 8, which reports some PSD for the different values of the Rice Factor. As K increases, the channels become flatter, resulting in a weaker entropy and hence, in a lower SKR.

Reducing the Alice–Bob correlation also impairs the SKR, as it means that the disagreements in the bit sequences harvested from the channel become more probable because of the lower reciprocity level. A further reduction in the SKR is triggered when Eve improves her correlation with respect to Alice/Bob, as she can then better infer some information about the key, thus reducing its overall secrecy. Since the SKR represents the total number of bits that can be extracted after the filterbank method, it is normal to observe higher values when four filters are employed (Figure 7).

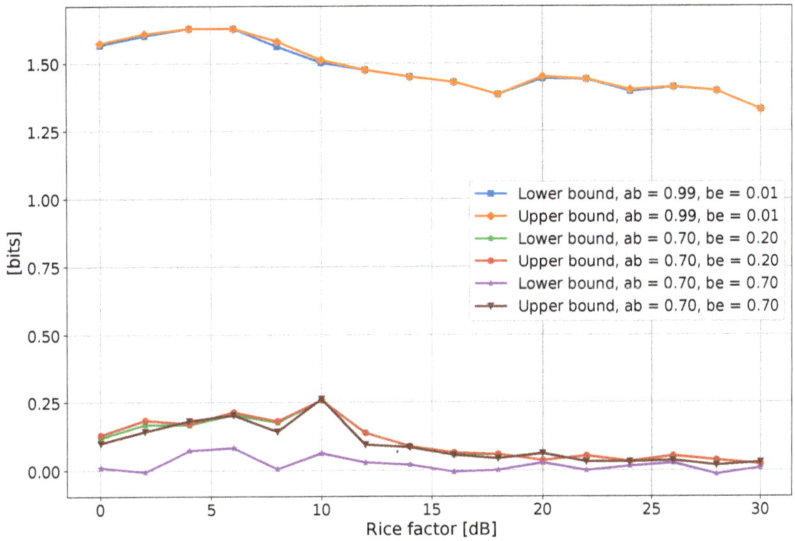

Figure 6. Secrecy key rate as a function of the Rice factor K, for different values of the correlation and with 1 filter. In the legend, "ab" and "be" stand for Alice–Bob correlation and Bob–Eve correlation.

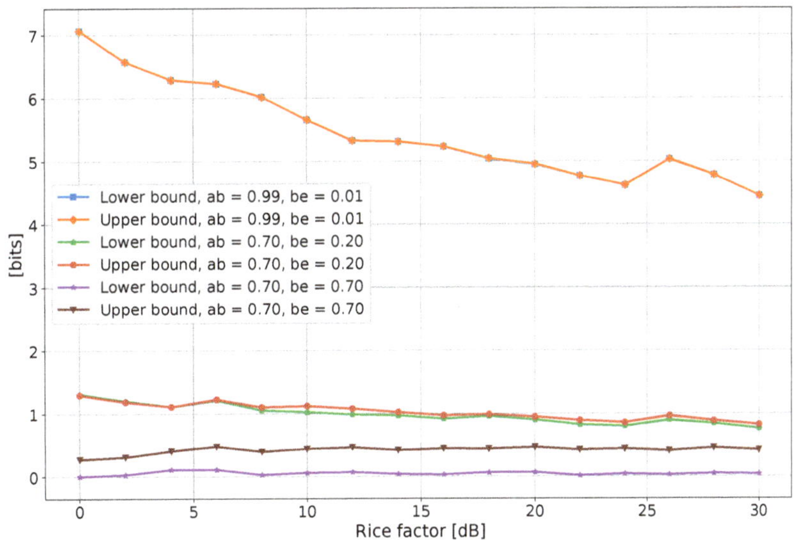

Figure 7. Secrecy key rate as a function of the Rice factor K, for different values of the correlation and with 4 filters. In the legend, "ab" and "be" stand for the Alice–Bob correlation and Bob–Eve correlation.

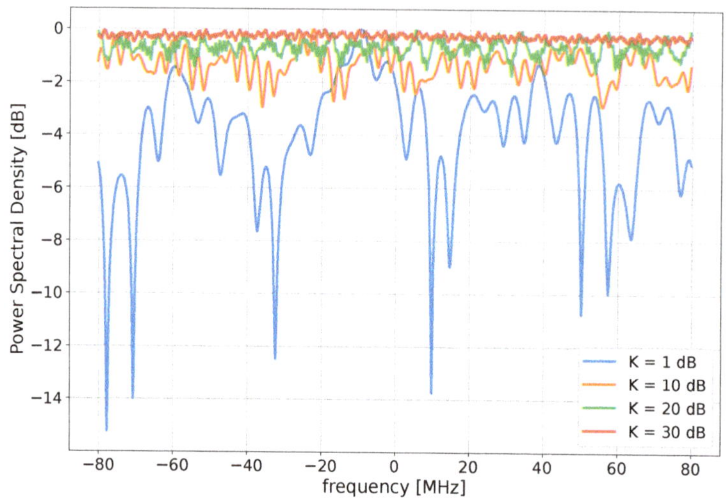

Figure 8. Power spectral densities for a different value of the Rice factor K.

5.2. SKR and SNR

The simulations were then performed with respect to the SNR experienced by Alice and Bob, whereas the SNR of Eve is always kept to 10 dB, the DS is 30 ns and the Rice factor was set to 10 dB. Once again, the simulations were repeated for different values of the correlation.

Figure 9 depicts the SKR as a function of the SNR with one filter, while Figure 10 shows the situation with four filters. In line with the Gaussian case described in Section 4, the SKR increases with the SNR, as a louder noise between Alice and Bob evidently affects the channel reciprocity, thus increasing the probability of disagreement between the key they finally receive from the channel observations. The sensitivity to the channels' correlation highlighted in Figures 9 and 10 is of course the same as that already discussed with reference to Figures 6 and 7.

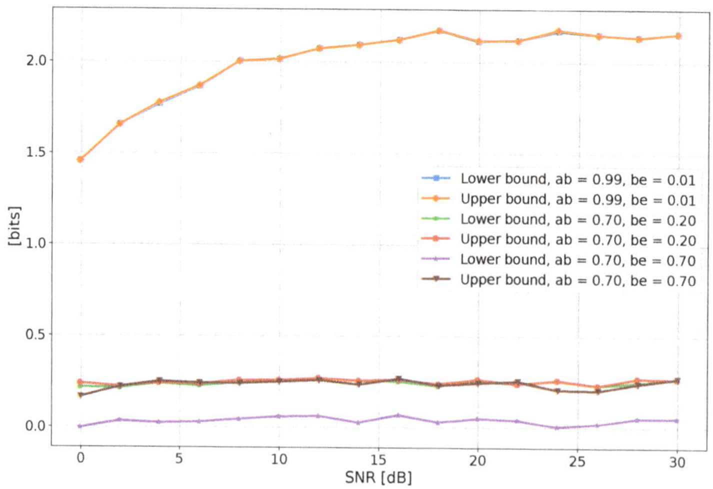

Figure 9. Secrecy key rate as a function of the SNR of Alice and Bob, for different values of the correlation and with 1 filter. In the legend, "ab" and "be" stand for Alice–Bob correlation and Bob–Eve correlation.

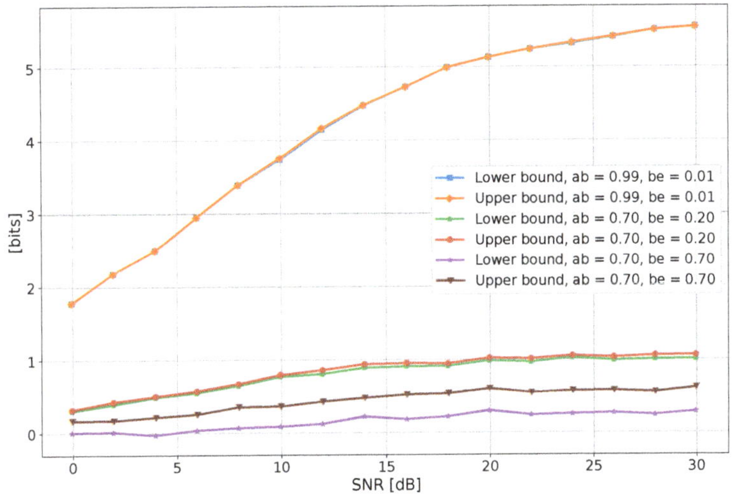

Figure 10. Secrecy key rate as a function of the SNR of Alice and Bob, for different values of the correlation and with 4 filters. In the legend, "ab" and "be" stand for Alice–Bob correlation and Bob–Eve correlation.

5.3. SKR and Delay Spread

As a last case, the simulations were performed to fix both the SNR and the Rice factor at 10 dB, but varying the DS of the channel. As in the previous cases, the simulations are repeated for different values of the correlation values.

As for the case with one filter, depicted in Figure 11, it is possible to notice that the DS does not seem to have a big impact on the SKR. Conversely, the SKR tends to decrease with the increasing DS, when multiple filters are employed (Figure 12). This trend is also in line with what has been reported in [13].

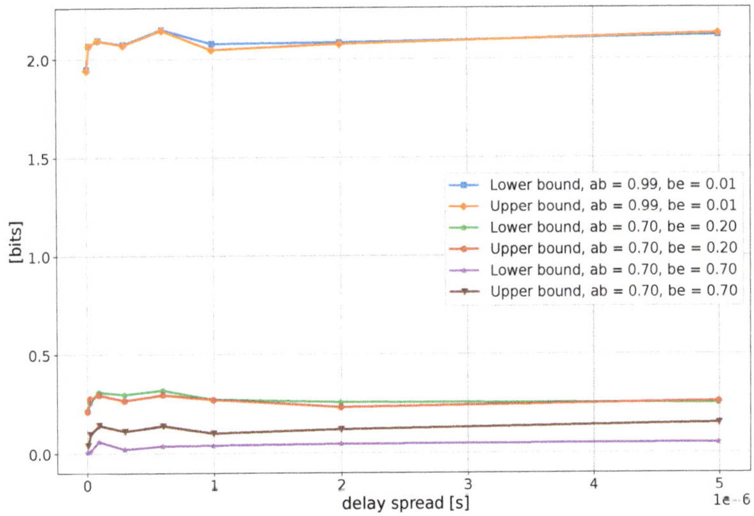

Figure 11. Secrecy key rate as a function of the delay spread of the channel of Alice and Bob, for different values of the correlation and with one filter. In the legend, "ab" and "be" stand for the Alice–Bob correlation and Bob–Eve correlation.

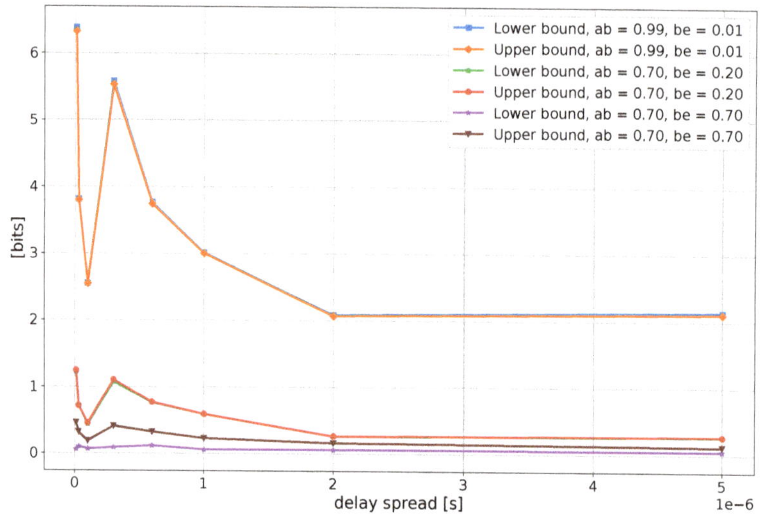

Figure 12. Secrecy key rate as a function of the delay spread of the channel of Alice and Bob, for different values of the correlation and with 1 filter. In the legend, "ab" and "be" stand for Alice–Bob correlation and Bob–Eve correlation.

The reason for this behavior can be understood by looking at Figure 13 and bearing in mind that the number of paths in the TDL is fixed: when the DS is low, there is a higher probability that the different paths cannot be resolved singularly; therefore, they might severely interfere and create a deep null in the PSD. In contrast, when the DS is larger, the different paths are spread over a wider delay range, and therefore they less frequently add up coherently inside the PSD, thus corresponding to a more oscillating PSD, but without deep fades.

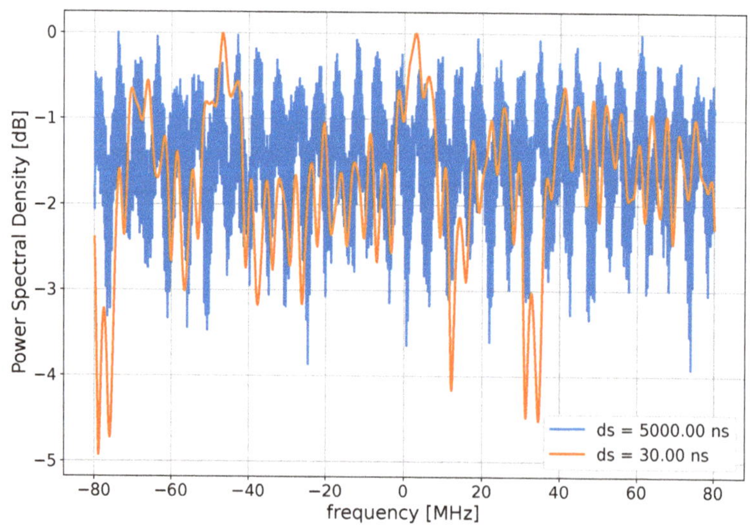

Figure 13. A realization of power spectral density with a different delay spread.

In terms of the entropy of the channel, and hence mutual information between Alice and Bob, having deep fades increases the randomness of the channel, translating into a higher SKR. Moreover, the effect of the deep fades is somehow mitigated in the case of one single filter, since it blunts the effects due to the presence of deep fades by averaging the PSD over the whole signal bandwidth. Instead, when four filters are employed, the deep fades in the case of low DS create more variability on the filter outputs, introducing more entropy.

6. Conclusions

In this work, a simulation framework for PLKG in the Rice channel was presented, in order to compute the SKR under different channel conditions. Moreover, the simulator is able to generate correlated wide-band channel in order to take into account the presence of an eavesdropper and the possible imperfections that lead to non-ideal channel reciprocity. The SKR was computed, showing a decreasing trend with respect to the Rice factor of the channel. Moreover, it was shown that a high correlation between the Alice and Bob channel samples is required in order to achieve a reasonable SKR. Finally, given the considered channel model, the DS has a detrimental effect on the SKR, since the higher DS situations lead to a lower SKR.

Author Contributions: Conceptualization, S.D.P.; methodology, S.D.P.; software, S.D.P.; validation, S.D.P., F.F. and M.B.; investigation, S.D.P.; writing—original draft preparation, S.D.P. and F.F.; writing—review and editing, M.B. All authors have read and agreed to the published version of the manuscript.

Funding: This research received no external funding.

Institutional Review Board Statement: Not applicable.

Informed Consent Statement: Not applicable.

Data Availability Statement: Data and software available under request.

Conflicts of Interest: The authors declare no conflict of interest.

Appendix A

Suppose we have two signals with AWGN:

$$A = s_a + n_a \tag{A1}$$
$$B = s_b + n_b \tag{A2}$$

where $s_a, s_b \sim \mathcal{N}(0,1)$, $n_a \sim \mathcal{N}(0,\sigma_a)$ and $n_b \sim \mathcal{N}(0,\sigma_b)$.

If SNR is the signal-to-noise ratio of the two users (assumed to be the same for simplicity):

$$\sigma_{n_a}^2 = \frac{1}{SNR}, \tag{A3}$$
$$\sigma_{n_b}^2 = \frac{1}{SNR}, \tag{A4}$$
$$\sigma_a^2 = 1 + \frac{1}{SNR}, \tag{A5}$$
$$\sigma_b^2 = 1 + \frac{1}{SNR}. \tag{A6}$$

Now, suppose that s_a and s_b have a covariance $\mathrm{cov}(s_a, s_b) = \eta$: since the variance of both s_a and s_b is equal to 1, then the covariance and the correlation are the same. The correlation between A and B can be computed as:

$$\mathrm{corr}(A,B) \equiv \rho = \frac{E[(A-\mu_A)(B-\mu_B)]}{\sigma_A\sigma_B}, \tag{A7}$$

but $\mu_A = \mu_B = 0$ since they are the sum of the zero mean Gaussian random variables, therefore

$$\text{corr}(A,B) \equiv \rho = \frac{E[(A-\mu_A)(B-\mu_B)]}{\sigma_A\sigma_B} = \frac{E[AB]}{\sigma_A\sigma_B} = \frac{E[(s_a+n_a)(s_b+n_b)]}{\sigma_A\sigma_B} = \\ = \frac{E[s_as_b]+E[s_an_b]+E[s_bn_a]+E[n_an_b]}{\sigma_A\sigma_B} = \frac{E[s_as_b]}{\sigma_A\sigma_B} = \\ = \frac{\text{cov}(s_a,s_b)}{\sigma_A\sigma_B} = \frac{\eta}{\sigma_A\sigma_B}. \tag{A8}$$

The mutual information of the two random variables can be rewritten as

$$\mathbb{I}(A;B) = h(A) + h(B) + h(A,B), \tag{A9}$$

where $h(A)$, $h(B)$ are the differential entropies of the two signals and $h(A,B)$ is the joint entropy, which in the Gaussian case, can be expressed as:

$$h(A) = \frac{1}{2}\log_2(2\pi e\sigma_A^2), \tag{A10}$$

$$h(B) = \frac{1}{2}\log_2(2\pi e\sigma_B^2), \tag{A11}$$

$$h(A,B) = \frac{1}{2}\log_2\Big((2\pi e)^2(\sigma_A^2\sigma_B^2 - \text{cov}^2(A,B))\Big) = \\ = \frac{1}{2}\log_2\Big((2\pi e)^2(\sigma_A^2\sigma_B^2 - \sigma_A^2\sigma_B^2\rho^2)\Big). \tag{A12}$$

Hence, the mutual information can be written as:

$$\mathbb{I}(A;B) = h(A) + h(B) + h(A,B) = \\ = \frac{1}{2}\log_2(2\pi e\sigma_A^2) + \frac{1}{2}\log_2(2\pi e\sigma_B^2) - \frac{1}{2}\log_2\Big((2\pi e)^2(\sigma_A^2\sigma_B^2 - \sigma_A^2\sigma_B^2\rho^2)\Big) = \\ = \frac{1}{2}\log_2\left(\frac{(2\pi e)^2\sigma_A^2\sigma_B^2}{(2\pi e)^2(\sigma_A^2\sigma_B^2 - \sigma_A^2\sigma_B^2\rho^2)}\right) = \frac{1}{2}\log_2\left(\frac{\sigma_A^2\sigma_B^2}{\sigma_A^2\sigma_B^2 - \sigma_A^2\sigma_B^2\rho^2}\right) = \\ = \frac{1}{2}\log_2\left(\frac{1}{1-\rho^2}\right) = \frac{1}{2}\log_2\left(\frac{\sigma_A^2\sigma_B^2}{\sigma_A^2\sigma_B^2 - \eta^2}\right). \tag{A13}$$

Appendix B

Let's start from the Lower Bound, assuming that Alice and Bob share highly correlated channel samples and that the correlation between Alice and Eve channel samples is the same as that between the Bob and Eve samples. Moreover, all the links share the same channel condition in terms of SNR and Rice factor K. The lower bound of the SKR can be reduced to:

$$\mathbb{R}(X^A, X^B \parallel X^E) \geq \max[\mathbb{I}(X^A;X^B) - \mathbb{I}(X^A;X^E), \mathbb{I}(X^A;X^B) - \mathbb{I}(X^B;X^E)] \\ = \mathbb{I}(X^A;X^B) - \mathbb{I}(X^A;X^E) \\ = h(X^A) - h(X^A|X^B) - h(X^A) + h(X^A|X^E) \simeq h(X^A|X^E). \tag{A14}$$

The conditioned entropy $h(X^A|X^B)$ is almost zero since the Alice and Bob channel observations are highly correlated, and therefore, the residual uncertainty on X^A by knowing X^B is almost null.

In the same way as before, the upper bound can be reduced to:

$$\mathbb{R}(X^A, X^B \parallel X^E) \leq \min[\mathbb{I}(X^A;X^B), \mathbb{I}(X^A;X^B \mid X^E)] = \\ = \min[h(X^A) - h(X^A|X^B), h(X^A|X^E) - h(X^A|X^B, X^E)] \\ \simeq \min[h(X^A), h(X^A|X^E)] = h(X^A|X^E). \tag{A15}$$

The term $h(X^A|X^B)$ is zero for the reasons explained before, and then the term $h(X^A|X^B, X^E)$ is almost zero since the conditioning happens on both X^B and X^E, but X^B is highly correlated with X^A, and hence, the residual uncertainty is almost zero.

Since the upper bound and the lower bound are equal, when Alice and Bob share highly correlated samples that the SKR reduces to $h(X^A|X^E)$; therefore, it is expected that for a high correlation, similar values for the upper and lower bounds should be achieved.

References

1. Zhang, J.; Li, G.; Marshall, A.; Hu, A.; Hanzo, L. A New Frontier for IoT Security Emerging from Three Decades of Key Generation Relying on Wireless Channels. *IEEE Access* **2020**, *8*, 138406–138446. [CrossRef]
2. Bhatia, V.; Ramkumar, K. An Efficient Quantum Computing technique for cracking RSA using Shor's Algorithm. In Proceedings of the 2020 IEEE 5th International Conference on Computing Communication and Automation (ICCCA), Greater Noida, India, 30–31 October 2020; pp. 89–94. [CrossRef]
3. Rao, S.; Mahto, D.; Yadav, D.; Khan, D. The AES-256 cryptosystem resists quantum attacks. *Int. J. Adv. Res. Comput. Sci.* **2017**, *8*, 404–408.
4. Bonnetain, X.; Naya-Plasencia, M.; Schrottenloher, A. Quantum Security Analysis of AES. *IACR Trans. Symmetric Cryptol.* **2019**, *2019*, 55–93. [CrossRef]
5. Shannon, C.E. Communication theory of secrecy systems. *Bell Syst. Tech. J.* **1949**, *28*, 656–715. [CrossRef]
6. Yang, P.; Xiao, Y.; Xiao, M.; Li, S. 6G Wireless Communications: Vision and Potential Techniques. *IEEE Netw.* **2019**, *33*, 70–75. [CrossRef]
7. Mucchi, L.; Jayousi, S.; Caputo, S.; Panayirci, E.; Shahabuddin, S.; Bechtold, J.; Morales, I.; Stoica, R.A.; Abreu, G.; Haas, H. Physical-Layer Security in 6G Networks. *IEEE Open J. Commun. Soc.* **2021**, *2*, 1901–1914. [CrossRef]
8. Shiu, Y.; Chang, S.Y.; Wu, H.; Huang, S.C.; Chen, H. Physical layer security in wireless networks: A tutorial. *IEEE Wirel. Commun.* **2011**, *18*, 66–74. [CrossRef]
9. Zhang, J.; Duong, T.; Marshall, A.; Woods, R. Key Generation from Wireless Channels: A Review. *IEEE Access* **2016**, *4*, 1. [CrossRef]
10. Cheng, C.; Lu, R.; Petzoldt, A.; Takagi, T. Securing the Internet of Things in a Quantum World. *IEEE Commun. Mag.* **2017**, *55*, 116–120. [CrossRef]
11. Maurer, U. Secret key agreement by public discussion from common information. *IEEE Trans. Inf. Theory* **1993**, *39*, 733–742. [CrossRef]
12. Kim, Y.; Kim, Y.; Oh, J.; Ji, H.; Yeo, J.; Choi, S.; Ryu, H.; Noh, H.; Kim, T.; Sun, F.; et al. New Radio (NR) and its Evolution toward 5G-Advanced. *IEEE Wirel. Commun.* **2019**, *26*, 2–7. [CrossRef]
13. Zoli, M.; Mitev, M.; Barreto, A.N.; Fettweis, G. Estimation of the Secret Key Rate in Wideband Wireless Physical-Layer-Security. In Proceedings of the 2021 17th International Symposium on Wireless Communication Systems (ISWCS), Berlin, Germany, 6–9 September 2021; pp. 1–6. [CrossRef]
14. Zeng, K. Physical layer key generation in wireless networks: Challenges and opportunities. *IEEE Commun. Mag.* **2015**, *53*, 33–39. [CrossRef]
15. Furqan, H.M.; Hamamreh, J.M.; Arslan, H. New Physical Layer Key Generation Dimensions: Subcarrier Indices/Positions-Based Key Generation. *IEEE Commun. Lett.* **2021**, *25*, 59–63. [CrossRef]
16. Pilc, M.; Remlein, P. The impact of LOS component on information disclosed to eavesdroppers in wireless channels with PHY-based secret key generation. In Proceedings of the 2018 Baltic URSI Symposium (URSI), Poznan, Poland, 14–17 May 2018; pp. 65–68. [CrossRef]
17. Zenger, C.; Zimmer, J.; Paar, C. Security Analysis of Quantization Schemes for Channel-based Key Extraction. *EAI Endorsed Trans. Secur. Saf.* **2015**, *2*, 267–272. [CrossRef]
18. Huth, C.; Guillaume, R.; Strohm, T.; Duplys, P.; Samuel, I.A.; Güneysu, T. Information reconciliation schemes in physical-layer security: A survey. *Comput. Netw.* **2016**, *109*, 84–104. [CrossRef]
19. Jana, S.; Premnath, S.N.; Clark, M.; Kasera, S.K.; Patwari, N.; Krishnamurthy, S.V. *On the Effectiveness of Secret Key Extraction from Wireless Signal Strength in Real Environments*; Association for Computing Machinery: New York, NY, USA, 2009; pp. 321–332. [CrossRef]
20. Zoli, M.; Barreto, A.N.; Köpsell, S.; Sen, P.; Fettweis, G. Physical-Layer-Security Box: A concept for time-frequency channel-reciprocity key generation. *EURASIP J. Wirel. Commun. Netw.* **2020**, *2020*, 114. [CrossRef]
21. 3GPP TR 38.901 Version 16.1.0 Release 16: Study on Channel Model for Frequencies from 0.5 to 100 GHz. Available online: https://www.etsi.org/deliver/etsi_tr/138900_138999/138901/16.01.00_60/tr_138901v160100p.pdf (accessed on 10 October 2021).
22. Zhang, K.; Song, Z.; Guan, Y.L. Simulation of Nakagami fading channels with arbitrary cross-correlation and fading parameters. *IEEE Trans. Wirel. Commun.* **2004**, *3*, 1463–1468. [CrossRef]
23. Kouiroukidis, N.; Evangelidis, G. The Effects of Dimensionality Curse in High Dimensional kNN Search. In Proceedings of the 2011 15th Panhellenic Conference on Informatics, Kastoria, Greece, 30 September–2 October 2011; pp. 41–45. [CrossRef]

electronics

MDPI

Article

A Cube Attack on a Reduced-Round Sycon

Minjeong Cho , Hyejin Eom , Erzhena Tcydenova and Changhoon Lee *

Department of Computer Science and Engineering, Seoul National University of Science and Technology, Seoul 01811, Korea
* Correspondence: chlee@seoultech.ac.kr

Abstract: The cube attack was proposed at the 2009 Eurocrypt. The attack derives linear polynomials for specific output bits of a BlackBox cipher. Cube attacks target recovery keys or secret states. In this paper, we present a cube attack on a 5-round Sycon permutation and a 6-round Sycon permutation with a 320-bit state, whose rate occupies 96 bits, and whose capacity is 224 bits. We found cube variables related to a superpoly with a secret state. Within the cube variables, we recovered 32 bits of the secret state. The target algorithm was Sycon with 5-round and 6-round versions of permutation. For the 5-round Sycon, we found a cube variable and recovered a state with a total of 2^{192} Sycon computations and 2^{37} bits of memory. For the 6-round Sycon, we found cube variables and recovered a state with a total of 2^{192} Sycon computations and 2^{70} bits of memory. When using brute force in a 5-round attack, 2^{224} operations were required, but the cube attack proposed in this paper had 2^{48} offline operations, and 2^{32} operations were required. When using brute force in a 6-round attack, 2^{224} operations were required, but the cube attack proposed in this paper required 2^{95} offline operations, and 2^{63} operations were required. For both attacks, offline could be used continuously after performing only once. To the best of our knowledge, this is the first cube attack on Sycon.

Keywords: sycon; cube attack; state recovery

Citation: Cho, M.; Eom, H.; Tcydenova, E.; Lee, C. A Cube Attack on a Reduced-Round Sycon. *Electronics* **2022**, *11*, 3605. https://doi.org/10.3390/electronics11213605

Academic Editor: Rameez Asif

Received: 24 August 2022
Accepted: 1 November 2022
Published: 4 November 2022

1. Introduction

Currently, wireless communication technology supports the high-speed communication of various devices, such as cellular phones, lightweight devices, and industrial sensors. In addition, wireless communication makes smart factories, smart cities, and self-driving cars possible and provides many conveniences for human beings. However, the importance of information security must be emphasized, because there are risks of being exposed to cyber security threat, manipulation or leakage of data and invasion of privacy during the transmission of data by wireless communication [1–3]. In wireless communication, information security may be provided through cryptographic algorithms [4,5]. However, since the security strength of cryptographic algorithms does not provide immutability, it must be continuously re-evaluated and reviewed for proper use.

A cube attack is the first type of attack to utilize existing linear, logarithmic, and correlation at the same time [6]. A cube attack can apply to block ciphers, stream ciphers, and MACs. With a cube attack, key recovery is possible. A cube attack creates a polynomial in GF(2) using a set of variables defined as a cube. A cube is the set of all cases with the given variables. The polynomial is associated with a cryptographic algorithm's output treated as a black box, expressed with a quotient and a remainder. The goal is to find the coefficient of the quotient. Then, the secret data are recovered using the obtained coefficient of the quotient.

At Eurocrypt 2009, a cube attack against Trivium, a block cipher-based stream cipher suitable for wireless communication environments, was announced [6]. Trivium is a stream cipher using an 80-bit key and 1152 initializations [7]. Since then, it has been proven through several papers that a cube attack is possible for several cryptographic algorithms such as SIMON-64/96, Ascon, ACORN, MROUS, GILMI, and Keyak [8–13].

This paper proposes a cube attack on Sycon. Sycon is an AEAD cipher sponge construction with a 128-bit key. Sycon was submitted to the NIST Lightweight Cryptography Competition [14] and was selected among the ciphers for the first round of this project.

In this paper, Sycon using a rate of 96 was reduced to five rounds and six rounds. Based on the low algebraic degree of Sycon, we were able to construct a cube attack with a complexity of 2^{192} for the 5-round Sycon permutation, described in Section 4. A total 2^{48} operations and 2^{37} bits of memory were required to recover the 32-bit state in the 5-round Sycon. In Section 6, we describe the use of similar algebraic properties to construct a cube attack to obtain a state recovery attack for the 6-round Sycon permutation. We recovered the same 32-bit state on the 6-round Sycon with 2^{95} operations and 2^{70} bits of memory. To the best of our knowledge, this paper describes the first cube attack on Sycon. The main contributions of this paper are summarized as follows:

- The cube attack against Sycon shows the potential threat applicable to wireless communication. Sycon could be considered in wireless communication for confidentiality and integrity, since Sycon is a lightweight AEAD algorithm.
- This is the first known state recovery attack against Sycon. The time complexity to recover the state of the 5-round and 6-round Sycon was 2^{192}, faster than brute force. This paper shows the possibility of transmission data tampering or sniffing by an attacker.

This paper proceeds as follows. Section 3 introduces cube attacks and the Sycon cipher. Section 4 describes a cube attack on Sycon. The Results and Discussion section describes the complexity of the attacks. Conclusions are provided in Section 6 with a summary of the proposed attack and the results.

2. Related Work

A cube attack was proposed on Trivium with 672 initialization rounds with 2^{19} bits operations, 735 initialization rounds with 2^{30} bits operations, and 767 initialization rounds with 2^{45} bits operations. Since then, an improved attack with an MILP model on Trivium with 675/735/840/841/842 initialization rounds was proposed in 2021, and an attack on Trivium with 843 initialization rounds was proposed in 2022 [15,16].

Ascon was selected as a finalist in the NIST Lightweight AEAD Cryptography Contest in 2017, and the strongest attack at that time was a cube attack, which had the time complexity of 2^{97} under the nonce misuse condition in the 7-round Ascon initialization step [17]. However, for Ascon in 2022, an improved cube attack was also conducted under the condition of nonce misuse in the initialization phase [18]. The complexity of the key recovery for a full-round Ascon was 2^{130}.

In [19], Dinur et al. presented a cube-like attack against Keccak hash function-based message authentication codes, authenticated encryption, and stream cipher. The key recovery attack was performed for up to seven rounds. Key recovery and forgery attacks were proposed for AE based on the Keccak hash function. In the case of the key recovery, the attacks were performed up to six rounds under the nonce respected condition, and the attacks were performed up to seven rounds under the nonce reused condition. A key recovery attack and keystream prediction attack were proposed for the Keccak hash function-based stream cipher, and it was shown that six rounds of key recovery and key stream prediction could perform attacks for up to nine rounds.

In [20], Salam et al. proposed a cube attack against the authenticated encryption stream cipher ACORN. This attack recovered a 128-bit key with a complexity of 2^{35} in 477 initialization rounds in ACORN, a proposed candidate for NIST CAESAR (Competition for Authenticated Encryption: Security, Applicability, and Robustness). In addition, full-round ACORN showed that a state recovery attack could be performed with a complexity of $2^{72.8}$ using a linear equation associated with the initial state. In [12], Yang et al. proposed a method of measuring the algebraic order and numerical mapping in NFSR-based ciphers and a method of finding a cube based on a greedy algorithm. It was shown that a key could be recovered with a complexity of $2^{127.46}$ with cube variables using 123 variables for the 772 reduced-round ACORN.

In [21], Huang et al. introduced an efficient key recovery attack for the Keccak hash function-based MAC or Keccak hash function-based AE algorithm Keyak using a conditional cube attack. In [19], a MAC-based 7-round Keccak hash function was proposed . With 2^8 times more data, the time complexity could be decreased to 2^{72}. An attack against Keyak was feasible with the time complexity of 2^{74} and a data complexity of 2^{74} for eight rounds.

3. Preliminaries

In this section, we briefly introduce the necessary background for this paper. Firstly, we provide the notations used in this paper. Then, we provide a brief description of Sycon and the concept of a cube attack.

3.1. Abbreviations and Notations

The abbreviations used in this paper are listed in Table 1.

Table 1. Abbreviations.

Abbreviation	Full Word
AEAD	Authenticated Encryption with Associated Data
LSB	Least Significant Bit
MSB	Most Significant Bit
LFSR	Linear Feedback Shift Register
SB	S-box layer
SD	Subblock Diffusion
RC	Add Round Constant Layer

The symbol notations used in this paper are listed in Table 2.

Table 2. Notations.

Symbol	Meaning
$x \oplus y$	Bitwise XOR of x and y
$(x \lll n)$	Left circular shift by n-bits
rc_i	Round constant at round i
Π^{ρ}	An iterated permutation with ρ rounds over $(0,1)^{320}$
X||Y	Concatenate data X and Y
Si_0	Most Significant 32 bit of si
Si_1	Least Significant 32 bit of si
s^i	The i-th bit of s
0^n	n bit sequences of 0
X ∘ Y	Y(X(data)) where the data are input, and X and Y are functions.
Q	Quotient of a polynomial
R	Remainder of a polynomial

3.2. Sycon Authenticated Encryption with the Associated Data Algorithm Specification

Sycon is an authenticated encryption with an associated data (AEAD) cipher [22]. AEAD is an encryption algorithm with a built-in integrity process using a secret key [23]. AEAD usually performs better than using two separate cryptographic processes with two different secret keys. Sycon provides two authenticated encryption algorithms with associated data and one hash algorithm in a sponge structure. In this section, we specify the Sycon whose rate is 96.

Sycon consists of initialization, related data processing, encryption/decryption, and finalization. The initialization phase loads 128-bit keys, a 128-bit nonce, and a 64-bit initialization vector into the 320-bit state variable. Then, it conducts two permutation calls, truncating the key by 64 bits and XORing it. The relevant data processing is applied after the initialization phase, if the related data are not empty. Relevant data processing performs the permutation with the associated data (AD) and the current state as input. The relevant data processing will not perform if the relevant data are empty. In encryption/decryption, the encryption algorithm generates the ciphertext with the same length as the input plaintext. In this case, the size of the plaintext is a multiple of 96, and padding is performed if it is less than 96 bits. Then, we conduct the permutation to update the state. This process is repeated until all 96 bits of plaintext are processed. The finalization absorbs the key back into the state via a ratio of two permutation calls, and a 128-bit tag is output. A tag is a value that concatenates the contents of S2 and S3 among the state variables.

The state is XORed with a key or plaintext after permutation as shown in Figure 1. The LSB 224 bits of the state are XORed with a domain separator. The domain separator of Sycon is as follows: 0^{224} for initialization, $100 \parallel 0^{221}$ for AD processing, $010 \parallel 0^{221}$ for massage, and $001 \parallel 0^{221}$ for tag generations. If the additional data are empty, $001 \parallel 0^{221}$ is replaced by $010 \parallel 0^{221}$.

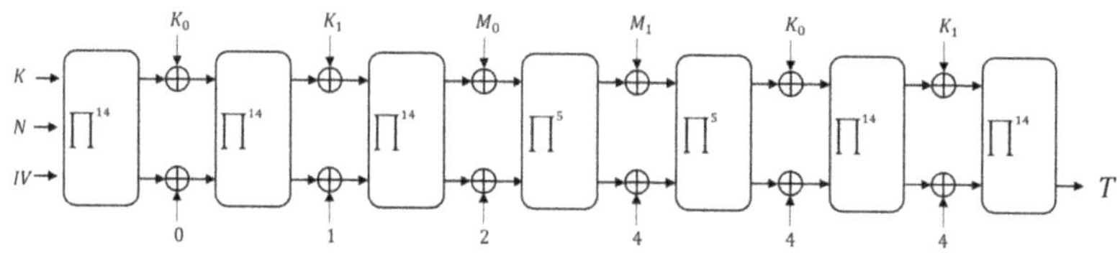

Figure 1. Sycon -AEAD-96 (when the length of the associated data is 0).

Sycon permutation is an iterative computation in a round function. In the round function, Sycon uses a 320-bit state. In the state, the first 64/96 bits are user message bits along the rate. The round function (R) of the Sycon permutation consists of a sequence of three distinct transformations: SBox (SB), SubBlockDiffusion (SD), and AddRoundConstant (RC), i.e., $R = RC \circ SD \circ SB$. The ρ-round permutation, denoted by Π^{ρ}, is constructed as $\Pi^{\rho} = R \circ \ldots \circ R$.

The first layer is a nonlinear computation. Sycon's round function is SPN. Thus, for nonlinear computation, Sycon uses 64 S-boxes. The process of the S-boxes in the equation is as follows:

$$\begin{aligned} y0 &= x0 \oplus x1x3 \oplus x2x3 \oplus x3x4 \oplus x4 \\ y1 &= x0x1 \oplus x0x3 \oplus x0 \oplus x1x3 \oplus x1 \oplus x2 \oplus x3 \oplus x4 \\ y2 &= x0x2 \oplus x1 \oplus x2x4 \oplus x2 \oplus x3 \\ y3 &= x0 \oplus x2x3 \oplus x2 \oplus x3x4 \oplus x3 \oplus 1 \\ y4 &= x0x4 \oplus x0 \oplus x1 \oplus x3 \end{aligned} \tag{1}$$

The second layer is a diffusion layer that performs linear transformation on five 64-bit sub-blocks. The diffusion layer uses the following linear transformation:

$$\begin{aligned} S0 &\leftarrow (S0 \oplus (S0 \lll 59) \oplus (S0 \lll 54)) \lll 40 \\ S1 &\leftarrow (S1 \oplus (S1 \lll 55) \oplus (S1 \lll 46)) \lll 32 \\ S2 &\leftarrow (S2 \oplus (S2 \lll 33) \oplus (S2 \lll 02)) \lll 16 \\ S3 &\leftarrow (S3 \oplus (S3 \lll 21) \oplus (S3 \lll 42)) \lll 56 \\ S4 &\leftarrow (S4 \oplus (S4 \lll 13) \oplus (S4 \lll 26)) \end{aligned} \tag{2}$$

The third layer is the add round constant layer. Round constants use a four-bit LFSR defined by the polynomial $x^4 + x + 1$ *over* $\mathbb{F}_2$. The LFSR status is expressed as $rc = (rc_{i+3}, rc_{i+2}, rc_{i+1}, rc_i)$, where $rc_{i+4} = rc_i \oplus rc_{i+1}$. Starting from the initial state $rc = (0,1,0,1)$, we generate a $\rho = 12$ round constant, where each state of the LFSR is given as a unique constant. The four-bit LFSR with status (rc_3, rc_2, rc_1, rc_0) is converted to a byte equal to $(0,0,0,0, rc_{i+3}, rc_{i+2}, rc_{i+1}, rc_i)$. The round constants are given in Table 3.

Table 3. The round constants rc_i.

Round	Constants	Round	Constants
0	0xaaaaaaaaaaaaaa15	4	0xaaaaaaaaaaaaaa17
1	0xaaaaaaaaaaaaaa1a	5	0xaaaaaaaaaaaaaa1b
2	0xaaaaaaaaaaaaaa1d	6	0xaaaaaaaaaaaaaa0d
3	0xaaaaaaaaaaaaaa0e	7	0xaaaaaaaaaaaaaa06

3.3. Cube Attack

Let the cryptography algorithms be expressed with a polynomial f. The input of the cryptographic algorithm (e.g., plaintext, initial vector, nonce, associated authentication data) will be f's input parameter, and the output of the cryptography algorithm (e.g., ciphertext, tag) will be the value of f's computed result. If the block cipher has input as plaintext P, initial vector IV, and key K, and the output is ciphertext C, we can express the block cipher $f(P, IV, K) = c$ with $P = p_0p_1p_2p_3 \ldots p_n$, $IV = iv_0iv_1iv_2iv_3 \ldots iv_n$, $K = k_0k_1k_2k_3 \ldots k_m$, and $C = c_0c_1c_2c_3 \ldots c_n$, where p_i, k_i, and c_i is a bit representation, respectively.

- **Degree** The dense polynomial f of degree d has $\sum_{i=0}^{d} \binom{2n+m}{i}$ possible polynomials over GF(2). To eliminate the nonlinear terms on the polynomials, the attack needs to eliminate $\sum_{i=2}^{d} \binom{2n+m}{i}$. Thus, when the degree becomes higher, it is hard to eliminate the nonlinear terms.
- **Cube Variables** To eliminate the nonlinear terms from the polynomial f, an attacker needs to divide the polynomials f by the other polynomial t, whose degree is $d-1$. If $f(P, IV, K)$ divided by $t = p_0p_1 \ldots p_{n-1}$, then f can be expressed as $f(P, IV, K) = tQ(p_n, IV, K) + R(P, IV, K)$. In this term, t is $n-1$ bit cube variables.
- **Superpolys** We assume that dense polynomial f divided by $t = p_0p_1 \ldots p_{n-1}$, $f(P, IV, K) = tQ(p_n, IV, K) + R(P, IV, K)$, as above. In this term, $Q(p_n, IV, K)$ is a superpoly with degree 1. In order to obtain a superpoly, the attacker can compute $\sum_{t \in C_t} f(P, IV, K) = Q(p_n, IV, K)$.

In a cube attack, finding cube variables is important, because when the degree of the polynomial becomes higher, an attacker needs more polynomials to use Gaussian elimination. When the cube variables are larger, the attacker breaks more rounds. Moreover, f can be divided by $m-1$ variables, and the quotient will be a degree 1 polynomial. A cube attack should first formulate the polynomial f. If f is not a dense polynomial, superpoly Q's degree will be changed along the chosen cube variables. For example, when we define cube variables as $t = p_0p_1 \ldots p_{n-1}$, and if A is bits that multiplied with $l, l <= n$ bits, then f could be $f(P, IV, K) = tQ(p_n, A) + R(P, IV, K)$. That is, the polynomial Q degree is $l-m$. Thus, in order to obtain Q, the attacker needs to use fewer cube variables $l-m-1$.

To make m independent polynomials, the attacker needs to analyze where there are no multiplication values between the target bits and the input bits that the attacker can control. The attacker will select the control bits that do not have multiplication with the target bits. The attacker can know whether multiplication will be computed by analyzing the cryptographic algorithms' process. A typical example that has a multiplication step in a cryptography algorithm is an S-box. After the attacker finds the proper cube variables from the polynomials, a cube attack can decide the round that an attacker can use. Then, the cube attack is presented as follows:

1. **Offline Phase** The attacker computes and stores the $Q(p_n, IV, K)$. The targeted data bits can be expressed as linear polynomials $Q(p_n, A)$. The attacker computes a linear polynomial from the Q values. The attacker assigns 0 except for the cube variables. Then, the attacker sets bit by bit on x. From the data, the attacker computes each coefficient of x.
2. **Online Phase** Considering the oracle as given, the attacker derives the cube sum of the oracle query results $Q(p_n, A)$. From the polynomials saved in the offline phase, the attacker recovers target bits A. If A has $l-m-1$ variables, the attacker needs an

$l - m - 1$ cube sum result. The attacker computes the Gaussian elimination to obtain the recovered target bits.

3. **Brute Force Phase** If the cube variables are not enough to obtain all rounds or all target bits, the attacker performs a brute force attack on the remaining bits. For example, if the recovered bits are l bits and the targeted bits are m bits, the attacker performs a 2^{m-l} exhaustive search.

4. State Recovery Attack on a Reduced-Round Sycon

In this section, we focus on state recovery attacks. First, we analyze the round-reduced Sycon. Then, in the later part of the section, the cube attacks on five-round Sycon and six-round Sycon are described.

4.1. Idea and Scenario

We propose an attack idea and scenario to recover the secret state of Sycon. Sycon has a 320-bit state variable, as described in Section 3. In this paper, the state variable $\mathbb{S}$ is expressed as a truncated form to 32-bit units as follows:

$$\mathbb{S} = S0_0||S0_1||S1_0||S1_1||S2_0||S2_1||S3_0||S3_1||S4_0||S4_1 \quad (3)$$

The polynomial of the output from the S-box has degree 2. After five rounds of the permutation are performed, the result $\mathbb{S}$ has a polynomial of degree 2^4. Therefore, we require $2^4 - 1$ variables in the cube for an attack against the 5-round Sycon. In the same way, after six rounds, the degree of $\mathbb{S}$ is 2^5. Thus, we need $2^5 - 1$ variables in the cube. We choose bits from the state variable that we have control over. We select $S0_0$, $S0_1$, and $S1_0$ as the cube variables. The cube attack is feasible if the variables are uniquely multiplied by $S0_0$, $S0_1$, and $S1_0$. The variable with this characteristic in the S-box is $S3_0$. $S1_0$ is only multiplied by $S0_0$ and $S3_0$. Choosing $S1_0$ as the cube variable allows us to recover $S3_0$. With the cube variables, we obtain the linear equations as follows:

$$L_i(S) = a_i^0 S1_0^0 + a_i^1 S1_0^1 + a_i^2 S1_0^2 + \ldots + a_i^{31} S1_0^{31} + c_i, i = 0, 1, 2, \ldots, 31 \quad (4)$$

To construct L_i, we obtain the result of the permutation by $M \oplus C$, and we choose the message bits that construct the cube variables. Figure 2 shows the attack scenario.

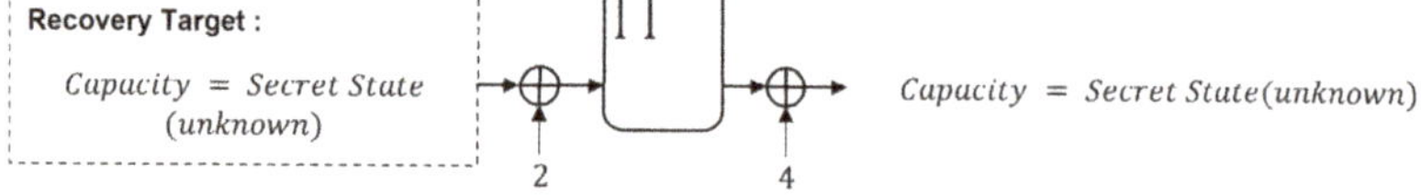

Figure 2. State recovery attack scenario on the reduced-round Sycon.

4.2. Attack on the Five-Round Sycon

4.2.1. Offline Phase

Sycon encrypts plaintext by truncating it in units of 96 bits. We chose 96 bits of plaintext. The state variable S was concatenated as $S0_0$, $S0_1$, $S1_0$, $S1_1$, $S2_0$, $S2_1$, $S3_0$, and $S3_1$, where each size was 32 bits.

$$\mathbb{S} = S0_0||S0_1||S1_0||S1_1||S2_0||S2_1||S3_0||S3_1||S4_0||S4_1 \quad (5)$$

P_0, P_1, and P_2 were the plaintext values chosen. Ciphertext C_0, C_1, and C_2 were computed by performing XOR of the current state variables MSB 96 bits $S0_0$, $S0_1$, and $S1_0$ and the plaintext P_0, P_1, and P_2.

$$\begin{aligned} C_0 &= S0_0 \oplus P_0 \\ C_1 &= S0_1 \oplus P_1 \\ C_2 &= S1_0 \oplus P_2 \end{aligned} \tag{6}$$

Since the ciphertext and the plaintext were known, we computed the XORing of the ciphertext and the plaintext to obtain $S0_0$, $S0_1$, $S1_0$.

$$\begin{aligned} S0_0 &= C_0 \oplus P_0 \\ S0_1 &= C_1 \oplus P_1 \\ S1_0 &= C_2 \oplus P_2 \end{aligned} \tag{7}$$

We set the cube variables in $S0_1$. Since the degree of the S-box was 2, the result after five rounds had a polynomial of degree 16. The attacker chose a 15-bit cube in the 5-round Sycon state variable to obtain a polynomial of degree 1 and the rest of the values as constants. We recovered 16 bits of MSB and 16 bits of LSB of $S3_0$, respectively. Therefore, two cube variables were required, and each cube variable set had 2^{15} elements. The cube variable could be $C_{S1_0} := \{0x0, 0x1, \ldots, 0xffff\}$ and $C_{S1_0} := \{0x00000, 0x10000, \ldots, 0xffff0000\}$. We assigned all the other variables to 0 except for the cube variable $S0_1$ and used $S0_0$ in the attack. $S1_0$ was used in the attack as it was multiplied only by $S0_0$ and $S3_0$, as described in Section 2.

Since $S3_0$ depends only on $S0_0$, after five rounds, the polynomial could be expressed as a polynomial for the quotient of $S0_0$ and $S3_0$.

$$\begin{aligned} \mathbb{S} &= S3_0 Q(S0_0, S3_1) + R(S0, S1, S2, S3, S4) \\ Q(S) &= \sum_{S1_0 \in C_{S1_0}} f(\mathbb{S}) \\ &= a^0 S1^0 + a^1 S1^1 \ldots + a^{31} S1^{31} + c \textit{ where } a^i \in \{0,1\} \end{aligned} \tag{8}$$

We used the process shown in Table 4 to find the variable coefficients of a polynomial for $Q(S)$. As a result, we obtained the following values. The results obtained were stored in the memory and used in the online phase.

$$\begin{aligned} &\sum_{(v_1,\ldots,v_{15}) \in C_{S1_0}} f(S0_0, S0_1, S1_0, S1_1, S2_0, S2_1, S3_0, S3_1, S4_0, S4_1) = Q(S0_0, S3_0), \\ &\quad \textit{with } S0_0 = 0x00000000, 0x00000001, 0x00000002, \ldots, 0x0000ffff, \\ &\quad S3_0 = 0x00000000, 0x00000001, 0x00000002, \ldots, 0x0000ffff \end{aligned} \tag{9}$$

$$\begin{aligned} &\sum_{(v_{16},\ldots,v_{31}) \in C_{S1_0}} f(S0_0, S0_1, S1_0, S1_1, S2_0, S2_1, S3_0, S3_1, S4_0, S4_1) = Q(S0_0, S3_0), \\ &\quad \textit{with } S0_0 = 0x00000000, 0x00010000, 0x00020000, \ldots, 0xffff0000, \\ &\quad S3_0 = 0x00000000, 0x00010000, 0x00020000, \ldots, 0xffff0000 \end{aligned} \tag{10}$$

Table 4. Attack on the five-round Sycon: Offline phase process.

Offline Pseudo Code

```
Input : State 𝕊 = S0₀S0₁S1₀S1₁S2₀S2₁S3₀S3₁S4₀S4₁, Cube Set C_{S1₀}
Output : CubeSum CS[2^16] = {0^510, 0^510, ..., 0^510}
Algorithm :
    𝕊 := 0^320
    For i in 0 to 0x0000ffff:
        CS[i] := 0^510
        S3₀ := i
        For j in 0 to 0x0000fffe:
            S0₀ := j
            For cube value in C_{S1₀}:
                𝕊 = Sycon(S)
                CStmp := CS[i] ⊕ S1₀
                S := 0^320
            CS[i] := CS[i]||CStmp
    For i in 0 to 0xffff0000:
        CS[i] := 0^510
        S3₀ := i
        For j in 0 to 0xfffe0000:
            S0₀ := j
            For cube value in C_{S1₀}:
                S = Sycon(S)
                CStmp := CS[i] ⊕ S1₀
                𝕊 := 0^320
            CS[i] := CS[i]||CStmp
    return CS
```

4.2.2. Online Phase

In the online phase, we implemented an oracle. The oracle allowed the choice of arbitrary plaintext $S0_1$ and $S1_0$, and the cube variables C_{S0_0} were defined internally as a fixed set. The oracle computed $Q(S0_0, S3_0)$ for the plaintext chosen. We computed two times for the MSB 32 bits and LSB 32 bits of $S3_0$. The oracle calculated the superpolys as follows:

$$\sum_{(v_1,\ldots,v_{15})\in C_{S1_0}} f(S0_0, S0_1, S1_0, S1_1, S2_0, S2_1, S3_0, S3_1, S4_0, S4_1) = Q(S0_0, S3_0), \quad with\ S0_0 = 0x00000000, 0x00000001, 0x00000002, \ldots, 0x0000ffff \tag{11}$$

$$\sum_{(v_16,\ldots,v_{31})\in C_{S1_0}} f(S0_0, S0_1, S1_0, S1_1, S2_0, S2_1, S3_0, S3_1, S4_0, S4_1) = Q(S0_0, S3_0), \quad with\ S0_0 = 0x00000000, 0x00010000, 0x00020000, \ldots, 0xffff0000. \tag{12}$$

We explored the superpoly $Q(S0_0, S3_0)$ obtained by querying the oracle in the memory space stored during the offline phase. If the same value as $Q(S0_0, S3_0)$ was in the memory, we determined $S3_0$, because the memory stored $Q(S0_0, S3_0)$ for every $S3_0$. The $Q(S0_0, S3_0)$ that we explored in the memory was as follows:

$$Q(0x00000000, S3_0), Q(0x00000001, S3_0), Q(0x00000002, S3_0), \ldots, Q(0x0000ffff, S3_0) \tag{13}$$

$$Q(0x00000000, S3_0), Q(0x00010000, S3_0), Q(0x00020000, S3_0), \ldots, Q(0xffff0000, s3_0) \tag{14}$$

4.2.3. Brute Force Phase

We recovered the remaining 192 bits of state by brute force. There was no memory required, but we needed 2^{192} computations.

4.3. Attack on the Six-Round Sycon

4.3.1. Offline Phase

We chose the 32-bit $S0_1$. With the chosen plaintext, we assigned the LSB 31 bits of $S1_0$ as the cube variable $C_{S1_0} := \{0^{32}, \ldots, 0xffffffff\}$. The cube variable 2^{32} of $S0_1$, $S2_0$, $S2_1$, $S3_1$, $S4_0$, and $S4_1$ was fixed as 0^{32}, because $S1_0$ was only multiplied with $S0_0$ and $S3_0$. We used the process shown in Table 5 to find the variable coefficients of a polynomial for Q(S). $S3_0$ could be rephrased as $Sycon(\mathbb{S}) = S3_0 Q(S0_0, S3_1) + R(S0, S1, S2, S3, S4)$. $S3_0 Q(S)$ was $\sum_{S1_0 \in C_{S1_0}} Sycon(\mathbb{S})$. For each $j \in \{0,1\}^{64}$, we computed the following:

$$\sum_{(v_1,\ldots,v_{31}) \in C_{S1_0}} f(S0_0, S0_1, S1_0, S1_1, S2_0, S2_1, S3_0, S3_1, S4_0, S4_1) = Q(S0_0, S3_0),$$
$$S0_0 = 0x00000000, 0x00000001, 0x00000002, \ldots, 0xffffffff, \quad (15)$$
$$S3_0 = 0x00000000, 0x00000001, 0x00000002, \ldots, 0xffffffff.$$

Table 5. Attack on the six-round Sycon: Offline phase process.

Offline Pseudo Code

Input: State $\mathbb{S} = S0_0 S0_1 S1_0 S1_1 S2_0 s2_1 S3_0 S3_1 S4_0 S4_1$, Cube Set C_{S1_0}
 Output: CubeSum $CS[2^{32}] = \{0^{510}, 0^{510}, \ldots, 0^{510}\}$
 Algorithm:
   $S := 0^{320}$
   For i in 0 to $0xffffffff$:
     $CS[i] := 0^{1020}$
     $S3_0 := i$
     For j in 0 to $0xfffffffe$:
       $s0_0 := j$
       For cube value in C_{S1_0}:
         $S = Sycon(S)$
         $CStmp := CS[i] \oplus S1_0$
         $\mathbb{S} := 0^{320}$
       $CS[i] := CS[i] \| CStmp$
   return CS

4.3.2. Online Phase

We implemented an oracle, where the cube variables C_{S0_0} were defined internally as a fixed set. The oracle computed $Q(S0_0, S3_0)$ for the plaintext chosen. The results calculated by the oracle were as follows for the online cube-sum:

$$\sum_{(v_1,\ldots,v_{31}) \in C_{S1_0}} f(S0_0, S0_1, S1_0, S1_1, S2_0, S2_1, S3_0, S3_1, S4_0, S4_1) = Q(S0_0, S3_0), \quad (16)$$
$$S0_0 = 0x00000000, 0x00000001, 0x00000002, \ldots, 0xffffffff.$$

We explored the value $Q(S0_0, S3_0)$ obtained through the oracle in the memory space stored during the offline phase. If the same value as $Q(S0_0, S3_0)$ was in the memory, we determined $S3_0$, because the memory stored $Q(S0_0, S3_0)$ for every $S3_0$. The $Q(S0_0, S3_0)$ that we explored in the memory was as follows:

$$Q(0x00000000, S3_0), Q(0x00000001, S3_0), Q(0x00000002, S3_0), \ldots, Q(0xffffffff, S3_0) \quad (17)$$

4.3.3. Brute Force Phase

We recovered the remaining 192-bit state bits by brute force.

5. Results and Discussion

First, we analyze the complexity in the offline phase. Computing the equations for just one case in the memory required 2^{15} 5-round Sycon and 16 bits of memory. We performed the same process 2^{32} times repeatedly for each $S3_0$ and $S0_0$. Therefore, in the offline phase, we required 2^{48} Sycon computations and 2^{37} bits of memory. In the online phase, we required 2^{32} computations to recover the target bits. The remaining 192 bits of state were recovered by brute force, so we required 2^{192}. In total, the attack needed a computational complexity of $2^{48} + 2^{32} + 2^{192} \approx 2^{192}$.

To recover the six-round Sycon permutation state, we needed an offline phase, an online phase, and a brute-force phase. In the offline phase, we precalculated the cube sum, and it needed a computation complexity of 2^{31} and 64 bits of memory. We performed this process 2^{64} times repeatedly for all cases in $S3_0$ and $S0_0$. In total, in the offline phase, 2^{95} 6-round Sycon computations and 2^{70} bits of memory were required. In the online phase, we only needed to perform 2^{64} of six-round Sycon permutations. The remaining 192 bits of state were recovered by brute force. In the brute-force phase, there was no memory required, but we needed 2^{192} computations. Thus, in total, the attack required a computational complexity of $2^{95} + 2^{70} + 2^{192} \approx 2^{192}$ (Table 6).

Table 6. Complexity of the 224-bit secret state recovery attack.

Round	Memory				Computation			
	This Paper			Brute Force *	This Paper			Brute Force *
	Offline	Online	Brute Force		Offline	Online	Brute Force	
5	2^{37}	-	-	-	2^{48}	2^{32}	2^{192}	2^{224}
6	2^{70}	-	-	-	2^{95}	2^{63}	2^{192}	2^{224}

* This was the first state recovery attack against Sycon. So the attack complexity based on brute force is the best result up to now.

The AEAD cipher encrypts the data to be transmitted and creates a tag for data integrity [24]. In order to give functions, Sycon has the following four phases: Initialization, associated data processing, encryption/decryption, and finalization. Each step should be analyzed in different ways [25]. In the future, we can extend the cube attack to the initialization or finalization phase of Sycon to recover the secret key.

6. Conclusions

In this paper, we proposed a state recovery attack against five-round and six-round Sycon. Sycon keeps 224 bits in secret and 96 bits as plaintext/ciphertext. With no information, we needed 2^{224} computations. From the attack we proposed, 32 bits of the state $S3_1$ could be recovered against the 5-round Sycon, with a time complexity of 2^{192}. We also proposed an attack against the six-round Sycon. The attack recovered the same 32 bits with the time complexity of 2^{95} and 2^{70} of memory in the offline phase. The time complexity was 2^{192}. This was faster than brute force over the 2^{224} possible states by a factor of about 2^{32}.

Author Contributions: Methodology, M.C.; Writing—original draft, M.C.; Formal analysis, H.E. and E.T.; Writing—review and editing, H.E. and E.T.; Supervision, C.L.; All authors read and agreed to the published version of the manuscript.

Funding: This work was supported as part of the Military Crypto Research Center (UD210027XD) funded by the Defense Acquisition Program Administration (DAPA) and the Agency for Defense Development (ADD).

Data Availability Statement: Not applicable.

Conflicts of Interest: The authors declare no conflict of interest.

References

1. Abdalzaher, M.S.; Samy, L.; Muta, O. Non-zero-sum game-based trust model to enhance wireless sensor networks security for IoT applications. *IET Wirel. Sens. Syst.* **2019**, *9*, 218–226. [CrossRef]
2. Abdalzaher, M.S.; Muta, O. A game-theoretic approach for enhancing security and data trustworthiness in IoT applications. *IEEE Internet Things J.* **2020**, *7*, 11250–11261. [CrossRef]
3. Abdalzaher, M.S.; Seddik, K.; Muta, O. Using repeated game for maximizing high priority data trustworthiness in wireless sensor networks. In Proceedings of the 2017 IEEE Symposium on Computers and Communications (ISCC), Heraklion, Greece, 3–6 July 2017 ; pp. 552–557.
4. Phuc, T.S.D.; Xiong, N.N.; Lee, C. Cryptanalysis of the XO-64 Suitable for Wireless Systems. *Wirel. Pers. Commun.* **2017**, *93*, 589–600. [CrossRef]
5. Phuc, T.S.D.; Lee, C. Cryptanalysis on SDDO-Based BM123-64 Designs Suitable for Various IoT Application Targets. *Symmetry* **2018**, *10*, 353. [CrossRef]
6. Dinur, I.; Shamir, A. Cube attacks on tweakable black box polynomials. In Proceedings of the Annual International Conference on the Theory and Applications of Cryptographic Techniques, Cologne, Germany, 26–30 April, 2009; pp. 278–299.
7. De Cannière, C.T.; Preneel, B. A stream cipher construction inspired by block cipher design principles. In *Information Security;* Katsikas, S.K., López, J., Backes, M., Gritzalis, S., Preneel, B., Eds.; Springer: Berlin/Heidelberg, Germany; pp. 171–186.
8. Mroczkowski, P.; Szmidt, J. The cube attack on stream cipher Trivium and quadraticity tests. *Fundam. Informaticae* **2012**, *114*, 309–318. [CrossRef]
9. Li, Z.; Dong, X.; Wang, X. Conditional Cube Attack on Round-Reduced ASCON. Cryptology ePrint Archive 2017. Available online: https://eprint.iacr.org/2017/160.pdf (accessed on 24 July 2022).
10. Bi, W.; Li, Z.; Dong, X.; Li, L.; Wang, X. Conditional cube attack on round-reduced River Keyak. *Des. Codes Cryptogr.* **2018**, *86*, 1295–1310.
11. Chen, S.; Xiang, Z.; Zeng, X.; Zhang, S. Cube attacks on round-reduced MORUS and GIMLI. *Sci. China Inf. Sci.* **2022**, *65*, 1–3. [CrossRef]
12. Yang, J.; Liu, M.; Lin, D. Cube cryptanalysis of round-reduced ACORN. In Proceedings of the International Conference on Information Security, New York, NY, USA, 16–18 September 2019; pp. 44–64.
13. He, Y.; Wang, G.; Li, W.; Ren, Y. Improved cube attacks on some authenticated encryption ciphers and stream ciphers in the Internet of Things. *IEEE Access* **2020**, *8*, 20920–20930. [CrossRef]
14. Lightweight Cryptography Round 1. Available online: https://csrc.nist.gov/Projects/lightweight-cryptography/round-1-candidates (accessed on 22 August 2022).
15. Sun, Y. Cube Attack against 843-Round Trivium. Cryptology ePrint Archive 2021. Available online: https://eprint.iacr.org/2021/547 (accessed on 24 July 2022).
16. Delaune, S.; Derbez, P.; Gontier, A.; Prud'Homme, C. A Simpler Model for Recovering Superpoly on Trivium. In *Proceedings of the International Conference on Selected Areas in Cryptography;* Springer: Cham, Switzerland, 2022; pp. 266–285.
17. Dobraunig, C.; Eichlseder, M.; Mendel, F.; Schläffer, M. Ascon v1. 2: Lightweight authenticated encryption and hashing. *J. Cryptol.* **2021**, *34*, 1–42. [CrossRef]
18. Chang, D.; Hong, D.; Kang, J. Conditional Cube Attacks on Ascon-128 and Ascon-80pq in a Nonce-Misuse Setting. Cryptology ePrint Archive 2022. Available online: https://eprint.iacr.org/2022/544 (accessed on 24 July 2022).
19. Dinur, I.; Morawiecki, P.; Pieprzyk, J.; Srebrny, M.; Straus, M. Cube attacks and cube-attack-like cryptanalysis on the round-reduced Keccak sponge function. In Proceedings of the Annual International Conference on the Theory and Applications of Cryptographic Techniques, Sofia, Bulgaria, 26–30 April 2015; pp. 733–761.
20. Salam, M.I.; Bartlett, H.; Dawson, E.; Pieprzyk, J.; Simpson, L.; Wong, K.K.H. Investigating cube attacks on the authenticated encryption stream cipher ACORN. In Proceedings of the International Conference on Applications and Techniques in Information Security, Cairns, QLD, Australia, 26–28 October 2016; pp. 15–26.
21. Huang, S.; Wang, X.; Xu, G.; Wang, M.; Zhao, J. Conditional cube attack on reduced-round Keccak sponge function. In Proceedings of the Annual International Conference on the Theory and Applications of Cryptographic Techniques, Paris, France, 30 April–4 May 2017; pp. 259–288.
22. Mandal, K.; Saha, D.; Sarkar, S.; Todo, Y. Sycon: A new milestone in designing ASCON-like permutations. *J. Cryptogr. Eng.* **2022**, 12, 305–327. [CrossRef]
23. Frankel, S.E.; Kent, K.; Lewkowski, R.; Orebaugh, A.D.; Ritchey, R.W.; Sharma, S.R. SP 800-77. Guide to IPsec VPNs. 2005. Available online: https://csrc.nist.rip/library/alt-SP800-77.pdf (accessed on 24 July 2022).
24. Seok, B.; Sicato, J.C.S.; Erzhena, T.; Xuan, C.; Pan, Y.; Park, J.H. Secure D2D communication for 5G IoT network based on lightweight cryptography. *Appl. Sci.* **2019**, *10*, 217. [CrossRef]
25. Teng, W.L.; Salam, I.; Yau, W.C.; Pieprzyk, J.; Phan, R.C.W. Cube attacks on round-reduced TinyJAMBU. *Sci. Rep.* **2022**, *12*, 1–13.

electronics

MDPI

Article

Provably Secure PUF-Based Lightweight Mutual Authentication Scheme for Wireless Body Area Networks

SangCheol Lee [1], SuHwan Kim [1], SungJin Yu [2], NamSu Jho [2] and YoHan Park [1,*]

1 College of Engineering, Department of Computer Engineering, Keimyung University, Daegu 42601, Republic of Korea
2 Electronics and Telecommunications Research Institute, Daejeon 34129, Republic of Korea
* Correspondence: yhpark@kmu.ac.kr; Tel.: +82-53-580-5229

Abstract: Wireless body area networks (WBANs) are used in modern medical service environments for the convenience of patients and medical professionals. Owing to the recent COVID-19 pandemic and an aging society, WBANs are attracting attention. In a WBAN environment, the patient has a sensor node attached to him/her that collects patient status information, such as blood pressure, blood glucose, and pulse; this information is simultaneously transmitted to his/her respective medical professional through a gateway. The medical professional receives and checks the patient's status information and provides a diagnosis. However, sensitive information, including the patient's personal and status data, are transmitted via a public channel, causing security concerns. If an adversary intercepts this information, it could threaten the patient's well-being. Therefore, a secure authentication scheme is essential for WBAN environments. Recently, Chen et al. proposed a two-factor authentication scheme for WBANs. However, we found out Chen et al.'s scheme is vulnerable to a privileged insider, physical cloning, verification leakage, impersonation, and session key disclosure attacks. We also propose a secure physical-unclonable-function (PUF)-based lightweight mutual authentication scheme for WBANs. Through informal security analysis, we demonstrate that the proposed scheme using biometrics and the PUF is safe against various security attacks. In addition, we verify the security features of our scheme through formal security analyses using Burrows–Abadi–Needham (BAN) logic, the real-or-random (RoR) model, and the Automated Validation of Internet Security Protocols and Applications (AVISPA). Furthermore, we evaluate the security features, communication costs, and computational costs of our proposed scheme and compare them with those of other related schemes. Consequently, our scheme is more suitable for WBAN environments than the other related schemes.

Keywords: wireless body area networks; authentication; biometric; physical unclonable function; BAN logic; RoR model; AVISPA

Citation: Lee, S.; Kim, S.; Yu, S.; Jho, N.; Park, Y. Provably Secure PUF-Based Lightweight Mutual Authentication Scheme for Wireless Body Area Networks. *Electronics* **2022**, *11*, 3868. https://doi.org/10.3390/electronics11233868

Academic Editor: Raed A. Abd-Alhameed

Received: 15 October 2022
Accepted: 18 November 2022
Published: 23 November 2022

1. Introduction

Recently, with the increasing number of elderly people in society, the demand for medical services is increasing, owing to the health problems of the aging society [1]. In addition, the emergence and spread of infectious diseases such as COVID-19 has accelerated this demand [2]. Therefore, solving the problem of meeting the supply and demand for healthcare has emerged as a challenge for governments in various countries. Many attempts have been made to use wireless sensor networks (WSNs) to address this problem. Because of sensor miniaturization and improved wireless communication technology, WSNs are widely used in various environments, such as the Industrial Internet of Things [3], smart homes [4], and healthcare [5]. A method was thus proposed that comprises a wireless body area network (WBAN) that incorporates WSNs into the medical field [6]. The WBAN framework includes medical professionals, gateways, and sensor nodes. Through a gateway, a medical professional receives information concerning a patient's condition from sensors attached to the patient or elderly person's body [7]. Medical services that use WBANs are more efficient for both medical professionals and patients. Using them, medical professionals can conveniently treat more patients than before, and patients can receive treatment regardless of location. This approach also limited the spread of infectious diseases by reducing contact between medical professionals and patients during the COVID-19 pandemic. Therefore, research on WBANs has been conducted continuously.

In a WBAN, sensitive information, such as patient status and personal information, is transmitted to medical professionals using insecure channels. Thus, an adversary could steal information from these public channels and attempt security breaches, including replay, impersonation, and man-in-the-middle (MITM) attacks [8]. In addition, a medical professional's mobile device could be stolen, and an adversary could attempt to impersonate the rightful owner using the parameters extracted from the device through power analysis attacks. Furthermore, an adversary could physically capture the sensor node, extract the secret parameters, and impersonate it. If a malicious adversary succeeds in any of the aforementioned attacks and gains sensitive patient information, this may have a significant adverse effect on the patient, such as a misdiagnosis [9]. Therefore, the security of authentication schemes for WBANs is directly related to the well-being of the patient [10].

In 2021, Chen et al. [11] proposed a two-factor authentication scheme for related existing WBAN schemes. They asserted that their scheme, which uses a single hash, is lightweight, heterogeneous, and allows joint operations to prevent various security threats, such as sensor node capture, privileged insider, and stolen verifier attacks. However, we demonstrate that Chen et al.'s scheme cannot resist physical cloning, privileged insiders, verification table leakage, impersonation, and session key disclosure attacks. To overcome the security issues in Chen et al.'s scheme, we designed a secure physical-unclonable-function (PUF)-based three-factor mutual authentication scheme, which we use with a fuzzy extractor [12] to increase security.

1.1. Research Contributions

The contributions of this paper are as follows:

- We review Chen et al.'s scheme to demonstrate that it cannot prevent physical cloning, privileged insider, verification table leakage, impersonation, and session key disclosure attacks.
- We propose a secure PUF-based three-factor mutual authentication scheme to remedy the security vulnerabilities in Chen et al.'s scheme.
- We conducted an informal security analysis to demonstrate that our scheme is secure against various security hazards, including stolen/lost mobile devices, privileged insiders, physical cloning, and stolen verifier attacks.
- We analyzed the security features of the proposed scheme using the well-known Burrows–Abadi–Needham (BAN) logic and real-or-random (RoR) model, which improve the mutual authentication and session key security, respectively. Furthermore, we utilized the Automated Verification of Internet Security Protocols and Applications (AVISPA) simulation tool to prove that the proposed scheme is resistant to replay and man-in-the-middle attacks.
- We evaluated the communication costs, computational costs, and security features of our scheme. Consequently, our scheme provides lower communication and computational costs and higher security levels compared with the existing schemes.

1.2. Organization

In Section 2, we introduce related works for WMSNs. We describe the system model, adversary model, PUF, and fuzzy extractor in Section 3. We provide a review of Chen et al.'s scheme and cryptanalysis of their scheme in Sections 4 and 5. Then, we propose the secure authentication scheme on WBANs in Section 6. The security and performance analyses of our scheme are shown in Sections 7 and 8. Lastly, we present the paper's conclusion in Section 9.

2. Related Works

Various authentication schemes have been proposed for wireless medical sensor networks (WMSNs). Kumar et al. [13] (2012) presented an authentication scheme for healthcare applications using WMSNs. This scheme provides a secure session key establishment between users and medical sensor nodes and allows the users to change their passwords. However, in 2013, He et al. [14] demonstrated that Kumar et al.'s scheme could not withstand attacks such as offline password guessing and privileged insider attacks. In addition, they proved that Kumar et al.'s scheme did not guarantee anonymity. Accordingly, He et al. proposed a more secure scheme and asserted that their scheme is robust against various attacks. Unfortunately, in 2015, Wu et al. [15] demonstrated that He et al.'s scheme was vulnerable to offline password guessing, user impersonation, and sensor node capture attacks. Accordingly, they proposed an authentication scheme using a smart card to store sensitive information from medical professionals, which provides a higher level of security in the WMSN environment. In 2017, Li et al. [16] proposed an anonymous mutual authentication and key agreement scheme for WMSNs using hash operations and XOR operations, which was more efficient than previous related schemes. Unfortunately, in 2020, Gupta et al. [17] demonstrated that Li et al.'s scheme could not prevent intermediate node capture, sensor node impersonation, and hub node impersonation attacks. They also proved that Li et al.'s scheme was vulnerable to linkable sessions and traceability. Therefore, they proposed an authentication scheme in the WBAN environments that overcomes the security vulnerabilities of Li et al.'s scheme. In 2019, Ostad–Sharif et al. [18] proposed an authentication key agreement scheme consisting of three tiers for WBANs. Their scheme ensured anonymity to protect users' sensitive information. However, in 2020, Alzahrani et al. [19] claimed that Ostad et al.'s scheme is vulnerable to brute-force guessing attacks, and it is possible to compute all previous session keys. Subsequently, they presented an anonymous authenticated key exchange scheme with better security and efficiency to demonstrate the known weaknesses of Ostad et al.'s scheme.

Recently, PUF-based authentication schemes have been proposed for various environments to prevent attacks. In 2018, Mahalat et al. [20] proposed a PUF-based scheme that secures WiFi authentication for Internet of Things (IoT) devices and protects them against invasive, semi-invasive, or tampering attacks. In 2019, Zhu et al. [21] proposed a lightweight RFID mutual authentication scheme using a PUF. Their scheme provides secure authentication between the server and a tag. They asserted that their scheme could prevent clone attacks because a PUF cannot be duplicated. In 2021, Mahmood et al. [22] suggested a mutual authentication and key exchange scheme for multiserver-based device-to-device (D2D) communication. The entire process of Mahmood et al.'s scheme uses only XOR operations and hash functions, and PUF is introduced to protect against physical capture attacks. In the same year, Chuang et al. [23] proposed a PUF-based authenticated key exchange scheme for IoT environments. Their scheme did not require verifiers or explicit challenge–response pairs (CRPs). Therefore, IoT nodes can freely authenticate each other and generate a session key without the assistance of any verifier or server. Kwon et al. [24] proposed a three-factor-based mutual authentication and key agreement scheme with a PUF for WMSNs. They proved that their scheme could protect against physical cloning attacks using a PUF.

In 2020, Fotouhi et al. [25] proposed a two-factor authentication scheme for WBANs and asserted that it was safe against sensor node capture attacks. Unfortunately, in 2021, Chen et al. [11] demonstrated that the aforementioned scheme is vulnerable to sensor node attacks and proposed an improved security-enhanced two-factor authentication scheme for WBANs. However, we discovered that their scheme is insecure against privileged insider attacks, physical cloning attacks, verification table leakage attacks, etc. Therefore, we propose a secure PUF-based lightweight mutual authentication scheme for WBANs that resolves these security issues.

3. Preliminaries

This section introduces the general system model, the threat model, and relevant mathematical preliminaries including the PUF and fuzzy extractor, which can improve our scheme's security.

3.1. System Model

Figure 1 shows the general system model of a WBAN, which consists of medical professionals such as doctors and nurses, sensor nodes, and a gateway. The details are as follows:

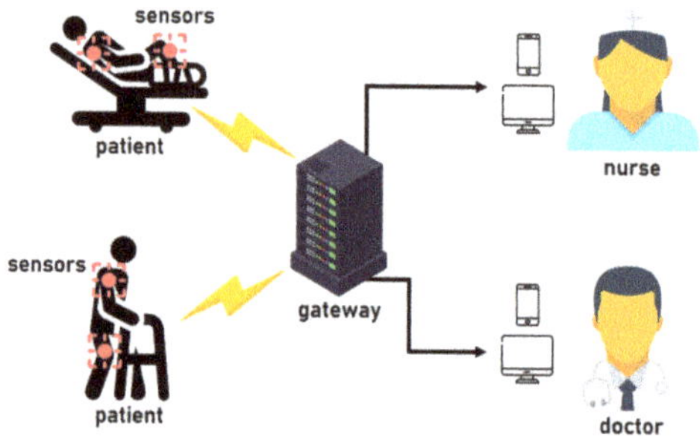

Figure 1. The general system model of WBANs.

- User (U_i): A user who wants to use the WBAN services receives a smart card from the gateway. After registration, the user can receive information from the sensor node attached to the patient's body.
- Gateway (GW_j): The gateway acts as a relay that connects patients with medical professionals. The gateway stores the value required for authentication.
- Sensor node (SN_k): The sensor node must be authenticated by the gateway. The authenticated sensor node is attached to the patient's body and transmits information to the medical professionals.

3.2. Adversary Model

To analyze the security of the proposed scheme, we applied the widely used Dolev–Yao (DY) adversary model. Under the DY model, a malicious adversary can inject, eavesdrop, modify, or delete messages transmitted using public channels. We also adopted the Canetti and Krawczyk (CK) adversary model to analyze the proposed scheme. The CK model is relatively strong compared with the DY model and is widely used to analyze scheme security. In the CK model, the adversary can intercept a random value and generate the master key of a gateway:

- An adversary can steal a medical professional's smart device and use a power analysis attack to extract sensitive information inside the cell phone.
- An adversary can obtain a patient's sensor node and extract important information within the sensor node through a physical cloning attack.
- An adversary can be a privileged insider, so it can also obtain a registration message from medical professionals
- An adversary can perform various attacks, such as password guessing, stolen verifier, and man-in-the-middle attacks.

3.3. Physical Unclonable Function

PUFs are physical circuits that operate using only a one-way function. The PUF circuit uses an input–output bit-string pair termed the "challenge–response pair". Even if numerous challenges are encountered in a PUF circuit, each has a unique output response.

In this paper. We express this process as $R = PUF(C)$, where R and C are a response and a challenge. The PUF's properties are as follows:

- The PUF is an unclonable circuit.
- The circuit of the PUF is easy to implement.
- The output of the PUF is unpredictable.
- The output of the PUF depends only on a physical circuit.

If the same challenge is entered into the PUF circuit of the same device, the same output response is printed. However, if a challenge is introduced into the PUF from different devices, different output responses are printed. Thus, the PUF provides a unique one-way function that cannot be replicated. The ability of the PUF to resist replication makes it impossible for adversaries to succeed with various attacks, such as physical cloning attacks.

3.4. Fuzzy Extractor

In this section, the purpose and basic concepts of the fuzzy extractor are discussed. However, biometric information is vulnerable to noise. Therefore, it is difficult to obtain a constant response value. Consequently, before users can utilize their biometrics, the biometric noise must be eliminated, for which we used a fuzzy extractor. The details are given below:

- $Gen(Bio_i) = <\sigma_i, \tau_i>$: This algorithm is intended to generate keys using biometric information. It receives biometric information as a parameter and returns the secret key data R_i and a public reproduction P_i as a helper value.
- $Rep(Bio_i^*, \tau_i) = \sigma_i$: This algorithm is for reproducing secret data R_i. The input of this algorithm is biometric information Bio_i^* and P_i. The algorithm returns the secret key R_i as a result.

4. Review of Chen et al.'s Scheme

In 2021, Chen et al. [11] proposed a two-factor authentication scheme for WBANs. Their scheme provides sensor node registration, user registration and mutual authentication, and a key exchange phase. The notations used in the Chen et al.s scheme are also presented in Table 1.

Table 1. Notations and definitions of Chen et al.'s scheme.

Notation	Definition
U_i	i-th user
ID_i, PW_i	identity of U_i, password of U_i
GW_j	j-th gateway
GID_j, G_j	identity of GW_j, secret key of GW_j
SN_k, SID_k	k-th sensor, its identity
CID_i, QID_k	Temporary pseudoidentity of U_i and SN_k
N_l	Network identifier of sensor set
M_i	i-th message
SG_k	Shared key between sensor and gateway
SK_u	Session key generated by user
SK_g	Session key generated by gateway
SK_s	Session key generated by sensor node
$R_s, R_0, R_u, R_g, R_x, R_y, R_z$	Temporary random number
$Gen(.)$	Fuzzy biometric generator
$Rep(.)$	Fuzzy biometric reproduction
BIO_i	Biometric template of the user
$h(.)$	Hash function
$\|\|$	Concatenation operator
$\oplus$	Exclusive-OR operator

4.1. User Registration Phase

A medical professional such as a doctor or nurse must register in the gateway to use this network system. We describe the sensor node registration phase below:

Step 1: The user enters her/his own ID_i, PW_i and imprints Bio_i into the mobile device. Then, U_i calculates $Gen(Bio_i) = \; < \sigma_i, \tau_i >$, $HPW_i = h(PW_i||\sigma_i)$ and sends ID_i, HPW_i as a registration request to the gateway through a secure channel.

Step 2: Upon receiving ID_i, PW_i determines whether the identity is new. If it is new, GW_j calculates $CID_i = h(ID_i)$ and stores CID_i, HPW_i. Then, GW_j selects a secret random number R_0. After that, GW_j computes $A_1 = h(CID_i||GID_j||R_0 \oplus G_j) \oplus HPW_i$ and $A_2 = h(GID_j||HPW_i) \oplus (R_0 \oplus G_j)$ and stores A_1 in memory. Finally, GW_j sends $\{A_2, GID_i\}$ to U_i via a secure channel.

Step 3: U_i computes $A_3 = h(ID_i||HPW_i)$. Then, U_i stores $\{A_2, A_3, GID_j, Gen(.), Rep(.), \tau_i\}$.

4.2. Sensor Node Registration Phase

The sensor node must be registered with the gateway to transmit the health information of the patient. We show the sensor node registration phase of Chen et al.'s scheme as follows:

Step 1: SN_k sends SID_k and N_l over a secure channel.

Step 2: GW_j determines whether SID_k is a new identity and generates a new pseudoidentity QID_k. GW_j computes $SG_k = h(SID_k||G_j \oplus N_l)$ and stores $\{QID_k, N_l\}$ in the memory. Then, GW_j sends $\{SG_k, QID_k\}$ to SN_k via a secure channel.

Step 3: SN_k computes $RSG_k = SG_k \oplus SID_k$ and saves $\{RSG_k, QID_k\}$ in the memory.

4.3. Login Phase

A medical professional must log in to the mobile device to use this network system. The detailed steps are illustrated in Figure 2:

User U_i		**Mobile Device**
U_i enters ID_i^* and PW_i^* Imprints BIO_i^*		
	input to mobile device $\longrightarrow$	
		$Rep(BIO_i^*, \tau_i) = \sigma_i^*$ $HPW_i^* = h(PW_i^*||\sigma_i^*)$ $A_3^* = h(ID_i^*||HPW_i)$ Verifies $A_3 \equiv A_3^*$ If true, user authentication passed

Figure 2. Login phase of Chen et al.'s scheme.

Step 1: U_i enters his/her own ID_i^*, PW_i^* and imprints $Bio_i^{'}$ into the mobile device.

Step 2: The mobile device computes $Rep(BIO_i^*, \tau_i) = \sigma_i^*$, $HPW_i^* = h(PW_i^*||\sigma_i^*)$, and $A_3^* = h(ID_i^*||HPW_i)$. Then, the mobile device verifies A_3 by comparison. If $A_3 = A_3^*$, the mobile device allows U_i to log in.

4.4. Authentication and Key Agreement Phase

In this phase, the medical professionals and the sensor node conduct a mutual authentication and key agreement phase to authenticate each other and establish a session key. Figure 3 shows the authentication and key agreement phase of Chen et al.'s scheme, and the details are as follows:

User U_i	Gateway GW_j	Sensor Node SN_k
Selects SID_k, R_u, T_1 Computes $(R_0 \oplus G_j) = A_2 \oplus h(GID_j\|\|HPW_i)$ $B_1 = SID_k \oplus h(GID_j\|\|HPW_i)$ $B_2 = R_u \oplus h(GID_j\|\|HPW_i \oplus SID_k)$ $B_3 = (R_0 \oplus G_j) \oplus h(GID_j\|\|R_u)$		
$M_1 = \{CID_i, GID_j, B_1, B_2, B_3, T_1\}$ →		
	Verifies $\|T_1 - T_c\| \leq \Delta T$ Gets HPW_i, QID_k Computes $SID_k = B_1 \oplus h(GID_j\|\|HPW_i)$ $R_u = B_2 \oplus h(GID_j\|\|HPW_i \oplus SID_k)$ $(R_0 \oplus G_j) = B_3 \oplus h(GID_j\|\|R_u)$ $A_1^* = h(CID_i\|\|GID_j\|\|R_0 \oplus G_j) \oplus HPW_i$ Checks $A_1 \equiv A_1^*$ Selects R_g, T_2 $SG_k = h(SID_k\|\|G_j \oplus N_l)$ $B_4 = R_u \oplus HPW_i \oplus SG_k$ $B_5 = R_g \oplus h(SG_k\|\|SID_k)$ $B_6 = h(QID_k\|\|B_4\|\|B_5\|\|SG_k\|\|R_u \oplus HPW_i\|\|R_g)$	
	$M_2 = \{QID_k, B_4, B_5, B_6, T_2\}$ →	
		Verifies $\|T_2 - T_c\| \leq \Delta T$ Gets RSG_k based on QID_k $SG_k = RSG_k \oplus SID_k$ $(R_u \oplus HPW_i) = B_4 \oplus SG_k$ $R_g = B_5 \oplus h(SG_k\|\|SID_k)$ $B_6^* = h(QID_k\|\|B_4\|\|B_5\|\|SG_k\|\|R_u \oplus HPW_i\|\|R_g)$ Verifies $B_6^* \equiv B_6$ Selects R_s, T_3 Computes $SK_s = h(R_u \oplus HPW_i\|\|R_g\|\|R_s)$ $B_7 = h(SG_k\|\|R_g) \oplus R_s$ $B_8 = h(R_g\|\|R_s\|\|SG_k\|\|T_3)$
		← $M_3 = \{B_7, B_8, T_3\}$
	Verifies $\|T_3 - T_c\| \leq \Delta T$ Computes $R_s = h(SG_k\|\|R_g) \oplus B_7$ $B_8^* = h(R_g\|\|R_s\|\|SG_k\|\|T_3)$ Checks $B_8^* \equiv B_8$ Selects T_4 $SK_g = h(R_u \oplus HPW_i\|\|R_g\|\|R_s)$ $B_9 = h(R_u \oplus GID_j\|\|HPW_i) \oplus (R_g\|\|R_s)$ $B_{10} = h(R_0 \oplus G_j\|\|SK_g\|\|R_u)$	
	← $M_4 = \{B_9, B_{10}, T_4\}$	
$\|T_4 - T_c\| \leq \Delta T$ Computes $(R_g\|\|R_s) = B_9 \oplus h(R_u \oplus GID_j\|\|HPW_i)$ $SK_u = h(R_u \oplus HPW_i\|\|R_g\|\|R_s)$ $B_{10}^* = h(R_0 \oplus G_j\|\|SK_u\|\|R_u)$ Checks $B_{10}^* \equiv B_{10}$ If true, communication is possible		

Figure 3. Authentication and key agreement phase of Chen et al.'s scheme.

Step 1: U_i selects the SID_k of the sensor to be accessed, generates a random number R_u, and creates a timestamp T_1. Then, U_i calculates $(R_0 \oplus G_j) = A_2 \oplus h(GID_j||HPW_i)$, $B_1 = SID_k \oplus h(GID_j||HPW_i)$, $B_2 = R_u \oplus h(GID_j||HPW_i \oplus SID_k)$, and $B_3 = (R_0 \oplus G_j) \oplus h(GID_j||R_u)$. Finally, U_i sends message $M_1\{CID_i, GID_j, B_1, B_2, B_3, T_1\}$ to GW_j via a public channel.

Step 2: GW_j receives the message M_1 and verifies the legitimacy of T_1 by determining whether it matches $|T_1 - T_c| \leq \Delta T$. GW_j retrieves the memory and obtains the HPW_i, QID_k that matches CID_i in M_1. $(SID_m||\alpha_m) = Dec_{MSK}(MID_m)$. Then, GW_j computes $SID_k = B_1 \oplus h(GID_j||HPW_i)$, $R_u = B_2 \oplus h(GID_j||HPW_i \oplus SID_k)$, $(R_0 \oplus G_j) = B_3 \oplus h(GID_j||R_u)$, and $A_1^* = h(CID_i||GID_j||R_0 \oplus G_j) \oplus HPW_i$. GW_j verifies $A_1 \equiv A_1^*$. If the verification is false, GW_j stops the conversation. Otherwise, GW_j confirms the justification of the identity of U_i, and it generates a random number R_g and a new timestamp T_2. Then, GW_j computes $SG_k = h(SID_k||G_j \oplus N_l)$, $B_4 = R_u \oplus HPW_i \oplus SG_k$, $B_5 = R_g \oplus h(SG_k||SID_k)$, and $B_6 = h(QID_k||B_4||B_5||SG_k||R_u \oplus HPW_i||R_g)$. Finally, GW_j sends $M_2\{QID_k, B_4, B_5, B_6, T_2\}$ to SN_k via a public channel.

Step 3: SN_k receives the message M_2 and verifies that $|T_2 - T_c| \leq \Delta T$. The message is fresh if the verification is true. Then, SN_k obtains the corresponding RSG_k in storage based on QID_k. SN_k computes $SG_k = RSG_k \oplus SID_k$, $(R_u \oplus HPW_i) = B_4 \oplus SG_k$, and $B_6^* = h(QID_k||B_4||B_5||SG_k||R_u \oplus HPW_i||R_g)$. Afterward, GW_j verifies whether $B_6^* \equiv B_6$. If it is true, SN_k generates a random number R_s and a timestamp T_3. SN_k calculates the keys $SK_s = h(R_u \oplus HPW_i||R_g||R_s)$, $B_7 = h(SG_k||R_g \oplus R_s)$, and $B_8 = h(R_g||R_s||SG_k||T_3)$. Then, SN_k sends message $M_3\{B_7, B_8, T_3\}$ to GW_j via a public channel.

Step 4: GW_j receives the message M_3 and verifies the freshness of timestamp T_3 using $|T_3 - T_c| \leq \Delta T$. If the verification passes, GW_j generates timestamp T_4 and calculates $R_s = h(SG_k||R_g) \oplus B_7$ and $B_8^* = h(R_g||R_s||SG_k||T_3)$, then verifies whether $B_8^* \equiv B_8$. If the verification is correct, GW_j generates T_4 and calculates $SK_s = h(R_u \oplus HPW_i||R_g||R_s)$, $B_9 = h(R_u \oplus GID_j||HPW_i) \oplus (R_g||R_s)$, and $B_{10} = h(R_0 \oplus G_j||SK_g||R_u)$. After that, GW_j sends message $M_4\{B_9, B_{10}, T_4\}$ to U_i via a public channel.

Step 5: U_i receives the message M_4 and verifies that $|T_2 - T_c| \leq \Delta T$. If the verification is true, the message is fresh. Then, U_i computes $(R_g||R_s) = B_9 \oplus h(R_u \oplus GID_j||HPW_i)$, $SK_u = h(R_u \oplus HPW_i||R_g||R_s)$, and $B_{10}^* = h(R_0 \oplus G_j||SK_u||R_u)$. Finally, U_i verifies whether $B_{10}^* \equiv B_{10}$, and if this is true, the verification and key exchange are a success.

5. Cryptanalysis of Chen et al.'s Scheme

In this section, we analyze the security defects of Chen et al.'s scheme. Our analysis shows that their scheme is vulnerable to privileged insider attacks, physical cloning attacks, and verification table leakage attacks. In addition, malicious adversary $\mathcal{A}$ can impersonate the user, sensor node, and gateway and disclose a session key.

5.1. Privileged Insider Attack

A privileged insider can support $\mathcal{A}$ by giving various important information such as registration message and values stored on the mobile device of the user. We describe the procedures are as follows:

Step 1: $\mathcal{A}$ can obtain a registration request message $\{ID_i, HPW_i\}$ and the secret parameter $\{A_2, A_3, GID_j, Gen(.), Rep(.), \tau_i\}$ extracted from the smart device of the user.

Step 2: The adversary $\mathcal{A}$ intercepts $M_1\{CID_i, GID_j, B_1, B_2, B_3, T_1\}$, and $M_3\{B_7, B_8, T_3\}$ transmitted by the public channel.

Step 3: $\mathcal{A}$ calculates $(R_0 \oplus G_j)^* = A_2 \oplus h(GID_j||HPW_i)$, $SID_k^* = B_1 \oplus h(GID_j||HPW_i)$, $R_u^* = B_2 \oplus h(GID_j||HPW_i \oplus SID_k)$, and $(R_g||R_s)^* = B_9 \oplus h(R_u \oplus GID_j||HPW_i)$. Then, $\mathcal{A}$ can extract the parameters $(R_0 \oplus G_j)^*$, SID_k^*, R_u^*, and $(R_g||R_s)^*$.

Step 4: $\mathcal{A}$ calculates $B_1^* = SID_k \oplus h(GID_j||HPW_i)$, $B_2^* = R_u \oplus h(GID_j||HPW_i \oplus SID_k)$, $B_3^* = (R_0 \oplus G_j) \oplus h(GID_j||R_u)$, and $SK_u = h(R_u \oplus HPW_i||R_g||R_s)$. Thereafter, $\mathcal{A}$ can generate $M_1\{CID_i, GID_j, B_1^*, B_2^*, B_3^*, T_1^*\}$ and send it to GW_j by impersonating legitimate user U_i. In addition, $\mathcal{A}$ can calculate $SK_u^* = h(R_u \oplus HPW_i||(R_g||R_s)^*)$ to generate session key SK_u^*. Thus, $\mathcal{A}$ can disclose or exploit the session key.

Thus, Chen et al.'s scheme is insecure against privileged insider attacks.

5.2. Physical Cloning Attack

In this attack, we assume that $\mathcal{A}$ can clone sensor node SN_k physically and extract the sensitive value $\{RSG_k, QID_k\}$ stored in the memory of SN_k. In order to be able to forward message $\{B_7, B_8, T_3\}$ on behalf of the legitimate GW_j and generate session key SK_s, then $\mathcal{A}$ has to calculate the value of $B_7 = h(SG_k||R_g \oplus R_s)$, $B_8 = h(R_g||R_s||SG_k||T_3)$, and $SK_s = h(R_u \oplus HPW_i||R_g||R_s)$ through the following steps:

Step 1: The adversary $\mathcal{A}$ can obtain the messages $M_2\{QID_k, B_4, B_5, B_6, T_2\}$ and $M_3\{B_7, B_8, T_3\}$ by the eavesdropping attack.

Step 2: $\mathcal{A}$ computes SG_k^* through $SG_k^* = RSG_k \oplus SID_k$.

Step 3: $\mathcal{A}$ calculates $(R_u \oplus HPW_i)^* = B_4 \oplus SG_k$, $R_g^* = B_5 \oplus h(SG_k||SID_k)$, and $R_s^* = h(SG_k||R_g) \oplus B_7$. Afterward, $\mathcal{A}$ obtains the parameters $(R_u \oplus HPW_i)^*$, R_g^*, and R_s^*.

Step 4: $\mathcal{A}$ can successfully compute $B_7^* = h(SG_k^*||R_g^*) \oplus R_s^*$, $B_8^* = h(R_g^*||R_s^*||SG_k^*||T_3^*)$, and $SK_s^* = h((R_u \oplus HPW_i)^*||R_g^*||R_s^*)$. Finally, $\mathcal{A}$ can generate authentication message $M_3^*\{B_7^*, B_8^*, T_3^*\}$ and session key SK_s.

Therefore, the scheme of Chen et al. cannot resist thephysical cloning attack.

5.3. Verification Table Leakage Attack

If $\mathcal{A}$ extracts the verification table $\{QID_k, N_l, CID_i, HPW_i, A_1\}$ of GW_j, $\mathcal{A}$ attempts to impersonate GW_j and generate a session key. The details are described below:

Step 1: The malicious adversary $\mathcal{A}$ can obtain the messages $M_1\{CID_i, GID_j, B_1, B_2, B_3, T_1\}$, $M_2\{QID_k, B_4, B_5, B_6, T_2\}$, and $M_3\{B_7, B_8, T_3\}$ transmitted by the public channel.

Step 2: $\mathcal{A}$ computes $SID_k^* = B_1 \oplus h(GID_j||HPW_i)$, $R_u^* = B_2 \oplus h(GID_j||HPW_i \oplus SID_k^*)$, $(R_0 \oplus G_j)^* = B_3 \oplus h(GID_j||R_u^*)$, $SG_k^* = R_u^* \oplus HPW_i \oplus B_4$, $R_g^* = B_5 \oplus h(SG_k^*||SID_k^*)$, and $R_s^* = h(SG_k^*||R_g^*) \oplus B_7$ to generate parameters $SID_k^*, R_u^*, (R_0 \oplus G_j)^*, SG_k^*, R_g^*, R_s^*$.

Step 3: $\mathcal{A}$ calculates $B_4 = R_u \oplus HPW_i \oplus SG_k$, $B_5 = R_g \oplus h(SG_k||SID_k)$, $B_6 = h(QID_k||B_4||B_5||SG_k||R_u \oplus HPW_i||R_g)$, $SK_g^* = h(R_u^* \oplus HPW_i||R_g^*||R_s^*)$, $B_9^* = h(R_u^* \oplus GID_j||HPW_i) \oplus (R_g^*||R_s^*)$, and $B_{10}^* = h((R_0 \oplus G_j)^*||SK_g^*||R_u^*)$.

Step 4: Eventually, $\mathcal{A}$ can generate authentication messages $M_2^*\{QID_k, B_4^*, B_5^*, B_6^*, T_2^*\}$ and $M_4^*\{B_9^*, B_{10}^*, T_4^*\}$ and send them to the user and gateway disguised as a legal GW_j. Furthermore, $\mathcal{A}$ can generate session key SK_g^* of GW_j and adversely affect the system by exposing SK_g^*.

Therefore, Chen et al.'s scheme cannot withstand verification table leakage attacks.

5.4. Impersonation Attack

(1) User impersonation attack: In the previous privileged insider attack in Section 5.1, $\mathcal{A}$ can generate authentication message $M_1\{CID_i, GID_j, B_1^*, B_2^*, B_3^*, T_1^*\}$ and send it to the gateway to impersonate a legitimate user. Therefore, the scheme of Chen et al. is vulnerable to the user impersonation attack.

(2) Gateway impersonation attack: In the previous verification table attack in Section 5.3, $\mathcal{A}$ can calculate authentication messages $M_2^*\{QID_k, B_4^*, B_5^*, B_6^*, T_2^*\}$ and $M_4^*\{B_9^*, B_{10}^*, T_4^*\}$ and send them to the sensor node and user. However, the sensor node and gateway cannot recognize that the message transmitted from a gateway was not legal. Therefore, the scheme of Chen et al. cannot resist the gateway impersonation attack.

(3) Sensor node impersonation attack: In the previous physical cloning attack in Section 5.2, a malicious adversary $\mathcal{A}$ can compute message $M_3^*\{B_7^*, B_8^*, T_3^*\}$ to be sent to the gateway. However, the gateway recognizes that the message was transmitted from a legitimate sensor node. Therefore, Chen et al.'s scheme cannot withstand sensor node impersonation attacks.

5.5. Session Key Disclosure Attack

In the previous attacks, privileged insider in Section 5.1, physical cloning in Section 5.2, and verification table leakage in Section 5.3, $\mathcal{A}$ can generate session keys SK_u, SK_k, and SK_g. $\mathcal{A}$ attempts to exploit the generated session key to adversely affect the system and disclose it to the outside. Thus, the scheme of Chen et al. cannot prevent session key disclosure attacks.

6. Proposed Scheme

In this section, we propose a secure three-factor mutual authentication scheme for WBANs to overcome the security weaknesses of Chen et al.'s scheme. Our scheme also considers the efficiency of the authentication process. Our scheme consists of user registration, sensor node registration, mutual authentication and key agreement, and password change phases. The notations and definitions used in the proposed scheme are explained in Table 2.

Table 2. Notations and definitionsof the proposed scheme.

Notation	Definition
U_i	i-th user
ID_i, PW_i	identity of U_i, password of U_i
GW_j	j-th gateway
GID_j, G_j	identity of GW_j, secret key of GW_j
SN_k, SID_k	k-th sensor, its identity
CID_i	Temporary pseudoidentity of U_i
M_i	i-th message
SG_k	Shared key between sensor and gateway
SK_u	Session key generated by user
SK_g	Session key generated by gateway
SK_s	Session key generated by sensor node
$R_u, R_g, R_s, R_0, R_1, R_2$	Temporary random number
$Gen(.)$	Fuzzy biometric generator
$Rep(.)$	Fuzzy biometric reproduction
BIO_i	Biometric template of the user
$h(.)$	Hash function
$\|\|$	Concatenation operator
$\oplus$	Exclusive-OR operator

6.1. User Registration Phase

In order for a medical professional to receive patient information from the sensor node, he/she must be registered with the gateway in advance. The details are shown in Figure 4:

User U_i | **Gateway GW_j**

Enters ID_i, PW_i, BIO_i
$< \sigma_i, \tau_i >= Gen(BIO_i)$
Calculates
$HID_i = h(ID_i||\sigma_i)$
$HPW_i = h(PW_i||\sigma_i)$

$\{HID_i\}$ $\longrightarrow$

Whether the identity is new
Generates random number R_0, R_1
Calculates
$CID_i = h(HID_i||R_0)$
$ER_j = R_1 \oplus G_j$
Stores CID_i, HID_i, ER_j
Computes
$A_0 = R_0 \oplus G_j$
$A_1 = h(HID_i||A_0) \oplus G_j$
Stores A_1 into memory

$\longleftarrow$ $\{A_0, R_1, CID_i\}$

Computes
$A_2 = A_0 \oplus R_1 \oplus \sigma_i$
$A_3 = h(ID_i||HPW_i)$
$ER_i = h(ID_i||PW_i) \oplus R_1$
Stores
$\{A_2, A_3, Gen(.), Rep(.), \tau_i, ER_i, CID_i\}$

Figure 4. User Registration of the proposed scheme.

Step 1: U_i inputs an identity ID_i, a password PW_i, and biometric template BIO_i into the mobile device. Then, the mobile device computes $Gen(BIO_i) =< \sigma_i, \tau_i >$, $HID_i = h(ID_i||\sigma_i)$, and $HPW_i = h(PW_i||\sigma_i)$. U_i sends HID_i to the gateway through a secure channel.

Step 2: GW_j receives HID_i from U_i and checks whether HID_i is new. If it is new, GW_j generates random numbers R_0 and R_1. Then, GW_j calculates $CID_i = h(HID_i||R_0)$ and $ER_j = R_1 \oplus G_j$ and stores CID_i, HID_i, ER_j. Afterward, GW_j computes $A_0 = R_0 \oplus G_j$ and $A_1 = h(HID_i||A_0) \oplus G_j$ and stores A_1 into memory. Finally, GW_j sends message $\{A_0, R_1, CID_i\}$ to U_i via a secure channel.

Step 3: U_i receives message A_0, R_1, CID_i from GW_j and computes $A_2 = A_0 \oplus R_1 \oplus \sigma_i$, $A_3 = h(ID_i||HPW_i)$, and $ER_i = h(ID_i||PW_i) \oplus R_1$. Then, GW_j stores $\{A_2, A_3, Gen(.), Rep(.), \tau_i, ER_i, CID_i\}$ in the mobile device.

6.2. Sensor Node Registration Phase

A sensor node must register with the gateway in order to transmit patient information to the medical professional. The sensor node registration phase is shown in Figure 5, and the detailed steps are as follows:

Sensor Node SN_k		**Gateway GW_j**
Generates a challenge CH_1	$\{SID_k, CH_1\}$ $\longrightarrow$	
		Whether SID_k is new Computes $SG_k = h(SID_k\|\|G_j)$ Stores SID_k, CH_1 into memory
	$\{SG_k\}$ $\longleftarrow$	
Computes $RE_1 = PUF(CH_1)$ $RSG_k = SG_k \oplus SID_k \oplus RE_1$ Stores $\{RSG_k, CH_1\}$		

Figure 5. Sensor node registration of the proposed scheme.

Step 1: SN_k generates a challenge CH_1 and sends identity SID_k and CH_1 to GW_j over a secure channel.

Step 2: GW_j receives SID_k and CH_1 from SN_k and determines whether SID_k is a new identity. If it is new, GW_j computes $SG_k = h(SID_k||G_j)$ and stores SID_k and CH_1 into memory. Then, GW_j sends SG_k to SN_k through a secure channel.

Step 3: SN_k receives SG_k from GW_j. Then, SN_k computes $RE_1 = PUF(CH_1)$ and $RSG_k = SG_k \oplus SID_k \oplus RE_1$ and saves $\{RSG_k, CH_1\}$ in the memory.

6.3. Login Phase

A medical professional must log in to the mobile device to utilize this WBAN system. The details are shown in Figure 6:

User U_i		**Mobile Device**
U_i enters ID_i^* and PW_i^* imprints BIO_i^*	*input to mobile device* $\longrightarrow$	
		$Rep(BIO_i^*, \tau_i) = \sigma_i^*$ $HPW_i^* = h(PW_i^*\|\|\sigma_i^*)$ $A_3^* = h(ID_i^*\|\|HPW_i^*)$ Verify $A_3 \equiv A_3^*$ If true, user authentication passed

Figure 6. Login phase of the proposed scheme.

Step 1: U_i enters ID_i^* and PW_i^* and imprints BIO_i^* into the mobile device.

Step 2: The mobile device calculates $Rep(BIO_i^*, \tau_i) = \sigma_i^*$, $HPW_i^* = h(PW_i^*||\sigma_i^*)$, and $A_3^* = h(ID_i^*||HPW_i^*)$. Then, the mobile device verifies A_3 by comparison. If $A_3 = A_3^*$, U_i logs in successfully.

6.4. Mutual Authentication and Key Agreement Phase

The medical professional sends an authentication message to the gateway and generates a session key among the medical professional, the sensor node, and the gateway. After that, the medical professionals can receive the patient's information from the sensor node. In Figure 7, we show the mutual authentication and key agreement phase of our scheme, and the details are given below:

Step 1: U_i selects SID_k, R_u, T_1 and computes $R_1 = ER_i \oplus h(ID_i||PW_i)$ and $A_0 = A_2 \oplus R_1 \oplus \sigma_i$. Then, U_i generates random nonce R_u and calculates $B_1 = R_u \oplus R_1$, $B_2 = A_0 \oplus R_u \oplus R_1 \oplus HID_i$. Finally, U_i sends $M_1\{SID_k, CID_i, B_1, B_2, T_1\}$ to GW_j through a public channel.

Step 2: GW_j receives message M_1 from U_i and verifies that $|T_1 - T_c| \leq \Delta T$. If the verification passes, GW_j checks whether $CID_i = CID_i^{old}$ or $CID_i = CID_i^{new}$. If $(CID_i == CID_i^{old})$, then it retrieves $\{HID_i^*, ER_j\}$ against CID_i^{old}, and if $(CID_i == CID_i^{new})$, it retrieves $\{HID_i^*, ER_j\}$ against CID_i^{new}. After that, GW_j computes $R_1 = ER_j \oplus G_j$, $R_u = B_1 \oplus R_1$, $A_0 = B_2 \oplus R_u \oplus R_1 \oplus HID_i$, and $A_1^* = h(HID_i||A_0) \oplus G_j$. If $A_1 \stackrel{?}{=} A_1^*$ is true, GW_j computes $CID_i^{new} = h(HID_i||R_u)$ and updates CID_i^{new}. Then, GW_j selects R_g, T_2 and calculates $SG_k = h(SID_k||G_j)$, $C_1 = Ru \oplus HID_i$, $B_3 = C_1 \oplus SG_k \oplus CH_1$, $B_4 = R_g \oplus h(SG_k||SID_k)$, and $B_5 = h(B_4||B_5||SG_k||C_1||R_g)$. Finally, GW_j sends $M_2\{B_3, B_4, B_5, T_2\}$ to SN_k via a public channel.

Step 3: SN_k receives the message $M_2\{B_3, B_4, B_5, T_2\}$ and verifies the freshness of timestamp T_2 using $|T_2 - T_c| \leq \Delta T$. If the verification is true, the message is fresh. Then, SN_k obtains the corresponding RSG_k, CH_1 and computes $RE_1 = PUF(CH_1)$, $SG_k = RSG_k \oplus SID_k \oplus RE_1$, $C_1 = B_3 \oplus SG_k \oplus CH_1$, $R_g = B_4 \oplus h(SG_k||SID_k)$, and $B_5^* = h(B_3||B_4||SG_k||C_1||R_g)$. SN_k verifies whether $B_5^* \stackrel{?}{=} B_5$. If verification is correct, SN_k selects R_s, T_3 and computes $SK_s = h(C_1||R_g||R_s)$, $B_6 = h(SG_k||R_g) \oplus R_s$, and $B_7 = h(R_g||R_s||SG_k||T_3||C_1)$. SN_k sends $M_3 = \{B_6, B_7, T_3\}$ to GW_j through a public channel.

Step 4: GW_j receives the message M_3 and verifies that $|T_3 - T_c| \leq \Delta T$. The message is fresh if the verification is true. Then, GW_j computes $R_s = h(SG_k||R_g) \oplus B_6$ and $B_7^* = h(R_g||R_s||SG_k||T_3||C_1)$. Afterward, GW_j verifies whether $B_7^* \stackrel{?}{=} B_7$. If it is true, GW_j selects T_4 and computes $SK_g = h(C_1||R_g||R_s)$, $B_8 = R_u \oplus (R_g||R_s)$, and $B_9 = h(A_0||SK_g||R_u)$. GW_j sends $M_4 = \{B_8, B_9, T_4\}$ to U_i via a public channel

Step 5: U_i receives the message M_4 and verifies the legitimacy of T_4 by determining whether it matches $|T_4 - T_c| \leq \Delta T$. U_i computes $(R_g||R_s) = B_8 \oplus R_u$, $SK_u = h(C_1||R_g||R_s)$, and $B_9^* = h(A_0||SK_u||R_u)$. Then, U_i verifies whether $B_9^* \stackrel{?}{=} B_9$. If the verification is true, U_i updates CID_i^{new}. Finally, the verification and key exchange are successful.

User U_i	**Gateway GW_j**	**Sensor Node SN_k**
Selects SID_k, R_u, T_1 Computes $R_1 = ER_i \oplus h(ID_i \| \| PW_i)$ $A_0 = A_2 \oplus R_1 \oplus \sigma_i$ Generates random nonce R_u $B_1 = R_u \oplus R_1$ $B_2 = A_0 \oplus R_u \oplus R_1 \oplus HID_i$		
$M_1 = \{SID_k, CID_i, B_1, B_2, T_1\}$ $\longrightarrow$		

	Verifies $\|T_1 - T_c\| \leq \Delta T$
	Checks whether
	$CID_i = CID_i^{old}$ or $CID_i = CID_i^{new}$
	if($CID_i == CID_i^{old}$)
	{Retrieves $\{HID_i^*, ER_j\}$ against CID_i^{old}}
	if($CID_i == CID_i^{new}$)
	{Retrieves $\{HID_i^*, ER_j\}$ against CID_i^{new}}
	Computes $R_1 = ER_j \oplus G_j$
	$R_u = B_1 \oplus R_1$
	$A_0 = B_2 \oplus R_u \oplus R_1 \oplus HID_i$
	$A_1^* = h(HID_i \| \| A_0) \oplus G_j$
	Check $A_1 \stackrel{?}{=} A_1^*$
	$CID_m^{new} = h(HID_i \| \| R_u)$
	Updates CID_i^{new}
	Selects R_g, T_2
	$SG_k = h(SID_k \| \| G_j)$
	$C_1 = R_u \oplus HID_i$
	$B_3 = C_1 \oplus SG_k \oplus CH_1$
	$B_4 = R_g \oplus h(SG_k \| \| SID_k)$
	$B_5 = h(B_4 \| \| B_5 \| \| SG_k \| \| C_1 \| \| R_g)$	

	$M_2 = \{B_3, B_4, B_5, T_2\}$ $\longrightarrow$	
		Verify $\|T_2 - T_c\| \leq \Delta T$ Gets RSG_k, CH_1 $RE_1 = PUF(CH_1)$ $SG_k = RSG_k \oplus SID_k \oplus RE_1$ $C_1 = B_3 \oplus SG_k \oplus CH_1$ $R_g = B_4 \oplus h(SG_k \| \| SID_k)$ $B_5^* = h(B_3 \| \| B_4 \| \| SG_k \| \| C_1 \| \| R_g)$ Verify $B_5^* \stackrel{?}{=} B_5$ Selects R_s, T_3 Computes $SK_s = h(C_1 \| \| R_g \| \| R_s)$ $B_6 = h(SG_k \| \| R_g) \oplus R_s$ $B_7 = h(R_g \| \| R_s \| \| SG_k \| \| T_3 \| \| C_1)$
	$\longleftarrow$ $M_3 = \{B_6, B_7, T_3\}$	
	Verify $\|T_3 - T_c\| \leq \Delta T$ Computes $R_s = h(SG_k \| \| R_g) \oplus B_6$ $B_7^* = h(R_g \| \| R_s \| \| SG_k \| \| T_3 \| \| C_1)$ Check $B_7^* \stackrel{?}{=} B_7$ Selects T_4 $SK_g = h(C_1 \| \| R_g \| \| R_s)$ $B_8 = R_u \oplus (R_g \| \| R_s)$ $B_9 = h(A_0 \| \| SK_g \| \| R_u)$	
$\longleftarrow$ $M_4 = \{B_8, B_9, T_4\}$		
$\|T_4 - T_c\| \leq \Delta T$ Computes $(R_g \| \| R_s) = B_8 \oplus R_u$ $C_1 = R_u \oplus HID_i$ $SK_u = h(C_1 \| \| R_g \| \| R_s)$ $B_9^* = h(A_0 \| \| SK_u \| \| R_u)$ Checks $B_9^* \stackrel{?}{=} B_9$ Updates CID_i^{new}		

Figure 7. Authentication and key agreement phase of the proposed scheme.

6.5. Password Update Phase

In our scheme, we provide an efficient password update process of the medical professional. We show the password update phase in Figure 8, and the detailed steps are as follows:

User U_i		Mobile Device
U_i enters ID_i^* and PW_i^* Imprints BIO_i^*		
	$\xrightarrow{\textit{input to mobile device}}$	
		$Rep(BIO_i^*, \tau_i) = \sigma_i^*$ $HPW_i^* = h(PW_i^* \| \| \sigma_i^*)$ $A_3^* = h(ID_i^* \| \| HPW_i^*)$ Verifies $A_3 \stackrel{?}{=} A_3^*$ If true, user authentication passed
	$\xleftarrow{Authenticate}$	
Inputs a new password PW_i^{new} and a new biometrics BIO_i^{new}		
	$\xrightarrow{\{PW_i^{new}, BIO_i^{new}\}}$	
		$< \sigma_i^{new}, \tau_i^{new} >= Gen(BIO_i^{new})$ $HPW_i^{new} = h(PW_i^{new} \| \| \sigma_i^{new})$ $R_1 = ER_i \oplus h(ID_i^* \| \| PW_i^*)$ $ER_i^{new} = h(ID_i^* \| \| PW_i^{new}) \oplus R_1$ $A_3^{new} = h(ID_i^* \| \| HPW_i^{new})$ Replaces $\{A_3, \tau_i, ER_i\}$ with $\{A_3^{new}, \tau_i^{new}, ER_i^{new}\}$

Figure 8. Password update phase of the proposed scheme.

Step 1: U_i enters ID_i^* and PW_i^* and imprints BIO_i^* to the mobile device.

Step 2: The mobile device calculates $Rep(BIO_i^*, \tau_i) = \sigma_i^*$, $HPW_i^* = h(PW_i^*||\sigma_i^*)$, and $A_3^* = h(ID_i^*||HPW_i^*)$ and verifies $A_3 \stackrel{?}{=} A_3^*$. If the equation is true, user authentication passes.

Step 3: U_i inputs a new password PW_i^{new} and a new biometric BIO_i^{new} to the mobile device.

Step 4: The mobile device computes $Gen(BIO_i^{new}) =< \sigma_i^{new}, \tau_i^{new} >$, $HPW_i^{new} = h(PW_i^{new}||\sigma_i^{new})$, $R_1 = ER_i \oplus h(ID_i^*||PW_i^*)$, $ER_i^{new} = h(ID_i^*||PW_i^{new}) \oplus R_1$, and $A_3^{new} = h(ID_i^*||HPW_i^{new})$. Finally, the mobile device replaces $\{A_3, \tau_i, ER_i\}$ with $\{A_3^{new}, \tau_i^{new}, ER_i^{new}\}$

7. Security Analysis

To prove the security features of the proposed scheme, we used BAN logic and the RoR model, which can prove the mutual authentication properties and session key security, respectively. Furthermore, we show that our scheme has resistance against man-in-the-middle and replay attacks using AVISPA. Furthermore, we claim that the proposed scheme can prevent various security attacks using informal analysis.

7.1. BAN Logic

In this section, BAN logic [26] is used to prove the mutual authentication of the proposed scheme. BAN logic uses a simple logic to explain the beliefs between the communication participants of authentication schemes. From that, many security schemes are proven by using BAN logic [27–29]. Table 3 shows the basic notation in BAN logic.

Table 3. Basic notations in BAN logic.

Notation	Definition
$\mathcal{C}_1, \mathcal{C}_2$	Principals
$\mathcal{T}_1, \mathcal{T}_2$	Statements
SK	Session key
$\mathcal{C}_1 \mid\equiv \mathcal{T}_1$	$\mathcal{C}_1$ **believes** $\mathcal{T}_1$
$\mathcal{C}_1 \mid\sim \mathcal{T}_1$	$\mathcal{C}_1$ once **said** $\mathcal{T}_1$
$\mathcal{C}_1 \Rightarrow \mathcal{T}_1$	$\mathcal{C}_1$ **controls** $\mathcal{T}_1$
$\mathcal{C}_1 \lhd \mathcal{T}_1$	$\mathcal{C}_1$ **receives** $\mathcal{T}_1$
$\#\mathcal{T}_1$	$\mathcal{T}_1$ is **fresh**
$(\mathcal{T}_1)_K$	$\mathcal{T}_1$ is **encrypted** with K
$\mathcal{C}_1 \overset{K}{\leftrightarrow} \mathcal{C}_2$	$\mathcal{C}_1$ and $\mathcal{C}_2$ have shared key K

7.1.1. Rules

We introduce five rules used in BAN logic:

1. Message meaning rule (MMR):

$$\frac{\mathcal{C}_1 \Big| \equiv \mathcal{C}_1 \overset{K}{\leftrightarrow} \mathcal{C}_2, \quad \mathcal{C}_1 \lhd (\mathcal{T}_1)_K}{\mathcal{C}_1 | \equiv \mathcal{C}_2 | \sim \mathcal{T}_1};$$

2. Nonce verification rule (NVR):

$$\frac{\mathcal{C}_1 | \equiv \#(\mathcal{T}_1), \quad \mathcal{C}_1 | \equiv \mathcal{C}_2 \Big| \sim \mathcal{T}_1}{\mathcal{C}_1 | \equiv \mathcal{C}_2 | \equiv \mathcal{T}_1};$$

3. Jurisdiction rule (JR):

$$\frac{\mathcal{C}_1 | \equiv \mathcal{C}_2 \Rightarrow \mathcal{T}_1, \quad \mathcal{C}_1 | \equiv \mathcal{C}_2 | \equiv \mathcal{T}_1}{\mathcal{C}_1 \Big| \equiv \mathcal{T}_1};$$

4. Belief rule (BR):

$$\frac{\mathcal{C}_1 \Big| \equiv (\mathcal{T}_1, \mathcal{T}_2)}{\mathcal{C}_1 \Big| \equiv \mathcal{T}_1};$$

5. Freshness rule (FR):

$$\frac{\mathcal{C}_1 \Big| \equiv \#(\mathcal{T}_1)}{\mathcal{C}_1 \Big| \equiv \#(\mathcal{T}_1, \mathcal{T}_2)}.$$

7.1.2. Goals

The final goal of BAN logic in the proposed scheme is to achieve mutual authentication by agreeing on the session key SK. We define U_i, GW_j, and SN_k as the user, gateway, and sensor node, respectively:

Goal 1: $U_i | \equiv GW_j \overset{SK}{\longleftrightarrow} U_i$;

Goal 2: $U_i | = GW_j | \equiv GW_j \overset{SK}{\longleftrightarrow} U_i$;

Goal 3: $GW_j | \equiv GW_j \overset{SK}{\longleftrightarrow} U_i$;

Goal 4: $GW_j | \equiv U_i | \equiv GW_j \overset{SK}{\longleftrightarrow} U_i$;

Goal 5: $SN_k| \equiv GW_j \stackrel{SK}{\longleftrightarrow} SN_k$;

Goal 6: $SN_k| \equiv GW_j| \equiv GW_j \stackrel{SK}{\longleftrightarrow} SN_k$;

Goal 7: $GW_j| \equiv GW_j \stackrel{SK}{\longleftrightarrow} SN_k$;

Goal 8: $GW_j| \equiv SN_k| \equiv GW_j \stackrel{SK}{\longleftrightarrow} SN_k$.

7.1.3. Idealized Forms

In the proposed scheme, $M_1 = \{SID_k, CID_i, B_1, B_2, T_1\}$, $M_2 = \{B_3, B_4, B_5, T_2\}$, $M_3 = \{B_6, B_7, T_3\}$, and $M_4 = \{B_8, B_9, T_4\}$ are transmitted through public channels. We restructure the messages to fit the BAN logic, named "idealized forms":

$\mathcal{T}_1: U_i \rightarrow GW_j : \{R_u, A_0, HID_i, T_1\}_{R_1}$;

$\mathcal{T}_2: GW_j \rightarrow SN_k : \{R_g, C_1, T_2\}_{SG_k}$;

$\mathcal{T}_3: SN_k \rightarrow GW_j : \{R_s, T_3\}_{SG_k}$;

$\mathcal{T}_4: GW_j \rightarrow U_i : \{R_g, R_s, T_4\}_{R_u}$.

7.1.4. Assumptions

The assumptions in the proposed scheme are shown as below:

$\mathcal{S}_1$: $GW_j| \equiv \#(T_1)$;

$\mathcal{S}_2$: $SN_k| \equiv \#(T_2)$;

$\mathcal{S}_3$: $GW_j| \equiv \#(T_3)$;

$\mathcal{S}_4$: $U_i| \equiv \#(T_4)$;

$\mathcal{S}_5$: $U_i| \equiv GW_j \Rightarrow (GW_j \stackrel{SK}{\longleftrightarrow} U_i)$;

$\mathcal{S}_6$: $GW_j| \equiv U_i \Rightarrow (GW_j \stackrel{SK}{\longleftrightarrow} U_i)$;

$\mathcal{S}_7$: $GW_j| \equiv SN_k \Rightarrow (GW_j \stackrel{SK}{\longleftrightarrow} SN_k)$;

$\mathcal{S}_8$: $SN_k| \equiv GW_j \Rightarrow (GW_j \stackrel{SK}{\longleftrightarrow} SN_k)$;

$\mathcal{S}_9$: $GW_j| \equiv GW_j \stackrel{R_1}{\longleftrightarrow} U_i$;

$\mathcal{S}_{10}$: $GW_j| \equiv GW_j \stackrel{SG_k}{\longleftrightarrow} SN_k$;

$\mathcal{S}_{11}$: $SN_k| \equiv GW_j \stackrel{SG_k}{\longleftrightarrow} SN_k$;

$\mathcal{S}_{12}$: $U_i| \equiv GW_j \stackrel{R_u}{\longleftrightarrow} U_i$.

7.1.5. BAN Logic Proof

Step 1: We can obtain PR_1 based on the first message T_1, and we obtain the following:

$$PR_1: GW_j \triangleleft \{R_u, A_0, HID_i, T_1\}_{R_1};$$

Step 2: Based on the message meaning rule, PR_1, and $\mathcal{S}_9$, we can obtain the following:

$$PR_2: GW_j| \equiv U_i| \sim (R_u, A_0, HID_i, T_1);$$

Step 3: Based on the freshness rule, PR_2, and $\mathcal{S}_1$, we can obtain the following:

$$PR_3\text{: } GW_j| \equiv \#(R_u, A_0, HID_i, T_1);$$

Step 4: Based on the nonce verification rule, PR_2, and PR_3, we obtain the following:

$$PR_4\text{: } GW_j| \equiv U_i| \equiv (R_u, A_0, HID_i, T_1);$$

Step 5: Based on the second message T_2, we obtain the following:

$$PR_5\text{: } SN_k \lhd \{R_g, C_1, T_2\}_{SG_k};$$

Step 6: Based on the message meaning rule, PR_5, and $\mathcal{S}_{11}$, we can obtain the following:

$$PR_6\text{: } SN_k| \equiv GW_j| \sim (R_g, C_1, T_2);$$

Step 7: Based on the freshness rule, PR_6, and $\mathcal{S}_2$, we can obtain the following:

$$PR_7\text{: } SN_k| \equiv \#(R_g, C_1, T_2);$$

Step 8: Based on the nonce verification rule, PR_6, and PR_7, we can obtain the following:

$$PR_8\text{: } SN_k| \equiv GW_j| \equiv (R_g, C_1, T_2);$$

Step 9: Based on the third message T_3, we can obtain the following:

$$PR_9\text{: } GW_j \lhd \{R_s, T_3\}_{SG_k};$$

Step 10: Based on the message meaning rule, PR_9, and $\mathcal{S}_{10}$, we can obtain the following:

$$PR_{10}\text{: } GW_j| \equiv SN_k| \sim (R_s, T_3);$$

Step 11: Based on the freshness rule, PR_{10}, and $\mathcal{S}_3$, we can obtain the following:

$$PR_{11}\text{: } GW_j| \equiv \#(R_s, T_3);$$

Step 12: Based on the nonce verification rule, PR_{10}, and PR_{11}, we can obtain the following:

$$PR_{12}\text{: } GW_j| \equiv SN_k| \equiv (R_s, T_3);$$

Step 13: Based on PR_8 and PR_{12}, SN_k and GW_j compute the session key $SK = h(C_1||R_g||R_s)$. Therefore, we can obtain the following goals:

$$PR_{13}\text{: } SN_k| \equiv GW_j| \equiv GW_j \overset{SK}{\longleftrightarrow} SN_k \quad \textbf{(Goal 6)}$$
$$PR_{14}\text{: } GW_j| \equiv SN_k| \equiv GW_j \overset{SK}{\longleftrightarrow} SN_k \quad \textbf{(Goal 8)};$$

Step 14: Based on the jurisdiction rule, PR_{13}, PR_{14}, $\mathcal{S}_7$, and $\mathcal{S}_8$, we can obtain the following goals:

$$PR_{15}\text{: } SN_k| \equiv GW_j \overset{SK}{\longleftrightarrow} SN_k \quad \textbf{(Goal 5)}$$
$$PR_{16}\text{: } GW_j| \equiv GW_j \overset{SK}{\longleftrightarrow} SN_k \quad \textbf{(Goal 7)};$$

Step 15: Based on the last message T_4, we can obtain the following:

$$PR_{17}\text{: } U_i \lhd \{R_g, R_s, T_4\}_{R_u};$$

Step 16: Based on the message meaning rule, PR_{17}, and $\mathcal{S}_{12}$, we can obtain the following:

$$PR_{18}\text{: } U_i \equiv SN_k| \sim (R_g, R_s, T_4);$$

Step 17: Based on the freshness rule, PR_{18}, and $\mathcal{S}_4$, we can obtain the following:

$$PR_{19}\text{: } U_i| \equiv \#(R_g, R_s, T_4);$$

Step 18: Based on the nonce verification rule, PR_{19}, and PR_{17}, we can obtain the following:

$$PR_{20}\colon U_i| \equiv GW_j| \equiv (R_g, R_s, T_4);$$

Step 19: Based on PR_4 and PR_{20}, U_i and GW_j compute the session key SK. Therefore, we can obtain the following goals:

$$PR_{21}\colon U_i| \equiv GW_j| \equiv GW_j \overset{SK}{\longleftrightarrow} U_i \quad \textbf{(Goal 2)}$$
$$PR_{22}\colon GW_j| \equiv U_i| \equiv GW_j \overset{SK}{\longleftrightarrow} U_i \quad \textbf{(Goal 4)};$$

Step 20: Based on the jurisdiction rule, PR_{21}, PR_{22}, $\mathcal{S}_5$, and $\mathcal{S}_6$, we can obtain the following goals:

$$PR_{23}\colon U_i| \equiv GW_j \overset{SK}{\longleftrightarrow} U_i \quad \textbf{(Goal 1)}$$
$$PR_{24}\colon GW_j| \equiv GW_j \overset{SK}{\longleftrightarrow} U_i \quad \textbf{(Goal 3)}.$$

7.2. RoR Model

To prove the security of the session key, we utilized a formal proof named the "real-or-random" (ROR) model [30]. Firstly, we define the participants, adversary, and queries. In the proposed scheme, there are three entities that perform the authentication phase to establish the session key. These entities are instantiated as participants and applied to the ROR model: EP^i_{US}, EP^j_{GW}, EP^k_{SN}. Note that i, j, and k are the instances of the user, gateway, and sensor node, respectively. Next, we define the adversary of the ROR model. The adversary can fully control the whole network, including modifying, deleting, hijacking, and intercepting messages. Moreover, we introduce queries that are utilized to reveal the session key security of the scheme. The details are as follows:

- $Exe(EP^i_{US}, EP^j_{GW}, EP^k_{SN})$: This is a passive attack, where the adversary obtain messages exchanged through public channels.
- $CorrD(EP^i_{US})$: The $CorrD$ query is an active attack. The adversary obtains secret parameters that are stored in the smart card of EP^i_{US} using power analysis attack.
- $Snd(EP)$: When the adversary uses the Snd query, the adversary transfers messages to EP^i_{US}, EP^j_{GW}, and EP^k_{SN}. Moreover, the adversary receives return messages from the participants.
- $Test(EP)$: An unbiased coin c is tossed, and the adversary obtains the result of this query. If the result value of c is 0, the session key is not fresh. If the result value of c is 1, we can demonstrate that the session key is fresh and secure. Otherwise, a null value $(\perp)$ is obtained.

Security Proof

Theorem 1. *We define the adversary and possibility of breaking the session key security as $\mathcal{M}$ and $\mathcal{A}_M(BP)$, respectively. In the ROR model, $\mathcal{M}$ tries to guess $SK = h(C_1||R_g||R_s)$ in polynomial time. To do this, we give a definition of hash and puf as the range space of the hash function and PUF, respectively. Moreover, q_{hash}, q_{puf}, and q_{snd} are the number of hash, puf, and Snd queries, respectively. We define C' and s' as Zipf's parameter [31], and the number of bits in the biometrics is BIO.*

$$\mathcal{A}_M(BP) \le \frac{q^2_{hash}}{|hash|} + \frac{q^2_{puf}}{|puf|} + 2max\{C'q^{s'}_{snd}, \frac{q_{snd}}{2^{BIO}}\}$$

Proof. In the proposed scheme, the ROR security proof consists of five games G_n $(0 \le n \le 4)$. $\mathcal{M}$ tries to compute the session key SK in each game G_k, and we define this winning possibility as WN_{G_k}. Our ROR security proof is performed according to the method of [32–34]:

G_0: $\mathcal{M}$ begins the real attack. Thus, $\mathcal{M}$ picks a random bit c. Therefore, we obtain Equation (1) as follows.

$$\mathcal{A}_M(BP) = |2\mathcal{M}[WN_{G_0}] - 1|. \tag{1}$$

G_1: As we mentioned before, $\mathcal{M}$ can obtain all of the messages in the proposed scheme using the query *Exe*. Thus, M_1, M_2, M_3, and M_4 can be intercepted and $\mathcal{M}$ executes the *Test* query as Equation (2). The session key SK is composed of $C_1 = R_u \oplus HID_i$, R_g, and R_s. Thus, $\mathcal{M}$ must know all of the random nonces and the secret parameter of US. This means that $\mathcal{M}$ cannot calculate SK.

$$|\mathcal{M}[WN_{G_1}]| = |\mathcal{M}[WN_{G_0}]|. \tag{2}$$

G_2: In this game, the *hash* and *Snd* queries are utilized. However, we used the "cryptographic hash function", which can overcome the hash collision problem in the proposed scheme. Thus, $\mathcal{M}$ has no advantage using the *hash* and *Snd* queries. We show the following inequation (3) by applying the birthday paradox [35].

$$|\mathcal{M}[WN_{G_2}] - \mathcal{M}[WN_{G_1}]| \leq \frac{q_{hash}^2}{|hash|}. \tag{3}$$

G_3: In G_3, $\mathcal{M}$ attempts to break the session key security using the puf query. However, it is impossible to guess or compute the PUF function according to Section 3.3. Therefore, we obtain the following Equation (4).

$$|\mathcal{M}[WN_{G_3}] - \mathcal{M}[WN_{G_2}]| \leq \frac{q_{puf}^2}{|puf|}. \tag{4}$$

G_4: In the final game G_4, $\mathcal{M}$ utilizes the *CorrD* query and obtains secret parameters $\{A_2, A_3, Gen(.), Rep(.), \tau_i, ER_i, CID_i\}$ from the smart card. In the proposed scheme, all of the parameters are masked in the user's identity, password, and biometrics. To calculate SK using the secret parameters, $\mathcal{M}$ must guess U_i, PW_i, and BIO_i at the same time. Since guessing them in polynomial time is obviously impossible, $\mathcal{M}$ cannot derive SK. We apply Zipf's law and obtain the following Equation (5).

$$|\mathcal{M}[WN_{G_4}] - \mathcal{M}[WN_{G_2}]| \leq max\{C' q_{snd}^{s'}, \frac{q_{snd}}{2^{BIO}}\} \tag{5}$$

After that, $\mathcal{M}$ obtains the result bits b. Moreover, we can set up the following Equation (6).

$$\mathcal{M}[WN_{G_4}] = \frac{1}{2} \tag{6}$$

Using (1) and (2), Equation (7) can be calculated.

$$\frac{1}{2}\mathcal{A}_M(BP) = |\mathcal{M}[WN_{G_0}] - \frac{1}{2}| = |\mathcal{M}[WN_{G_1}] - \frac{1}{2}| \tag{7}$$

From (6) and (7), Equation (8) can be calculated.

$$\frac{1}{2}\mathcal{A}_M(BP) = |\mathcal{M}[WN_{G_1}] - \mathcal{M}[WN_{G_4}]| \tag{8}$$

Using the triangular inequality, we can obtain the following Equation (9).

$$\begin{aligned} \frac{1}{2}\mathcal{A}_M(BP) &= |\mathcal{M}[WN_{G_1}] - \mathcal{M}[WN_{G_4}]| \\ &\leq |\mathcal{M}[WN_{G_1}] - \mathcal{M}[WN_{G_3}]| \\ &+|\mathcal{M}[WN_{G_3}] - \mathcal{M}[WN_{G_4}]| \\ &\leq |\mathcal{M}[WN_{G_1}] - \mathcal{M}[WN_{G_2}]| \\ &+|\mathcal{M}[WN_{G_2}] - \mathcal{M}[WN_{G_3}]| \\ &+|\mathcal{M}[WN_{G_3}] - \mathcal{M}[WN_{G_4}]| \end{aligned} \tag{9}$$

$$\leq \frac{q_{hash}^2}{2|hash|} + \frac{q_{puf}^2}{2|puf|} + max\{C' q_{snd}^{s'}, \frac{q_{snd}}{2^{BIO}}\} \quad (10)$$

We obtain the resulting inequation by multiplying (10) by two.

$$\mathcal{A}_M(BP) \leq \frac{q_{hash}^2}{|hash|} + \frac{q_{puf}^2}{|puf|} + 2max\{C' q_{snd}^{s'}, \frac{q_{snd}}{2^{BIO}}\}.$$

Thus, we prove the Theorem. □

7.3. AVISPA Simulation

In this section, we utilize the AVISPA simulation tool [36,37] to verify the resistance against the replay and man-in-the-middle attacks of the proposed scheme. The AVISPA simulation tool verifies the authentication scheme through a code called "High-Level scheme Specification Language (HLPSL)" on the Linux OS. Afterwards, the HLPSL code is converted to "Intermediate Format (IF)" to perform security verification on the four backends ("On-the-Fly Model Checker (OFMC)", "Three Automata based on Automatic Approximations for Analysis of Security Protocol (TA4SP)", "SAT-based Model Checker (SATMC)", and "Constraint Logic-based Attack Searcher (CL-AtSe)"). In this paper, we used the CL-AtSe and OFMC backends because these backends can support the XOR operator. Finally, the result window, i.e., "Output Format (OF)", is shown, and we can demonstrate that the proposed scheme can resist the replay and man-in-the-middle attacks if the OF summarizes the verification as "SAFE". We show the three basic roles of the proposed scheme: user UI, gateway GWJ, and sensor node SNK. The session, environment, and goals are shown in Figure 9. We also show the role of UI in Figure 10.

```
role session(UI, GWJ, SNK : agent, SKuigw, SKsngw : symmetric_key, PUF,H : hash_func)

def=
local SN1, SN2, SN3, RV1, RV2, RV3 : channel(dy)
composition
user(UI, GWJ, SNK, SKuigw, SKsngw, PUF, H, SN1, RV1)
/\ gateway(UI, GWJ, SNK, SKuigw, SKsngw, PUF, H, SN2, RV2)
/\ sensornode(UI, GWJ, SNK, SKuigw, SKsngw, PUF, H, SN3, RV3)

end role

%%%%%%%%%%%%%%%%%%%%%%%%%%%%%%%%%%%%%%%

role environment()
def=
const ui, gwj, snk : agent,
                puf, h : hash_func,
                skuigw, sksngw : symmetric_key,
                ui_gw_ru, ui_sn_ru, gw_sn_rg, gw_ui_rg, sn_gw_rs, sn_ui_rs : protocol_id,
                sp1, sp2, sp3, sp4 : protocol_id,
                idi, cidi, sidk : text
intruder_knowledge = {h, idi, cidi, sidk}
composition
session(ui, gwj, snk, skuigw, sksngw, puf, h)
/\session(i, gwj, snk, skuigw, sksngw, puf, h)
/\session(ui, i, snk, skuigw, sksngw, puf, h)
/\session(ui, gwj, i, skuigw, sksngw, puf, h)

end role

%%%%%%%%%%%%%%%%%%%%%%%%%%%%%%%%%%%%%%%

goal
secrecy_of sp1, sp2, sp3, sp4
authentication_on ui_gw_ru
authentication_on ui_sn_ru
authentication_on gw_sn_rg
authentication_on gw_ui_rg
authentication_on sn_gw_rs
authentication_on sn_ui_rs

end goal

environment()
```

Figure 9. Role specification for the session, environment, and goals.

```
%%%AVISPA Simulation
role user(UI, GWJ, SNK : agent, SKuigw, SKsngw : symmetric_key, PUF,H : hash_func, SND,RCV : channel(dy))

played_by UI
def=
local State : nat,
        IDi, PWi, HIDi, HPWi, BIOi, R0, R1, CIDi, ERj, A0, A1, Gj, A2, A3, ERi : text,
        SIDk, CH1, SGk, RE1, RSGk, Ru, Rs, Rg, T1, T2, T3, T4, C1, B1, B2, B3, B4, B5, B6, B7, B8, B9, SK : text
        const sp1, sp2, sp3, sp4, ui_gw_ru, ui_sn_ru, gw_sn_rg, gw_ui_rg, sn_gw_rs, sn_ui_rs : protocol_id

init State := 0
transition
%%%%%%%User registration phase
1. State = 0 /\ RCV(start) =|>
State' := 1
/\ HIDi' := H(IDi. BIOi)
/\ HPWi' := H(PWi. BIOi)
/\ SND({HIDi'}_SKuigw)
/\ secret({HIDi'}, sp1, {UI, GWJ})

2. State = 1 /\ RCV({A0. R1. H(H(IDi. BIOi). R0)}_SKuigw) =|>
State' := 2
/\ A2' := xor(xor(A0, R1), BIOi)
/\ A3' := H(IDi. H(PWi. BIOi))
/\ ERi' := xor(H(IDi. PWi), R1)
%login and authentication phase
/\ Ru' := new()
/\ T1' := new()
/\ B1' := xor(Ru', R1)
/\ B2' := xor(xor(xor(A0, Ru'), R1), H(IDi. BIOi))
/\ SND(SIDk. H(H(IDi. BIOi). R0). B1'. B2'. T1')
/\ witness(UI,GWJ,ui_gw_ru,Ru')
/\ witness(UI,SNK,ui_sn_ru,Ru')

3. State = 2 /\ RCV(xor(xor(Ru', Rs'), Rg'). H(A0. H(xor(Ru'. H(IDi. BIOi)). Rg'. Rs'). Ru'). T4') =|>
State' := 3
/\SK' := H(xor(Ru'. H(IDi. BIOi)). Rg'. Rs')
/\request(GWJ,UI,gw_ui_rg,Rg')
/\request(SNK,UI,sn_ui_rs,Rs')

end role
```

Figure 10. Role specification for the user.

In State 1, UI receives the start message and computes HID_i and HPW_i. Then, UI sends $\{HID_i\}$ to GWJ. GWJ registers UI and returns $\{A_0, R_1, CID_i\}$ through a secure channel. State 2 is the login and authentication phase, for which UI generates R_u, T_1 and computes the authentication request message $\{SID_k, CID_i, B_1, B_2, T_1\}$ to GWJ. At the same time, UI generates function $witness(UI, GWJ, ui_g w_r u, Ru')$ and $witness(UI, SNK, ui_s n_r u, Ru')$, which means the proof of random nonce R_u's freshness. Finally, UI receives $\{B_8, B_9, T_4\}$ and computes the session key $SK = h(C_1||R_g||R_s)$. We verified the proposed scheme in the CL-AtSe and OFMC backends, and the result window is shown in Figure 11. Therefore, the proposed scheme can resist the replay and man-in-the-middle attacks.

```
% OFMC
% Version of 2006/02/13
SUMMARY
  SAFE
DETAILS
  BOUNDED_NUMBER_OF_SESSIONS
PROTOCOL
  /home/span/span/testsuite/results/SANG.if
GOAL
  as_specified
BACKEND
  OFMC
COMMENTS
STATISTICS
  parseTime: 0.00s
  searchTime: 6.31s
  visitedNodes: 1480 nodes
  depth: 12 plies
```

```
SUMMARY
  SAFE

DETAILS
  BOUNDED_NUMBER_OF_SESSIONS
  TYPED_MODEL

PROTOCOL
  /home/span/span/testsuite/results/SANG.if

GOAL
  As Specified

BACKEND
  CL-AtSe

STATISTICS

  Analysed   : 0 states
  Reachable  : 0 states
  Translation: 0.10 seconds
  Computation: 0.00 seconds
```

Figure 11. The AVISPA simulation result of the proposed scheme.

7.4. Informal Analysis

In this section, we demonstrate the security features of our proposed scheme, including those that resist against privileged insider, insider, physical, cloning, verification table leakage, impersonation, session key disclosure, ephemeral secret leakage, replay, man-in-the-middle, stolen mobile device, offline password guessing, and denial-of-service attacks. Moreover, the proposed scheme can provide user anonymity and perfect forward secrecy.

7.4.1. User Anonymity

In our scheme, $\mathcal{A}$ cannot obtain the legitimate $U_i's$ identity ID_i, and even $\mathcal{A}$ extracts values $\{A_2, A_3, Gen(.), Rep(.), \tau_i, ER_i, CID_i\}$ inside $U_i's$ mobile device. ID_i is masked by a hash function with $U_i's$ biometric information or PW_i such that $HID_i = h(ID_i||\sigma_i)$, $A_3 = h(ID_i||HPW_i)$, and $ER_i = h(ID_i||PW_i) \oplus R_1$.

7.4.2. Privileged Insider Attack

We can assume privileged insider $\mathcal{A}$ obtains the registration request message $\{HID_i\}$ of the medical professional. Furthermore, $\mathcal{A}$ can extract the parameters $\{A_2, A_3, Gen(.), Rep(.), \tau_i, ER_i, CID_i\}$ from the stolen mobile device of the medical professional using power analysis attack. $\mathcal{A}$ can also intercept transmitted messages such as M_1 and $M4$ on a public channel. After that, $\mathcal{A}$ attempts to impersonate a medical professional. To calculate authentication message $M_1\{SID_k, CID_i, B_1, B_2, T_1\}$, $\mathcal{A}$ must compute parameters R_1 and A_0. However, $\mathcal{A}$ cannot compute $R_1 = ER_i \oplus h(ID_i||PW_i)$ and $A_0 = A_2 \oplus R_1 \oplus \sigma_i$ because $\mathcal{A}$ cannot generate the ID_i, PW_i and biometric information BIO_i of U_i. Therefore, it is difficult for $\mathcal{A}$ to calculate the authentication message M_1 to impersonate a medical professional. $\mathcal{A}$ can also attempt to compute $SK_u = h(C_1||R_g||R_s)$. However, $\mathcal{A}$ cannot generate a session key of U_i SK_u. $\mathcal{A}$ cannot calculate $(R_g||R_s) = B_8 \oplus R_u$ and $R_u = B1 \oplus R_1$. In conclusion, the proposed scheme can resist the privileged insider attack.

7.4.3. Insider Attack

Suppose that U_i registers with GW_j as a legal user and intercepts the transmitted messages such as M_2, M_3, and M_4. However, U_i cannot calculate important parameters such as the symmetric key SG_k shared by GW_j and SN_k. Thus, U_i cannot attempt various attacks, including the impersonate and session key disclosure attacks. As as result, our scheme can prevent the insider attack.

7.4.4. Physical Cloning Attack

Assume that an adversary $\mathcal{A}$ physically captures a sensor node SN_k and attempts to authenticate with GW_j by disguising it as SN_k. $\mathcal{A}$ physically clones SN_k to obtain a values $\{RSG_k, CH_1\}$ in the memory of SN_k and intercepts authentication request messages M_2 on the public channel. Then, $\mathcal{A}$ attempts to generate authenticate message $M_3\{B_6, B_7, T_3\}$. However, $\mathcal{A}$ cannot generate a message M_3 because he/she cannot calculate the parameter RE_1 necessary to generate message M_3. $\mathcal{A}$ can replicate the same CH_1 from SN_k, but cannot generate the same RE_1. The PUF circuit cannot be forged. Thus, our scheme can withstand the physical cloning attack.

7.4.5. Verification Table Leakage Attack

Suppose that $\mathcal{A}$ intercepts $\{CID_i, HID_i, ER_j, A_1, SID_k, CH_1\}$ in $GW_j's$ verification table of GW_j. Then, $\mathcal{A}$ eavesdrops the transmitted messages such as M_1, M_2, M_3 and intercepts message M_4 via an insecure channel. After that, $\mathcal{A}$ attempts to compute authentication request messages M_2 or $SK_g = h(C_1||R_g||R_s)$. However, $\mathcal{A}$ cannot calculate $SG_k = h(SID_k||G_j)$, which is essential for generating M_2 and SK_g, because $GW_j's$ secret key G_j is unknown. Therefore, $\mathcal{A}$ cannot generate both M_2 and SK_G. As a result, our scheme can protect against verification table leakage attack.

7.4.6. Impersonation Attack

(1) User impersonation attack: For this attack, suppose an adversary $\mathcal{A}$ attempts to impersonate U_i. $\mathcal{A}$ must generate a valid authentication request message $M_1\{SID_k, CID_i, B_1, B_2, T_1\}$. $\mathcal{A}$ can extract CID_i from $U_i's$ mobile device and intercept message $M_1\{SID_k, CID_i, B_1, B_2, T_1\}$ through a public channel, but cannot calculate the remaining values $\{B_1, B_2\}$ because $U_i's$ ID_i, PW_i, and BIO_i are essential for calculating the remaining values $\{B_1, B_2\}$. Therefore, the proposed scheme is resilient against the user impersonation attack.
(2) Gateway impersonation attack: Suppose malicious adversary $\mathcal{A}$ tries to impersonate GW_j and sends a authentication request message $M_2\{B_3, B_4, B_5, T_2\}$ to SN_k. To do this, $\mathcal{A}$ eavesdrops the transmitted messages M_1 and M_2. However, without having credentials SG_k, C_1, HID_i, CH_1, it is an impossible task for $\mathcal{A}$ to compute $M_2\{B_3, B_4, B_5, T_2\}$. Hence, the proposed scheme provides protection against the gateway impersonation attack.
(3) Sensor node impersonation attack: A malicious adversary $\mathcal{A}$ can try to impersonate SN_k. To do this, $\mathcal{A}$ intercepts transmitted messages M_2 and M_3 via an insecure channel and calculates the key agreement message $M_3\{B_6, B_7, T_3\}$. However, since $PUF(.)$ is a physically unclonable circuit, $\mathcal{A}$ cannot calculate $RE_1 = PUF(CH_1)$ and $SG_k = RSG_k \oplus SID_k \oplus RE_1$. Therefore, $\mathcal{A}$ cannot generate message $M_3\{B_6, B_7, T_3\}$. Thus, the proposed scheme prevents the sensor node impersonation attacks.

7.4.7. Session Key Disclosure Attack

If $\mathcal{A}$ tries to calculate a legitimate session key $SK = h(C_1||R_g||R_s)$, the adversary must obtain HID_i, R_u, R_g, R_s. However, $\mathcal{A}$ cannot obtain these values. R_u, R_g, and R_s are temporary random nonces used in a session, and HID_i is masked as the legitimate $U_i's$ biometric information BIO_i. Hence, the proposed scheme provides protection against the session key disclosure attacks.

7.4.8. Perfect Forward Secrecy

$\mathcal{A}$ obtains long-term secret keys $\{SG_k, G_j\}$ and intercepts transmitted message $\{M_1, M_2, M_3, M_4\}$ through a public channel. After that, $\mathcal{A}$ attempts to generate M_4 to impersonate GW_j or calculate $SK_g = h(C_1||R_g||R_s)$ to exploit the session key. However, $\mathcal{A}$ cannot compute the parameters C_1 without $U_i's$ identity HID_i and random nonce R_u. For these reasons, our scheme provides perfect forward secrecy.

7.4.9. Ephemeral Secret Leakage Attack

$\mathcal{A}$ obtains random numbers $\{R_u, R_g, R_s, R_0, R_1, R_2\}$. After that, $\mathcal{A}$ attempts to compute the session key $SK_G = h(C_1||R_g||R_s)$. Unfortunately, $\mathcal{A}$ cannot generate session key SK because $\mathcal{A}$ cannot calculate $C_1 = R_u \oplus HID_i$, which is essential for generating a session key SK. Thus, the proposed scheme can prevent the ESL attacks.

7.4.10. Replay and Man-in-the-Middle Attack

We assume that $\mathcal{A}$ eavesdrop transmitted message $\{M_1, M_2, M_3, M_4\}$ through a public channel. However, $\mathcal{A}$ cannot impersonate U_i, GW_j, and SN_k by sending a message again. Because timestamps and random numbers such as $\{T_1, T_2, T_3, R_u, R_g, R_s\}$ are essential to generate a message, and the transmitted message is verified by $\{T_1, T_2, T_3, R_u, R_g, R_s\}$. Therefore, our scheme can prevent replay and man-in-the-middle attack.

7.4.11. Stolen Mobile Device Attack

Suppose that $\mathcal{A}$ succeeds in extracting stored values $\{A_2, A_3, Gen(.), Rep(.), \tau_i, ER_i, CID_i\}$ from $U_i's$ stolen mobile device. However, $\mathcal{A}$ cannot compute any meaningful value from U_i. The values stored in the mobile device are masked with ID_i, PW_i, and BIO_i such as $A_2 = A_0 \oplus R_1 \oplus \sigma_i$, $A_3 = h(ID_i||HPW_i)$, $ER_i = h(ID_i||PW_i) \oplus R_1$. Therefore, $\mathcal{A}$ cannot attempt any attack. Thus, our scheme can resist the stolen mobile device attacks.

7.4.12. Offline Password Guessing Attack

$\mathcal{A}$ obtains $U_i's$ mobile device and extracts parameters $\{A_2, A_3, Gen(.), Rep(.), \tau_i, ER_i, CID_i\}$ using the power analysis attack. After that, $\mathcal{A}$ tries to guess the password of U_i using the extracted parameters. However, $\mathcal{A}$ cannot guess the $U_i's$ password PW_i because the password is masked by the $U_i's$ ID_i, BIO_i, or random nonce R_1 such as $HPW_i = h(PW_i||\sigma_i)$, $A_3 = h(ID_i||HPW_i)$, and $ER_i = h(ID_i||PW_i) \oplus R_1$. Therefore, the proposed scheme is secure against the offline password guessing attacks.

7.4.13. Denial-of-Service

Assume that malicious $\mathcal{A}$ attempts to send $M_1\{SID_k, CID_i, B_1, B_2, T_1\}$ to GW_j as a replay message. To do this, $\mathcal{A}$ must verify the value of $A_3 = h(ID_i||HPW_i)$ and pass the login phase. However, $\mathcal{A}$ cannot calculate a valid A_3 because $\mathcal{A}$ cannot obtain ID_i and HPW_i. Therefore, $\mathcal{A}$ cannot transmit a replay message M_1 to GW_j. Thus, the proposed scheme is secure against the denial-of-service attacks.

7.4.14. Untraceability

Suppose a malicious $\mathcal{A}$ obtains $U_i's$ pseudoidentity CID_i. However, $\mathcal{A}$ cannot attempt any attack with the obtained CID_i. Every session, GW_j updates the CID_i stored with a CID_i^{new} using random nonce R_u after verifying that it is a legitimate user through $A_1 \stackrel{?}{=} A_1^*$ verification. For this reason, the proposed scheme ensures untraceability.

7.4.15. Mutual Authentication

To ensure mutual authentication, our scheme verifies that each entity is justified by $A_1 \stackrel{?}{=} A_1^*$, $B_5 \stackrel{?}{=} B_5^*$, $B_7 \stackrel{?}{=} B_7^*$, and $B_9 \stackrel{?}{=} B_9^*$. Moreover, all entities have verified freshness of messages through random values R_u, R_g, and R_s generated by each entity. When the verification processes are passed, the entities are authenticated with each other. Therefore, our scheme achieves mutual authentication.

8. Performance

In this section, we evaluate the security features, communication costs, and computational costs of our scheme compared with the related schemes [11,38–41].

8.1. Security Features Comparison

We compared the performance of the proposed scheme with the related existing schemes [11,38–41]. As shown in Table 4, we considered various security functionalities and attacks, including "user anonymity", "privileged-insider attack", "offline password guessing attack", "stolen mobile device attack", "denial-of-service attack", "replay attack", "man-in-the-middle attack", "mutual authentication", "session key security", "known session specific temporary information attack", "untraceability property", "server-independent password update phase", "physical cloning attack", "perfect forward secrecy", "impersonation attack", "session-specific random number leakage attack", and "stolen verifier attack". Therefore, our scheme offers functional features and security in comparison with the related schemes [11,38–41].

8.2. Communication Cost Comparison

In this section, we demonstrate the comparison analysis for the communication cost of the proposed scheme with related existing schemes [11,38–41]. According to [42], we define that the bit lengths for the SHA-256 hash output, random number, identity, password, PUF challenge–response, timestamp, and ECC point are 256, 256, 128, 128, 128, 32, and 320 bits, respectively. Therefore, the communication costs of the proposed scheme can be described as below:

Table 4. Security and functionality features' comparison with related schemes.

Security Properties	[38]	[39]	[40]	[41]	[11]	Proposed
$SP1$	×	✓	✓	×	×	✓
$SP2$	×	✓	×	×	×	✓
$SP3$	✓	✓	✓	×	✓	✓
$SP4$	✓	✓	✓	×	✓	✓
$SP5$	✓	✓	✓	✓	✓	✓
$SP6$	×	✓	✓	✓	×	✓
$SP7$	×	✓	✓	✓	×	✓
$SP8$	✓	✓	✓	✓	✓	✓
$SP9$	✓	×	×	✓	✓	✓
$SP10$	✓	✓	✓	✓	✓	✓
$SP11$	✓	✓	×	✓	✓	✓
$SP12$	✓	✓	×	×	×	✓
$SP13$	×	×	✓	×	×	✓
$SP14$	×	✓	✓	✓	✓	✓
$SP15$	×	✓	✓	×	×	✓
$SP16$	×	✓	✓	✓	✓	✓
$SP17$	✓	✓	✓	×	×	✓

Note: $SP1$: user anonymity; $SP2$: privileged insider attack; $SP3$: offline password guessing attack; $SP4$: stolen mobile device attack; $SP5$: denial-of-service attack; $SP6$: replay attack; $SP7$: man-in-the-middle attack; $SP8$: mutual authentication; $SP9$: session key security; $SP10$: known session specific temporary information attack; $SP11$: untraceability property; $SP12$: server-independent password update phase; $SP13$: physical cloning attack; $SP14$: perfect forward secrecy; $SP15$: impersonation attack; $SP16$: session-specific random number leakage attack; $SP17$: stolen verifier attack; ✓: provides or supports the security/functionality feature. ×: does not provide or support the security/functionality feature.

- Message 1: The message $M_1 = \{SID_k, CID_i, B_1, B_2, T_1\}$ needs (128 + 256 + 256 + 256 + 32) = 928 bits;
- Message 2: The message $M_2 = \{B_3, B_4, B_5, T_2\}$ requires (256 + 256 + 256 + 32) = 800 bits;
- Message 3: The message $M_3 = \{B_6, B_7, T_3\}$ requires (256 + 256 + 32) = 544 bits;
- Message 4: The message $M_4 = \{B_8, B_9, T_4\}$ needs (256 + 256 + 32) = 544 bits.

Therefore, the total communication cost of our scheme is 928 + 800 + 544 + 544 = 2816 bits. We show the total communication cost of our scheme and other related scheme [11,38–41] in Table 5. As a result, Figure 12 illustrates that our scheme has more efficient communication costs than other related schemes.

Table 5. Comparison of communication costs required for AKA.

Schemes	Communication Costs	Messages
Li et al. [38]	3584 bits	4 messages
Shin et al. [39]	4480 bits	4 messages
Rangwani et al. [40]	2816 bits	4 messages
Masud et al. [41]	3200 bits	4 messages
Chen et al. [11]	3072 bits	4 messages
Proposed	2816 bits	4 messages

8.3. Computational Cost Comparison

We evaluated the computational costs of our scheme. According to [24], we determined the comparative analysis for the computational cost of the proposed scheme with [11,38–41] in the AKA phase. According to [24], we define T_H, T_{RNG}, T_{EM}, T_{EA}, T_F, and T_{PUF} as the hash function (≈0.00023 ms), random number generation (≈0.0539 ms), ECC multiplication (≈0.2226 ms), ECC addition (≈0.00288 ms), fuzzy extractor (≈0.268 ms), and PUF operation time (≈0.012 ms), respectively. Additional, we did not consider the execution time of Exclusive-OR (⊕) operations because it is computationally negligible. Table 6 shows the detail.

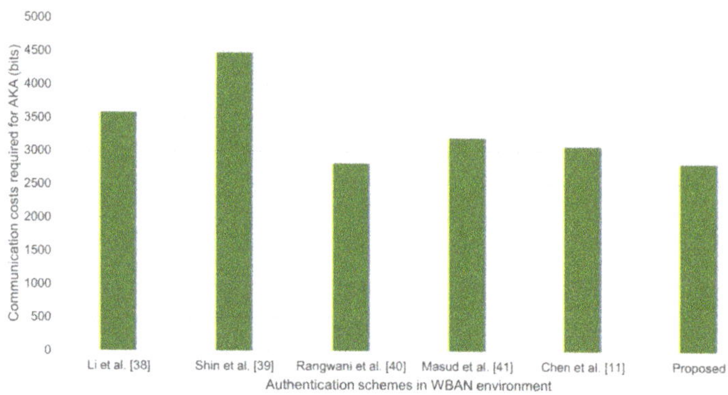

Figure 12. Communication cost comparison of related schemes [11,38–41].

The total computational costs of our scheme was estimated to be lower than other related schemes, except Masud et al.'s scheme. However, our scheme uses the fuzzy extractor and PUF to outperform Masud et al.'s scheme. Figure 13 shows that the computational cost (delay) increases with increasing numbers of users.

Table 6. Computational costs of each related scheme.

Scheme	User	Gateway	Sensor Node	Total	Total Cost (s)
Li et al. [38]	$1T_{RNG} + 9T_H + 3T_{EM}$	$1T_{RNG} + 8T_H + 1T_{EM}$	$1T_{RNG} + 4T_H + 2T_{EM}$	$3T_{RNG} + 21T_H + 6T_{EM}$	≈1.5021 ms
Shin et al. [39]	$1T_{RNG} + 1T_F + 14T_H + 2T_{EM}$	$12T_H + 1T_{EM}$	$1T_{RNG} + 5T_H + 1T_{EM}$	$2T_{RNG} + 1T_F + 31T_H + 4T_{EM}$	≈1.232 ms
Rangwani et al. [40]	$5T_H + 2T_{EM} + 3T_{EA}$	$4T_H + 2T_{EM} + 3T_{EA}$	$8T_H + 2T_{EM} + 4T_{EA}$	$17T_H + 6T_{EM} + 10T_{EA}$	≈1.36831 ms
Masud et al. [41]	$1T_{RNG} + 3T_H$	$4T_{RNG} + 3T_H$	$2T_{RNG} + 2T_H$	$7T_{RNG} + 8T_H$	≈0.379 ms
Chen et al. [11]	$9T_H$	$7T_H + 2T_{ENC}$	$7T_H$	$23T_H + 2T_{ENC}$	≈0.739 ms
Proposed	$5T_H + 1T_{RNG} + 1T_F$	$9T_H + 1T_{RNG}$	$5T_H + 1T_{RNG} + 1T_{PUF}$	$19T_H + 3T_{RNG} + 1T_F + 1T_{PUF}$	≈0.44607 ms

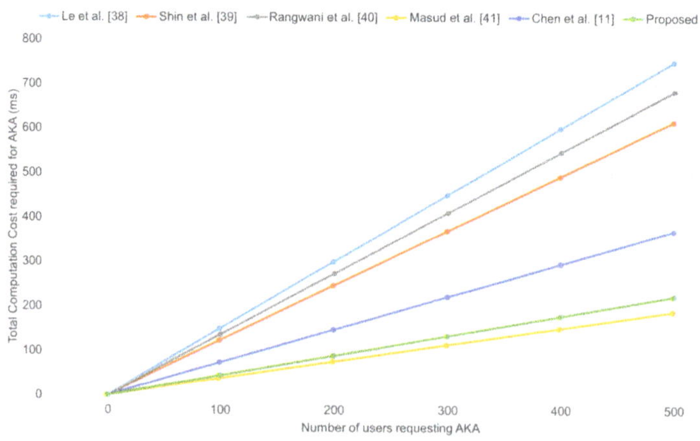

Figure 13. Total computation cost with increasing the AKA requests [11,38–41].

9. Conclusions

In this paper, we reviewed Chen et al.'s scheme and demonstrated that it is vulnerable to several attacks, such as privileged insider attacks, physical cloning attacks, verification leakage attacks, impersonation attacks, and session key disclosure attacks. Therefore, it is hard for Chen et al.'s scheme to be applied to WBANs properly, and a secure user authentication scheme should be presented for wireless medical environments. To enhance the security level of Chen et al.'s scheme, we proposed a secure three-factor mutual authentication and key agreement scheme using a secure PUF in the WBAN environment. Our scheme is lightweight because it uses only hash functions and Exclusive-OR operators and a fuzzy extractor to provide a secure login process. Moreover, our scheme resists physical cloning attacks using the PUF. The proposed scheme guarantees mutual authentication through BAN logic and utilizes the RoR model by which the session key is secured. Using the AVISPA simulation tool, we also demonstrated that our proposed scheme could withstand the replay and man-in-the-middle attacks. Moreover, we performed an informal security analysis to show that our proposed scheme provides protection against diverse hazards and attacks, including privileged insiders, physical cloning, verification table leakage, impersonation, session key disclosure, ephemeral secret leakage, replay, man-in-the-middle, stolen mobile device, offline password guessing, and denial-of-service attacks. We also proved that our scheme provides user anonymity, mutual authentication, and perfect forward secrecy. Finally, we compared the communication and computational costs of our scheme with those of related schemes after estimation. Based on the results, our scheme provides a lower communication cost and a higher security level compared to related existing schemes. Accordingly, we expect that our proposed scheme is to provide secure medical environments and to increase the use of the various healthcare applications.

Author Contributions: Conceptualization, S.L.; Formal analysis, S.Y. and Y.P.; Methodology, S.L. and S.K.; Software, S.Y.; Validation, N.J. and Y.P.; Formal Proof, Y.P.; Writing—original draft, S.L. and Y.P.; Writing—review and editing, S.K. and Y.P.; Supervision, N.J. and Y.P. All authors have read and agreed to the published version of the manuscript.

Funding: This work was supported by the Korean Government under Electronics and Telecommunications Research Institute (ETRI) Grant (20ZR1300, Core Technology Research on Trust Data Connectome).

Institutional Review Board Statement: Not applicable.

Informed Consent Statement: Not applicable.

Data Availability Statement: Not applicable.

Conflicts of Interest: The authors declare no conflict of interest.

References

1. Mastnak, T.; Maver, U.; Finšgar, M. Addressing the Needs of the Rapidly Aging Society through the Development of Multifunctional Bioactive Coatings for Orthopedic Applications. *Int. J. Mol. Sci.* **2022**, *23*, 2786. [CrossRef] [PubMed]
2. Abdulsalam, Y.; Hossain, M.S. COVID-19 networking demand: An auction-based mechanism for automated selection of edge computing services. *IEEE Trans. Netw. Sci. Eng.* **2020**, *9*, 308–318.
3. Lara, E.; Aguilar, L.; Sanchez, M.A.; García, J.A. Lightweight authentication protocol for M2M communications of resource-constrained devices in industrial Internet of Things. *Sensors* **2020**, *20*, 501. [CrossRef] [PubMed]
4. Oh, J.; Yu, S.; Lee, J.; Son, S.; Kim, M.; Park, Y. A secure and lightweight authentication protocol for IoT-based smart homes. *Sensors* **2021**, *21*, 1488. [CrossRef] [PubMed]
5. Park, K.; Noh, S.; Lee, H.; Das, A.K.; Kim, M.; Park, Y.; Wazid, M. LAKS-NVT: Provably secure and lightweight authentication and key agreement scheme without verification table in medical internet of things. *IEEE Access* **2020**, *8*, 119387–119404. [CrossRef]
6. Honeine, P.; Mourad, F.; Kallas, M.; Snoussi, H.; Amoud, H.; Francis, C. Wireless sensor networks in biomedical: Body area networks. In Proceedings of the International Workshop on Systems, Signal Processing and Their Applications, WOSSPA, Tipaza, Algeria, 9–11 May 2011; pp. 388–391.
7. Aileni, R.M.; Suciu, G. IoMT: A blockchain perspective. In *Decentralised Internet of Things*; Springer: Berlin/Heidelberg, Germany, 2020; pp. 199–215.

8. Rahman, M.; Jahankhani, H. Security vulnerabilities in existing security mechanisms for IoMT and potential solutions for mitigating cyber-attacks. In *Information Security Technologies for Controlling Pandemics*; Springer: Berlin/Heidelberg, Germany, 2021; pp. 307–334.
9. Hajar, M.S.; Al-Kadri, M.O.; Kalutarage, H.K. A survey on wireless body area networks: Architecture, security challenges and research opportunities. *Comput. Secur.* **2021**, *104*, 102211. [CrossRef]
10. Yaghoubi, M.; Ahmed, K.; Miao, Y. Wireless Body Area Network (WBAN): A Survey on Architecture, Technologies, Energy Consumption, and Security Challenges. *J. Sens. Actuator Netw.* **2022**, *11*, 67. [CrossRef]
11. Chen, C.M.; Li, Z.; Chaudhry, S.A.; Li, L. Attacks and solutions for a two-factor authentication protocol for wireless body area networks. *Secur. Commun. Netw.* **2021**, *2021*, 3116593. [CrossRef]
12. Dodis, Y.; Reyzin, L.; Smith, A. Fuzzy extractors: How to generate strong keys from biometrics and other noisy data. In *International Conference on the Theory and Applications of Cryptographic Techniques*; Springer: Berlin/Heidelberg, Germany, 2004; pp. 523–540.
13. Kumar, P.; Lee, S.G.; Lee, H.J. E-SAP: Efficient-strong authentication protocol for healthcare applications using wireless medical sensor networks. *Sensors* **2012**, *12*, 1625–1647. [CrossRef] [PubMed]
14. He, D.; Kumar, N.; Chen, J.; Lee, C.C.; Chilamkurti, N.; Yeo, S.S. Robust anonymous authentication protocol for health-care applications using wireless medical sensor networks. *Multimed. Syst.* **2015**, *21*, 49–60. [CrossRef]
15. Wu, F.; Xu, L.; Kumari, S.; Li, X. An improved and anonymous two-factor authentication protocol for health-care applications with wireless medical sensor networks. *Multimed. Syst.* **2017**, *23*, 195–205. [CrossRef]
16. Li, X.; Ibrahim, M.H.; Kumari, S.; Sangaiah, A.K.; Gupta, V.; Choo, K.K.R. Anonymous mutual authentication and key agreement scheme for wearable sensors in wireless body area networks. *Comput. Netw.* **2017**, *129*, 429–443. [CrossRef]
17. Gupta, A.; Tripathi, M.; Sharma, A. A provably secure and efficient anonymous mutual authentication and key agreement protocol for wearable devices in WBAN. *Comput. Commun.* **2020**, *160*, 311–325. [CrossRef]
18. Ostad-Sharif, A.; Nikooghadam, M.; Abbasinezhad-Mood, D. Design of a lightweight and anonymous authenticated key agreement protocol for wireless body area networks. *Int. J. Commun. Syst.* **2019**, *32*, e3974. [CrossRef]
19. Alzahrani, B.A.; Irshad, A.; Albeshri, A.; Alsubhi, K.; Shafiq, M. An improved lightweight authentication protocol for wireless body area networks. *IEEE Access* **2020**, *8*, 190855–190872. [CrossRef]
20. Mahalat, M.H.; Saha, S.; Mondal, A.; Sen, B. A PUF based light weight protocol for secure WiFi authentication of IoT devices. In Proceedings of the 2018 8th International Symposium on Embedded Computing and System Design (ISED), Cochin, India, 13–15 December 2018; pp. 183–187.
21. Zhu, F.; Li, P.; Xu, H.; Wang, R. A lightweight RFID mutual authentication protocol with PUF. *Sensors* **2019**, *19*, 2957. [CrossRef] [PubMed]
22. Mahmood, K.; Shamshad, S.; Rana, M.; Shafiq, A.; Ahmad, S.; Akram, M.A.; Amin, R. PUF enable lightweight key-exchange and mutual authentication protocol for multi-server based D2D communication. *J. Inf. Secur. Appl.* **2021**, *61*, 102900. [CrossRef]
23. Chuang, Y.H.; Lei, C.L. PUF Based Authenticated Key Exchange Protocol for IoT without Verifiers and Explicit CRPs. *IEEE Access* **2021**, *9*, 112733–112743. [CrossRef]
24. Kwon, D.; Park, Y.; Park, Y. Provably secure three-factor-based mutual authentication scheme with PUF for wireless medical sensor networks. *Sensors* **2021**, *21*, 6039. [CrossRef]
25. Fotouhi, M.; Bayat, M.; Das, A.K.; Far, H.A.N.; Pournaghi, S.M.; Doostari, M.A. A lightweight and secure two-factor authentication scheme for wireless body area networks in health-care IoT. *Comput. Netw.* **2020**, *177*, 107333. [CrossRef]
26. Burrows, M.; Abadi, M.; Needham, R. A logic of authentication. *ACM Trans. Comput. Syst.* **1990**, *8*, 18–36. [CrossRef]
27. Son, S.; Lee, J.; Kim, M.; Yu, S.; Das, A.K.; Park, Y. Design of secure authentication protocol for cloud-assisted telecare medical information system using blockchain. *IEEE Access* **2020**, *8*, 192177–192191. [CrossRef]
28. Kwon, D.K.; Yu, S.J.; Lee, J.Y.; Son, S.H.; Park, Y.H. WSN-SLAP: Secure and lightweight mutual authentication protocol for wireless sensor networks. *Sensors* **2021**, *21*, 936. [CrossRef]
29. Kim, M.; Lee, J.; Park, K.; Park, Y.; Park, K.H.; Park, Y. Design of secure decentralized car-sharing system using blockchain. *IEEE Access* **2021**, *9*, 54796–54810. [CrossRef]
30. Abdalla, M.; Fouque, P.A.; Pointcheval, D. Password-based authenticated key exchange in the three-party setting. In *International Workshop on Public Key Cryptography*; Springer: Berlin/Heidelberg, Germany, 2005; pp. 65–84.
31. Wang, D.; Cheng, H.; Wang, P.; Huang, X.; Jian, G. Zipf's law in passwords. *IEEE Trans. Inf. Forensics Secur.* **2017**, *12*, 2776–2791. [CrossRef]
32. Wazid, M.; Bagga, P.; Das, A.K.; Shetty, S.; Rodrigues, J.J.; Park, Y. AKM-IoV: Authenticated key management protocol in fog computing-based Internet of vehicles deployment. *IEEE Internet Things J.* **2019**, *6*, 8804–8817. [CrossRef]
33. Lee, J.; Yu, S.; Kim, M.; Park, Y.; Das, A.K. On the design of secure and efficient three-factor authentication protocol using honey list for wireless sensor networks. *IEEE Access* **2020**, *8*, 107046–107062. [CrossRef]
34. Kwon, D.; Son, S.; Park, Y.; Kim, H.; Park, Y.; Lee, S.; Jeon, Y. Design of Secure Handover Authentication Scheme for Urban Air Mobility Environments. *IEEE Access* **2022**, *10*, 42529–42541. [CrossRef]
35. Boyko, V.; MacKenzie, P.; Patel, S. Provably secure password-authenticated key exchange using Diffie-Hellman. In *International Conference on the Theory and Applications of Cryptographic Techniques*; Springer: Berlin/Heidelberg, Germany, 2000; pp. 156–171.

36. Armando, A.; Basin, D.; Cuellar, J.; Rusinowitch, M.; Viganò, L. The AVISPA tool for the automated validation of internet security protocols and applications. In *Lecture Notes in Computer Science*; Springer: Berlin/Heidelberg, Germany, 2005; pp. 281–285.
37. Glouche, Y.; Genet, T.; Houssay, E. *SPAN: A Security Protocol ANimator for AVISPA*; IRISA/Université de Rennes 1: Rennes, France, 2008.
38. Li, X.; Peng, J.; Obaidat, M.S.; Wu, F.; Khan, M.K.; Chen, C. A secure three-factor user authentication protocol with forward secrecy for wireless medical sensor network systems. *IEEE Syst. J.* **2020**, *14*, 39–50. [CrossRef]
39. Shin, S.; Kwon, T. A privacy-preserving authentication, authorization, and key agreement scheme for wireless sensor networks in 5G-integrated Internet of Things. *IEEE Access* **2020**, *8*, 67555–67571. [CrossRef]
40. Rangwani, D.; Om, H. A secure user authentication protocol based on ECC for cloud computing environment. *Arab. J. Sci. Eng.* **2021**, *46*, 3865–3888. [CrossRef]
41. Masud, M.; Gaba, G.S.; Choudhary, K.; Hossain, M.S.; Alhamid, M.F.; Muhammad, G. Lightweight and anonymity-preserving user authentication scheme for IoT-based healthcare. *IEEE Internet Things J.* **2021**, *9*, 2649–2656. [CrossRef]
42. Son, S.; Park, Y.; Park, Y. A Secure, Lightweight, and Anonymous User Authentication Protocol for IoT Environments. *Sustainability* **2021**, *13*, 9241. [CrossRef]

electronics

MDPI

Article

A Meta-Model to Predict and Detect Malicious Activities in 6G-Structured Wireless Communication Networks

Haider W. Oleiwi [1,*], Doaa N. Mhawi [2] and Hamed Al-Raweshidy [1]

1 Department of Electronic and Electrical Engineering, Brunel University London, London UB8 3PH, UK
2 Technical Institute for Administration, Middle Technical University, Baghdad 10010, Iraq
* Correspondence: haider.al-lami@brunel.ac.uk

Abstract: The rapid leap in wireless communication systems incorporated a plethora of new features and challenges that accompany the era of 6G and beyond being investigated and developed. Recently, machine learning techniques were widely deployed in many fields, especially wireless communications. It was used to improve network traffic performance regarding resource management, frequency spectrum optimization, latency, and security. The studies of modern wireless communications and anticipated features of ultra-densified ubiquitous wireless networks exposed a risky vulnerability and showed a necessity for developing a trustworthy intrusion detection system (IDS) with certain efficiency/standards that have not yet been achieved by current systems. IDSs lack acceptable immunity against repetitive, updatable, and intelligent attacks on wireless communication networks, significantly concerning the modern infrastructure of 6G communications, resulting in low accuracies/detection rates and high false-alarm/false-negative rates. For this objective principle, IDS system complexity was reduced by applying a unique meta-machine learning model for anomaly detection networks was developed in this paper. The five main stages of the proposed meta-model are as follows: the accumulated datasets (NSL KDD, UNSW NB15, CIC IDS17, and SCE CIC IDS18) comprise the initial stage. The second stage is preprocessing and feature selection, where preprocessing involves replacing missing values and eliminating duplicate values, leading to dimensionality minimization. The best-affected subset feature from datasets is selected using feature selection (i.e., Chi-Square). The third step is represented by the meta-model. In the training dataset, many classifiers are utilized (i.e., random forest, AdaBoosting, GradientBoost, XGBoost, CATBoost, and LightGBM). All the classifiers undergo the meta-model classifier (i.e., decision tree as the voting technique classifier) to select the best-predicted result. Finally, the classification and evaluation stage involves the experimental results of testing the meta-model on different datasets using binary-class and multi-class forms for classification. The results proved the proposed work's high efficiency and outperformance compared to existing IDSs.

Keywords: 6G wireless communications; chi-square; cybersecurity; intrusion detection system; machine learning techniques; meta-model; stacking ensemble learning; voting techniques

Citation: Oleiwi, H.W.; Mhawi, D.N.; Al-Raweshidy, H. A Meta-Model to Predict and Detect Malicious Activities in 6G-Structured Wireless Communication Networks. *Electronics* **2023**, *12*, 643. https://doi.org/10.3390/electronics12030643

Academic Editors: Shihao Yan, Guanglin Zhang, Li Sun, Tsz Hon Yuen, YoHan Park, Changhoon Lee and Tao Huang

Received: 3 January 2023
Revised: 20 January 2023
Accepted: 24 January 2023
Published: 28 January 2023

1. Introduction

The advancement of modernized wireless communication networks with their accompanying features, technologies, heterogeneously connected networks/gadgets, service demands, and the huge amount of data traffic has brought more complexity and sophistication to communication systems [1]. The 6G revolution and internet of everything (IoE) technology drive artificial intelligence (AI)-based incorporations (e.g., machine learning (ML)) in the ubiquitous connection of billions of sub-networks, users, and devices. Furthermore, the new features of 6G and beyond wireless communications, movable infrastructure, and the potential intelligent services add critical security risks to the network's core, edge, and associated devices [1–4]. Modern networks benefit significantly from AI and ML in various ways, such as intelligent communications, network optimization, and big data analytics. However, the threats of renewable intelligent attacks on the networks increase proportionally with the complexity increase (caused by heterogeneity, enormous scale, and variety of applications these networks serve) [5–10]. The difficulty of creating adequate security procedures to defend the network increases due to the possibility of attackers discovering network vulnerabilities utilizing AI techniques. Thus, it is highly necessary to build a robust intelligent intrusion detection system (IDS) to comply with the evolution of intelligent attacks and to secure future networks [11–15]. The new networks connect a variety of billions of users/devices to serve people, providing a plethora of services/applications via the network's main components, e.g., the base station (BS) using the edge of technologies, e.g., terahertz communications, non-orthogonal multiple access, and IoE [12,15,16]. In risk-sensitive systems safety, the realization of a zero-day attack is not an easy process, especially with the proliferation of numerous malicious activities. Figure 1 demonstrates a sample of the 6G general expected infrastructure with a number of nominated applications and media over different areas [17].

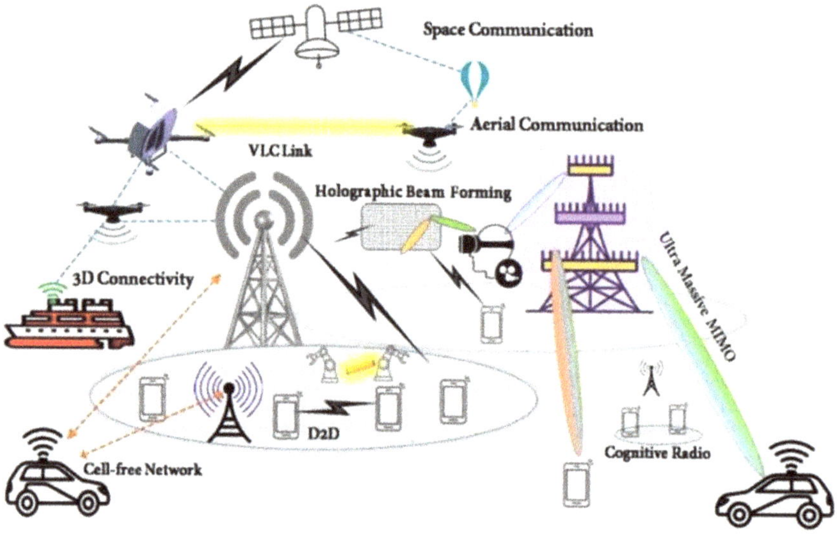

Figure 1. A sample of 6G expected infrastructure and applications.

IDSs send out notifications when discovering an unexpected activity or identified hazards. Any destructive behavior that interferes with the information system is considered an intrusion [18]. IDSs scan computers for unusual activities a conventional packet filter may fail. IDSs note any indicator for potentially dangerous action of network packets, as well as signals for highly resilient cyber defenses against disruptive activities and non-authorized access to a computer system. IDSs use two methods to detect intrusions (i.e., misuse and anomaly). A new IDS that includes these two methods was presented to overcome these limitations to increase accuracy and decrease FAR [11,19–25]. Furthermore; feature selection (FS) is a useful approach for IDSs to specify the significant features and cancel the useless features with less performance degradation [26–28]. IDSs require classifier methods to detect the final results and there are different AI methods for this task, e.g., ensemble learning (EL). EL techniques were used as building blocks for more complicated models by integrating many weak learners in EL methods, e.g., Bagging, boosting, AdaBoosting, and stacking (meta-model). These models of classifiers are used to reduce variance when using the bagging method, manipulated high bias to achieve strong classifiers inside these models when using the boosting, and the main session of the stacking (meta-model) is to combine the strengths of several effective models to provide predictions that perform better than any one model in EL [29].

However; IDSs still do not achieve the needed optimization for detection rate (DR), false alarm rate (FAR), or running time because of the high-dimensional dataset and abundant Zero-day attacks. Despite having a direct influence on resources, time complexity was not given as a significant consideration. Besides, the technological realm is envisioning IoE and 6G networks depending on the equipment that is programmed using lightweight algorithms.

This work targets initiating more sufficient/robust ML techniques-based attack-resistant detection to increase the IDSs' stability and accuracy by reducing the amount of computation/time needed by using four different datasets. The proposed model trains the FS method and ML algorithms to realize accurate/efficient IDs. Utilizing AI systems, the orientation of wireless communications must be thought about. Therefore; the contributions of this work are:

- In the context of FS and preprocessing, we used the Chi-square method for cleaning and preparing four different unbalanced datasets (NSL_KDD, UNSW _NB15, CIC_IDS17, and SCE_CIC_IDS18) to select the best subset features. Furthermore; enhancing the effectiveness of the training and testing stages is much more advantageous. These datasets undergo the cleansing and selection processes to select only the affected features to reduce time and achieve the best accuracy result.
- We enhance the performed effectiveness of the multiclass and binary class forms used with the four imbalanced datasets. Hence, the proposed work presents a novel meta-model that uses different ML techniques (i.e., random forest RF, Gradient Boost, AdaBoosting, LightGBM, XGBoosting, and CatBoosting) to work as a base classifier and then applies the meta-model technique using decision tree (DT) to select the best-affected result (prediction). The meta-model works as a prediction method to select only the classifiers with high accuracy and then enter the results into the testing part to achieve the final result.

The remaining sections of this paper are organized as follows:

Section 2 implies several similar works, while Section 3 provides a detailed definition of the proposed system's methodology and addresses the experimental findings. Furthermore, it illustrates how the proposed method was implemented with the applied datasets and addresses the technical constraints. Finally, the conclusions are stated in Section 4, which summarizes the results, directions for further investigation, and future suggestions.

2. Literature Review

In this section, the authors study the other related similar studies and demonstrate them in Table 1 for better understandable readability. Furthermore, to distinguish each of those related studies the main FS method with the number of FSs, type of the classification method, experimental results, and disadvantages.

Table 1. Similar related studies.

References/Authors	FS Methods and Number of Features	Classifiers Methods	Experimental Results	Cons
[11], Oleiwi et. al.	They used correlation FS combined with RF EL. This system selected 30, 35, and 40 FSs for (NSL, UNSW_NB, AND CIC_IDS) respectively.	Adopting two modified classifiers (RF and SVM) and applying the classifiers as AdaBoosting and bagging EL; then aggregating these classifiers by the voting average technique.	The experimental results are 99.6% accuracy with 0.004 FAR for NSL_KDD, 99.1% accuracy with 0.008 FAR for UNSW_NB2015, and 99.4% accuracy with 0.0012 FAR for CIC_IDS2017.	Complexity time measurement took too much time, due to the merging of two methods of EL techniques for splitting and disseminating normal or suspicious network traffic attacks.
[30], Gaikwad, D. and Thool, R.	N/A	DT and rule learner-based EL.	It shows that the classifiers methods of IDS exhibit the lowest false positive rate (FPR) with higher classification accuracy (i.e., 80%, 81%, 15.1%) for (accuracy, DR, FAR).	Not accurate results and undetected several attacks. Furthermore; A long time for searching with the lowest accuracy and false negative rate (FNR).
[31], Pajouh et. al.	linear discriminant analysis (They have chosen 16 features).	Two-tier anomaly-detection model using K-Nearest Neighbor KNN.	The experimental evaluation of 83.24% accuracy, 4.83% FAR, 82% true positive rate (TPR), and 5.43 FPR.	needed more execution time. Insufficient dealing with the network imbalance of anomaly datasets.

Table 1. *Cont.*

References/Authors	FS Methods and Number of Features	Classifiers Methods	Experimental Results	Cons
[32], Kanakarajan, N.K. and Muniasamy, K.	Information gain adopts 32 features for binary class and with 10-features for multiclass.	Hybrid RF with Adaptive Greedy randomized.	Accuracy is 85.0559% with information gain reaching an accuracy of 78.9035%.	Less accuracy and high FAR.
[33], Mittal, M. et. al.	DT for FS.	ML techniques for energy efficiency and anomaly detection in hybrid wireless sensor networks.	The experimental results showed that accuracy is 95%, where the precision is 94.00%, recall is 98.00%, and F1-Score is 96.00%.	Long time for searching and A high FAR.
[34], Jaw, E. and Wang, X.	The wrapper method is based on a genetic algorithm to select (11, 8, and 13) features.	Different classifiers are used for classification.	The results showed 98.99% for CIC_IDS17, 98.73% for NSL_KDD, 97.997% for UNSW_NB15 accuracy, with 98.75%, 96.64%, 98.93% DRs.	Not accurate results and undetected several attacks. A long time for searching. Furthermore; low FNR.
[35], Gupta, N. et. al.	RF was adopted to select the best subset features. By used NSL_KDD, CIDDS-001, and CIC_IDS2017.	The extreme gradient Boosting algorithm is used as a classifier with deep learning.	The experimental results are 99% for NSL, 96% for CICIDS-001%, and 92% for CIC_IDS2017.	Complexity time measurement has taken several hours, due to the deep learning techniques for splitting and disseminating normal or suspicious network traffic attacks.
[23], Mhawi. et. al.	Hybrid of Correlation FS coupled with Forest Panelized Attributes.	They used four different classifiers (i.e., SVM, RF, Naïve Bayes NB, and K-Nearest-Neighbor).	The experimental results are 99.7% for CIC_IDS17 of accuracy with 0.0053 FNR, and 0.004 FAR.	Complexity system in the FS stage and classification stage. It takes high time in the training part.

To the researchers' knowledge, the provided system outperforms the earlier systems in terms of performance and outcomes. Using numerous datasets, it considerably excels in literature performance and delivers the highest results.

3. Methodology

IDSs observe malicious or suspicious activities in the traffic across the whole communication network. They were presented to wireless communication networks to examine for any abnormal activity occurring throughout control/data communication. The hacker attempts to penetrate networks to stop communications or capture important data. By breaching networks' security and affecting the behaviors of sensors/networks, the attacker inserts bugs into a network. To solve this sensitive issue and protect the system from malicious actors, a properly secured framework is required. The proposal's main structure is shown in Figure 2.

Figure 2 shows different stages to detect suspicious/malicious activities (anomalies) over the communication network undergoing preprocessing. Before these stages, collecting different types of datasets and detecting the missing values are required, replacing the null values with some values, while average values are considered. After that, duplicate values are deleted from datasets (NSL_KDD, UNSW_NB15, CICI_IDS17, and SCE_CIC_IDS18).

Next step, data normalization and encoding processes are performed. Encoded data undergoes a dimensionality decrease to aid data handling. Accordingly, features are optimized to attain the optimal features out of the entire data. This is helpful to detect anomalies within data. After preprocessing, the cleansed data will transfer to the next level to utilize impacted features only to the finalized results by applying Chi-square. Ultimately, the proposed system uses meta-ML models as a classifier to detect and predict malicious activities in the network traffic. It includes a number of stages that include several steps with a dedicated task each. Each stage's outcome represents an input to its next stage. The stages are described in detail successively.

3.1. First Stage: Datasets Collection

The researchers' main problem is finding an appropriate dataset for evaluating IDSs. Therefore; there are different collected datasets used with different features (NSL_KDD, UNSW_NB15, CIC_IDS17, and SCE_CIC_IDS18). They were collected from different sites and contained different types of attacks. These datasets are used for experiments, and each dataset is briefly described as follows:

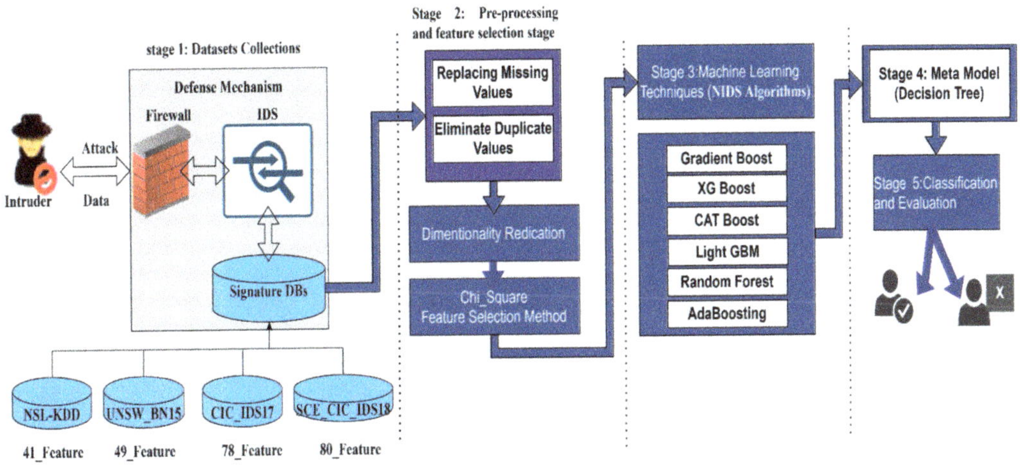

Figure 2. The proposed system's general structure.

3.1.1. First: NSL_KDD Dataset

NSL-KDD is a dataset suggested to solve some of the inherent problems of the KDD'99 dataset. Because of the scarcity of freely available datasets for networking-built IDSs, the new dataset's version is still in service as a high-impact benchmark dataset to help the researchers in comparison of multiple ID strategies, although they have technical issues noted by McHugh. NSL-KDD training set and testing set have a notable quantity of records. The achieved gain enables cost-effective experimentation on the entire set without arbitrary selection of a limited subset.

3.1.2. Second: UNSW_NB15 Dataset

It is a network intrusion dataset that is collected by the university of the new southern western network base in 2015. It contains nine types of attacks. Raw network packets are included in the dataset. There are 175,341 records in the train set and 82,332 records from various types of activities in the test set (attacks and normal activities).

3.1.3. Third: CIC_IDS17 Dataset

The CIC_IDS17 dataset (compiled in 2017) was released by the Canadian Institute for Cybersecurity (CIC). It offers positive information and the most current widespread attacks. The outcomes of the network traffic analysis using the CIC flow meter are also presented. Time-stamped flows exist for protocols, source/destination IPs, ports, and attacks. One of the most recent datasets is this one. Updated DDoS, Brute Force, XSS, SQL Injection, Infiltration, Port Scan, and Botnet assaults are among the things it contains. There are 2,830,743 records total in this dataset, which is divided into eight files. Each record comes with 78 unique characteristics and labels. In order to maintain the same magnitude order for each dataset when multi-classification is required.

3.1.4. Fourth: SCE_CIC_IDS18 Dataset

The University of New Brunswick created this dataset for analyzing DDoS data. It was sourced completely from 2018 and stopped updates. The dataset was built depending on the university's servers' logs, which have observed a variety of DoS attacks during the free availability era. When writing the dataset, ML notebooks observed that the label column is the precious portion, as it determines if the transmitted packets are malicious or benign. Data is divided into various files based on date. Each file is unbalanced, and it is up to the notebook creator to divide the dataset into a balanced form for higher-quality predictions. It has eighty columns, each of which corresponds to an entry in the IDS logging system the University of New Brunswick has. Given the system divides traffic into forward and backward. The most important columns within this dataset (i.e., Destination port, Protocol, Flow Duration, total forward packets (Tot Fwd Pkts), total backward packets (Tot Bwd Pkts), and label (Label).

3.2. Second stage: Preprocessing and FS

The datasets collected in the first stage undergo preprocessing and FS steps. The processing of these steps is demonstrated in Algorithm 1.

Algorithm 1. Preprocessing and FS.

Input: Reading Four different Datasets [] = [D1, D2, D3, and D4], N = sample size.
Output: *BestFeature.*
Begin
LOOP:
Repeat from 1 to N
1. Preprocessing steps:
(Filteration process):
Reading Datasets [i]
Repeat
  If Datasets [i] = np. information or -np. information then
  Datasets [i] = NAN */np,-np are negative, positive infinity */
  If Datasets [i] = Missing_values and duplicated_values then
  Datasets [i] = dropping values.
(Transformation process):
  If the Datasets [i] = nonnumerical_values then
  Call One_Hot_encoding function then return new datasets [i].
Normalization (Computing MinMax Scal function):
  Check Call Min_value [i] function for each dataset [i].
  Check Call Max_value [i] function for each dataset [i].
  $\mathrm{XiValue[i]} = \frac{\mathrm{XiValue-Min\_value[i]}}{\mathrm{Max_{value}[i]-Min}}$
Until Datasets [i] greater than N;
Return XiValue[i].
End Loop
2. Feature_Selection steps:
For each dataset [i] split XiValue[i] into two parts Training_part [i] and Testing_part [i]. */ 70%training_part and 30% testing_part */.
Repeat
DF = N − 1. (Freedom degrees (DF)) */It refers to the maximum number of logically independent values that can vary*/.
Compute each part Chi-square as follows: $x_2^c = \sum \frac{(Oi-Ci)^2}{Ei}$
End Loop
Return the best features Xi for each dataset [i].
End

In Algorithm 1, raw data in each dataset is passed into two main steps. Firstly, preprocessing to clean and prepared data (filtration process) then non-numerical values are converted into numerical using the one-hot encoding (transformation process) and then converted into the binary form using the Minimax scaling function (normalization). The outcome of this algorithm is to return the best subset features of each dataset. Therefore; the best subset features are (20, 30, 35, and 38) for NSL_KDD, UNSW_NB15, CIC_IDS17, and SCE_CIC_IDS18 datasets, respectively.

3.3. Third and Fourth Stages: ML Techniques for NIDS (Training Set) and Voting Techniques (Meta-Model) for the Testing Set

For the training stage, many different classifiers are used (i.e., XGBoosting, random forest (RF), AdaBoosting, GradientBoosting, LightGBM, and CatBoost) each of them considered as a base classifier. Each of these classifiers manipulates the training data independently by taking the Di of each dataset. Afterby, the results of each base classifier (predictions) are aggregated into the meta-model (DT), Figure 3 demonstrates the main idea of meta-model classifiers. Furthermore, the testing stage begins in the meta-model to get the prediction results to check the evaluation and performance of the proposed meta-model. Algorithm 2 illustrates this stage.

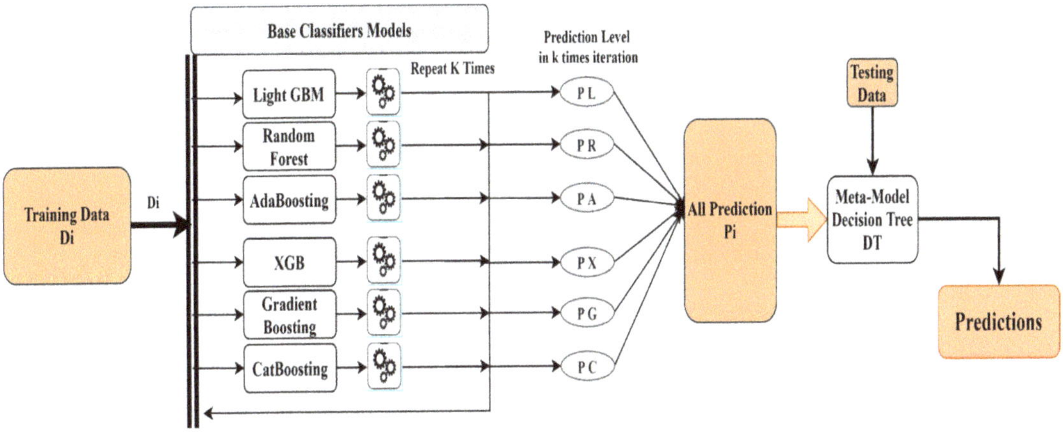

Figure 3. The Meta-model structure.

The meta-model working mechanism are demonstrated in detail in the following subsections.

3.3.1. The Datasets Partitioning Mechanism

It is necessary to aggregate the result of each classifier through the composite model and then send them to the stacking model to select the best result for voting. Furthermore, the voting technique is a type of EL methods that combines the predictions of several different models (classifiers) and selects the best prediction with the most votes.

As shown in Figure 3, the meta-model system has four traffic datasets, it uses three datasets as source datasets to train the meta-model, whereas the fourth dataset is used as a target to fine-tune it and then test the model performance. Each source dataset requires splitting into training and validation partitions. During training, it randomly selects two batches of samples from the training datasets, using one batch to compute the task-specific parameters and the other batch to compute the loss. Then repeat the same process with the validation dataset to be able to select the best prediction model. After the training, it is essential to fine-tune the model upon the target dataset.

3.3.2. Classifiers Work and Aggregation Techniques

In Algorithm 2 there are different classifiers, each of which performs a specific process and manipulates problems precisely. RF is a meta-estimator that fits several DT classifiers on different datasets' sub-samples, applying averaging to enhance predictive accuracy and controlling overfitting. Subsample and original input sample sizes are usually the same, however, samples are drawn with replacement if bootstrap = True. While XGBoost optimized gradient boosted DT. This classifier does not need normalized features and works well if the data is nonlinear, non monotonic, or with segregated clusters. Whereas the AdaBoosting classifier is to fit a sequence of weak-learners (e.g., models that are better than stochastic guessing, like small DTs) on repetitively modifying data versions. Consequently, the predictions get integrated by a weighted majority vote (or sum) to generate the final prediction. Data modifications at each so-called boosting iteration include applying weights $\omega_1, \omega_2, \omega_3, \ldots \ldots, \omega_N$ to every training sample.

Algorithm 2. ML and meta-model techniques.

Input: Xi for each dataset [i] from Algorithm 1;
K /* is the number of classifiers*/;
Learning_Rate (LR);
Random_state (RS);
M_i; /* Error rate of each classifier*/; (i.e., (Mi) $= \sum_{j=1}^{d} wj \times err(Xj)$);
Number of Estimators (NS); /* subset number*/;
Criterion; /* type of measure*/;
Machin learning classifiers (Bse classifiers); (i.e.,
RandomForest (C1),
XGBoosting (C2),
AdaBoost (C3),
GradientBoosting (C4),
LightGBM (C5),
and CatBoosting (C6)).
Meta-model classifier (i.e., DT (C8))
Output: A composite model.
Begin
1. ML techniques (base-classifiers):
Read a number of K.
Loop: from 1 to k
RandomForest (C1) Determine attribute:
RS = 1, NS = 10, LR = 0.01, max_features [integer] /*The number of features to consider when looking for the best split*/.
XGBoosting classifiers (C2) Determine attribute:
Determine attribute: LR = 0.01, RS = 1.
AdaBoosting classifiers (C3) Determine attribute:
Determine attribute: RS = 1, and NS = 10, wi = 1/N.
GradientBoosting (C4) Determine attribute: (Loss = 'deviance', LR = 0.1, number of estimators = 100, minimum split samples = 2, maximum depth = 3, fraction of validation = 0.1).
LightGBM (C5) Determine attribute:
RS = 1, and NS = 10.
CatBoosting (C6) Determine attribute:
RS = 1, and NS = 10.
Repeat
For i = 1 to 6 do
Mi for the prediction by applying:
$M_i = \sum_{j=1}^{d} wj \times err(Xj)$.
If M_i is larger than half then
$[\log(1 - (M_i))/(M_i)]$.
End if
Until the results of 6 Ci
End for
Return all Ci with minimum M_i.
2. Meta-model (DT) and compute (Voting techniques):
Repeat
Compute average weighting techniques for all Ci by $\frac{1}{mj} = l\sum_{i=1}^{l} pci(\frac{wi}{x})$.
Measurements of the binary and multi-class forms:
DR, FNR, FPR, TPR, TNR, accuracy, FAR, precision, and recall.
Until the result is the best.
Return composite-model.
End

Since all weights are initially set to $\omega i = 1/N$, the initial step trains a learning algorithm using initial data. The sample weights are individually adjusted for each further iteration, and the learning process is then performed once more on the reweighted data. Furthermore; to compute and adjust weight, it undergoes the following steps:

- Assigning equal weights to all the data points to find the stump that does the best job, classifying the new collection of samples by finding their Gini Index and selecting the sample's weight with the lowest Gini index.
- Calculating the "Amount of Say" and "Total error" to update the previous sample weights.
- Normalizing the new sample weights.

The consequences of training examples at a particular stage are changed to reflect whether or not the boosted model that was induced in the preceding step accurately predicted those training examples. Examples that are challenging to foresee get growing importance during the iterative process. As a result, each weak learner after them in the chain is compelled to focus on the instances that they missed before. Using gradient-boosting tree strategies has numerous benefits, which include:

- Generally, more accurate compared to other classifiers models.
- Train faster, especially on larger datasets.
- Most of them provide support handling categorical features.
- Some of them handle missing values natively.
- Often provides unbeatable predictive accuracy.
- Plenty of flexibility could optimize various loss functions.
- Provides multiple hyper-parameter setting options, making the function fit very flexibly.

LightGBM is a fast-distributed high-performance gradient-boosting framework based on DT algorithms, it is used for ranking, classification, and many other ML tasks. The CatBoost classifier is an algorithm for gradient boosting on DTs. It is used for search, recommendation systems, personal assistants, self-driving cars, weather prediction, and many other tasks in different companies.

3.4. Fifth Stage: Implementation and Evaluation

3.4.1. Implementation

It is carried out by applying four datasets (NSL_KDD, UNSW_NB15, CIC_IDS17, and SCI_CIC_IDS18). The train portion is 70% while the test portion is 30% to evaluate the proposal.

System Performance is evaluated by implementing the proposal using four various features selected using chi-square. The intrusion is detected by using different ML techniques with multiclass and binary-class forms of confusion matrices. Ultimately, performance evaluation is done by using multiple measurements; recall, precision, DR, FAR, and FNR. It is carried out by anaconda python 3.9 software and colab platform with Sklearn, Kearse, and Tensor Flow libraries with laptop hardware with the: CPU Core i7, generation 10th, and 11 windows operating system with 64-bit.

3.4.2. Evaluation and Experimental Results

1 Binary-Class and Multi-Class Confusion-Matrix forms

The experiment is conducted at this stage of the ML and meta-model (voting techniques) using four different datasets. Confusion-matrix is adopted in each class, which includes benign and attack network traffic. Furthermore, four Features are applied to detect suspicious activities on the network traffic. The proposed system uses binary and multi-class forms confusion matrices.

The distribution of the four states of true-positive (TP), false-positive (FP), true-negative (TN), and false-negative (FN) with different numbers of FSs and computing accuracy and FNR are explained in Table 2.

Table 2 explains the best features and results of accuracy and FNR (i.e., false negative detections are classified into FN and TP detections in the experiment) when using NSL_KDD, UNSW_NB15, CIC_IDS17, and SCE_CIC_IDS18 are (20, 30, 35, and 38), respectively. This measurement is significant to measure the efficiency and professionalism of the proposal due to calculating the total number of errors found in every attack diagnosed as normal. additionally, applying other features leads to an insufficiency of FNR and accuracy measures.

Table 2. Accuracy and FNR for (NSL_KDD, UNSW_NB15, CIC_IDS17, and SCE_CIC_IDS18) datasets when applied to different FSs.

Datasets	FS	TP	TN	FP	FN	Accuracy	FNR
NSL_KDD	10	9000	2280	715	605	9000 + 2280/12,600 = 0.89	605/(605 + 9000) = 0.06
	20	9714	2885	1	0	9714 + 2885/12,600 = 0.99	0/(0 + 9714)= 0
	30	9500	2480	215	405	9500 + 2480/12,600 = 0.95	405/(405 + 9500) = 0.04
	all	1525	630	144	201	1525 + 630/2470 = 0.87	201/201 + 1525 = 0
UNSW_NB15	10	1500	400	226	344	1500 + 400/2470 = 0.76	344/344 + 1500 = 0.19
	20	1525	630	144	201	1525 + 630/2470 = 0.87	201/201 + 1525 = 0.11
	30	1701	744	0	25	1701 + 744/2470 = 0.99	25/25 + 1701 = 0
	all	1000	400	226	844	1000 + 400/2470 = 0.56	844/844 + 1000 = 0.45
CIC_IDS17	10	443,615	48,561	10,650	62,736	492,176/565,562 = 0.87	62,736/443,615 + 62,736 = 0.123
	20	437,550	86,556	16,715	24,741	524,106/565,562 = 0.92	24,741/24,741 + 437,550 = 0.053
	30	453,916	10,928	1349	369	453,916/565,562 = 0.99	369/1369 + 453,916 = 0.0008
	35	453,916	110,928	349	369	564,844/565,562 = 0.99	369/369 + 453,916 = 0
	40	453,890	111,048	249	357	564,938/565,562 = 0.98	249/454,247 = 0.0005
	50	437,550	86,556	16,715	24,741	524,106/565,562 = 0.92	24,741/24,741 + 437,550 = 0.053
	all	443,615	48,561	10,650	62,736	492,176/565,562 = 0.87	62,736/443,615 + 62,736 = 0.123
SCE_CIC_IDS18	10	100,000	142,945	42,439	27,971	242,945/313,426 = 0.77	27,971/(100,000 + 27,971) = 0.218
	20	127,945	142,439	42,000	971	270,384/313,426 = 0.86	971/137,655 = 0.0705
	30	127,945	152,539	32,000	871	280,484/313,426 = 0.89	871/871 + 127,945 = 0.006,76,158
	38	142,439	170,916	0	71	313,355/313,426 = 0.99	71/(71 + 142,439) = 0.000,02,1821
	40	127,945	152,539	32,000	871	280,484/313,426 = 0.89	871/871 + 127,945 = 0.006,76,158
	50	127,945	142,439	42,000	971	270,384/313,426 = 0.86	971/137,655 = 0.0705
	60	100,000	142,945	42,439	27,971	242,945/313,426 = 0.77	27,971/(100,000 + 27,971) = 0.218
	all	100,045	102,000	43,339	67,971	202,045/313,426 = 0.64	67,971/(100,045 + 67,971) = 0.40

The core objective of utilizing different datasets is to train the proposed system for different types of attacks and make it more robust against suspicious traffic activities. Figures 4 and 5 demonstrate the final results of the binary form and multiclass form of the confusion matrix.

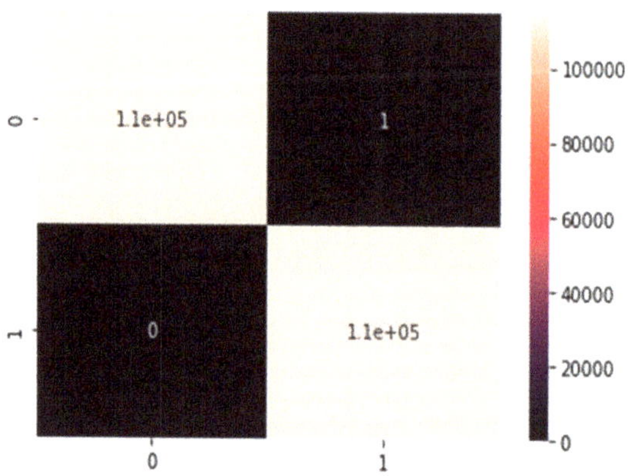

Figure 4. Binary-class confusion matrix.

Figure 4 shows that the proposed system achieves the best prediction results, it distinguishes benign activities and attacks precisely, and it can be noticed that only one percent of the benign activities is predicted as an attack; this result does not affect the final results.

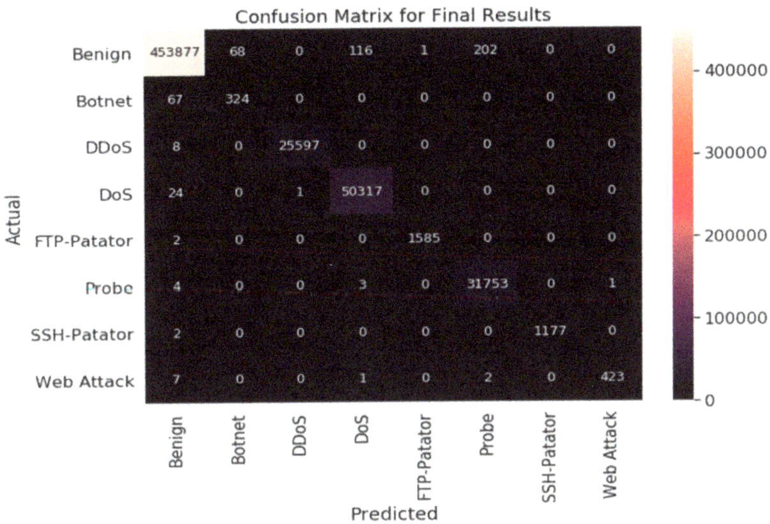

Figure 5. Multi-class confusion matrix.

In Figure 5, irrespective of the individual class's accuracy, the accuracy of the entire system (i.e., 99%) depends on the average accuracy of all the classes.

Furthermore; Figures 6 and 7 demonstrate the training and testing confusion matrix with the final measurements' results.

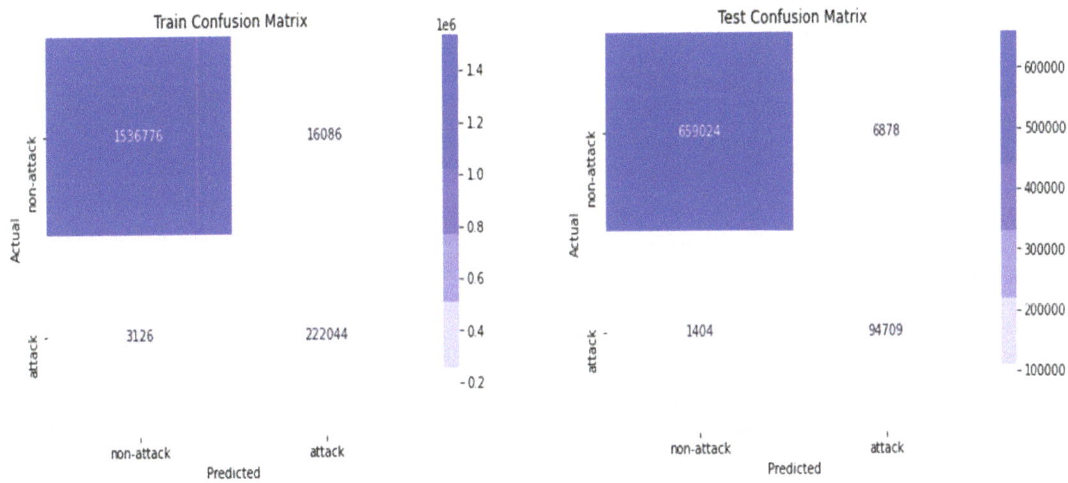

Figure 6. Train and Test confusion matrix.

2 BIG O Notation Measures

The complexity time of this proposed system is measured by applying the Big O notation (i.e., O (N^2)). It contains the calculations of complexity time. However, Figure 8 illustrates datasets classes with the required running time. Noticed the running time is increasing proportionally with input increase.

Figure 8 explains system complexity with respect to the applied datasets. The proposed meta-model reduces the number of features by selecting only the affected and sufficient features. In addition, in the training phase, the meta-model system selects the results of the best-predicted classifiers to be used in the testing phase.

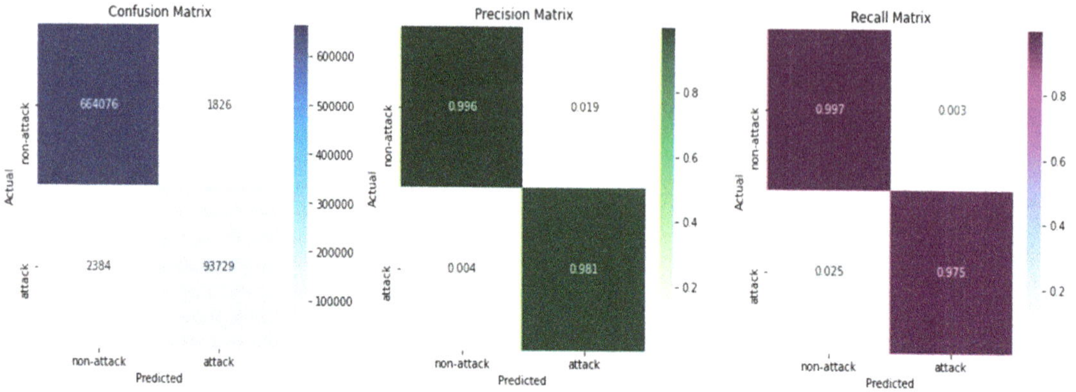

Figure 7. Final measurement matrix when applying meta-model system.

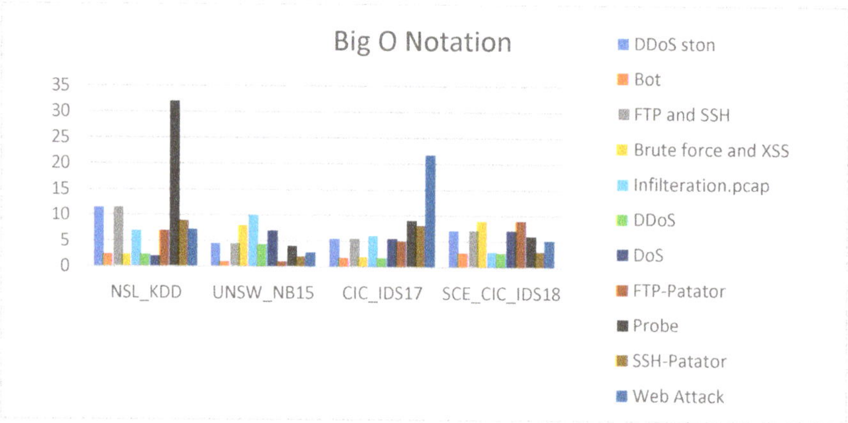

Figure 8. Big O notation idea for four datasets.

3 Analysis Results and Comparison with Other Related Studies

The first stage is very important to clear the datasets and process them from all problems, then pass to the FS stage (chi-square). In this stage, each dataset's class passes through an analysis procedure to check and choose the best effective features' subset to the final results and find the suitable subset feature of NSL_KDD is 20 features, 30 features of the UNSW_NB15, 35-features in CIC_IDS17, and 38-features in SCE_CIC_IDS18. Afterby, the ML and voting techniques stages begin to make each classifier work independently and aggregated applying the voting average technique to return the best result for the classifiers.

The proposal is assessed and compared to other previous systems by accuracy, FAR, DR, and a number of FS, Table 3 demonstrates the outperform of the meta-model is 99% for training and 90.1% for testing, as compared with other similar studies.

4 Challenges

Experimental results indicate that IDS based on a new NIDS is proposed using a meta-model (ML) with DT as a voting technique. The main objective is to build a secure system which able to distinguish malicious/suspicious traffic activities. The proposed meta-model proves sufficiency and effectiveness to detect intrusions and suspicious traffic activities, however, some limitations have come into view to be recommended to other researchers. It includes the following constraints:

- The accuracy of the entire system depends on the average accuracy of all the classes. Hence, for more efficient and accurate results, it is recommended to compute the accuracy of each class a side and accordingly the system average accuracy of all the classes for optimal performance.
- The meta-model system outperforms excellent performance when testing the system by four different datasets, however, it does not consider further attacks sourced by external networks.
- Analyzing data connections aids in the detection of non-detectable attacks throughout the application of IDS to each connection record separately. Thus, it always requires updated preprocessing and FS for accurate analyses.
- Deploying the proposed NIDS to the classified information servers of security establishments. Hence, this requires constant development for up-to-date NIDSs.

Table 3. Results comparison with other studies.

References/ Published Year	Dataset	FS Method	Number of FS	Classification Method	Accuracy %	DR %	FAR %
[30], 2016	NSL_KDD	DT	N/A	EL Methods (Rule base)	80	81	N/A
[31], 2017		KNN	16	NB	83	82	4.83
[32], 2021		symmetrical uncertainty, Information Gain and CFS	32	Gradient Adaptive Rate	85	N/A	15.00
			10		78	N/A	1.00
[33], 2021		Entropy	42	SVM	95	96	5.11
[34], 2021	UNSW_NB2015	Wrapper based GA	13	logistic regression as an EL algorithm	97.99	96.64	N/A
			8		98.73	98.93	N/A
	CIC_ID17		11		98.99	98.75	N/A
[35], 2022	NSL_KDD	Deep NN	N/A	Gradient Boosting algorithm	99	N/A	N/A
[35], 2022	CIC_ID17		N/A		92	N/A	N/A
[11], 2022	NSL_KDD	CFS-RF	30	Voting (RF, and SVM)	99.4	99.9	0.004
	UNSW_NB15		35		99.8	99.6	0.008
	CIC_ID17		40		99.7	99.4	0.0012
Meta-model	NSL_KDD	Chi-square	20	ML with meta-model classifiers (i.e., XGB (C1), Random Forest (C2), DT (C3), AdaBoost (C4), GradientBoosting (C5), LightGBM (C6), and CatBoost (C7)).	99.9	99	0.002
	UNSW_NB15		30		99.5	99	0.004
	CIC_ID17		35		99.8	99	0.0013
	SCE_CIC_IDS18		38		99.3	99	0.0021

4. Conclusions

In nutshell, it was discovered that the existing IDSs are still ineffectual despite having intentionally utilized a range of ML techniques to increase their performance, principally as a result susceptibility of to the anticipated 6G wireless paradigm and the rapidly evolving sophisticated threats. The meta-model system initiated a new IDS mechanism to apply to unbalanced/high dimensional network traffic having a low DR given the needed ML classifiers and voting mechanisms. The proposed meta-model system complexity was reduced while applying Chi-Square to present (20, 30, 35, and 38) features for NSL KDD, UNSW NB15, CIC IDS17, and SCI CIC IDS18, respectively to acquire the ideal subset of the best FS and dimensionality reduction. For each dataset, the experiment's results of the meta-model achieve high accuracies for all datasets reach 0.99% and low FAR values for NSL KDD, UNSW NB15, CIC IDS17, and SCI CIC IDS18 were 0.002, 0.004, 0.0013, and 0.0021, respectively. Other findings are concisely displayed within the results comparison table. The suggested method also outperformed current classification methods. As can be observed, this method significantly increased the IDS market's competitive edge over other strategies. Despite the system's benefits, further work is still required to make it capable of handling potential threats from future infrequent traffic.

Author Contributions: Conceptualization, H.W.O. and D.N.M.; methodology, D.N.M. and H.W.O.; software, D.N.M.; validation, H.W.O., D.N.M. and H.A.-R.; formal analysis, D.N.M. and H.W.O.; resources, D.N.M.; data curation, D.N.M.; writing—original draft preparation, H.W.O. and D.N.M.; writing—review and editing, H.W.O.; visualization, H.W.O. and D.N.M.; supervision, H.A.-R.; project administration, H.W.O. All authors have read and agreed to the published version of the manuscript.

Funding: This research received no external funding.

Data Availability Statement: NSL_KDD, and UNSW_NB15 Dataset free downloaded from the link: http://www.di.uniba.it/~andresini/datasets.html, accessed on 18 February 2022. CICIDS2017 Dataset free downloaded from the link: http://205.174.165.80/CICDataset/CIC-IDS-2017/Dataset/, accessed on 24 June 2022, and SCE_CIC_IDS18Dataset free downloaded from the link: https://www.unb.ca/cic/datasets/ids-2018.html, accessed on 12 January 2022.

Conflicts of Interest: The authors declare no conflict of interest.

References

1. Letaief, K.B.; Chen, W.; Shi, Y.; Zhang, J.; Zhang, Y.J.A. The Roadmap to 6G: AI Empowered Wireless Networks. *IEEE Commun. Mag.* **2019**, *57*, 84–90. [CrossRef]
2. Duan, Z.; Song, P.; Yang, C.; Deng, L.; Jiang, Y.; Deng, F.; Jiang, X.; Chen, Y.; Yang, G.; Ma, Y.; et al. The impact of hyperglycaemic crisis episodes on long-term outcomes for inpatients presenting with acute organ injury: A prospective, multicentre follow-up study. *Front. Endocrinol. (Lausanne)* **2022**, *13*, 1–11. [CrossRef] [PubMed]
3. Lin, Z.; An, K.; Niu, H.; Hu, Y.; Chatzinotas, S.; Zheng, G.; Wang, J. SLNR-based Secure Energy Efficient Beamforming in Multibeam Satellite Systems. *IEEE Trans. Aerosp. Electron. Syst.* **2022**, 1–4. [CrossRef]
4. Lin, Z.; Lin, M.; Wang, J.B.; De Cola, T.; Wang, J. Joint Beamforming and Power Allocation for Satellite-Terrestrial Integrated Networks with Non-Orthogonal Multiple Access. *IEEE J. Sel. Top. Signal Process.* **2019**, *13*, 657–670. [CrossRef]
5. Mokhtari, S.; Abbaspour, A.; Yen, K.K.; Sargolzaei, A. A machine learning approach for anomaly detection in industrial control systems based on measurement data. *Electron* **2021**, *10*, 407. [CrossRef]
6. Sommestad, T.; Holm, H.; Steinvall, D. Variables influencing the effectiveness of signature-based network intrusion detection systems. *Inf. Secur. J.* **2021**, *31*, 711–728. [CrossRef]
7. Winanto, E.A.; Idris, M.Y.; Stiawan, D.; Nurfatih, M.S. Designing consensus algorithm for collaborative signature-based intrusion detection system. *Indones J. Electr. Eng. Comput. Sci* **2021**, *22*, 485–496. [CrossRef]
8. Creech, G.; Hu, J. A semantic approach to host-based intrusion detection systems using contiguousand discontiguous system call patterns. *IEEE Trans. Comput.* **2014**, *63*, 807–819. [CrossRef]
9. Sahu, K.K.; Nayak, S.C.; Behera, H.S. Multi-step-ahead exchange rate forecasting for South Asian countries using multi-verse optimized multiplicative functional link neural networks. *Karbala. Int. J. Mod. Sci.* **2021**, *7*, 7. [CrossRef]
10. Jabardi, M.; Hadi, A.S. Twitter fake account detection and classification using ontological engineering and semantic web rule language. *Karbala. Int. J. Mod. Sci.* **2020**, *6*, 404–413. [CrossRef]
11. Oleiwi, H.W.; Mhawi, D.N.; Al-Raweshidy, H. MLTs-ADCNs: Machine Learning Techniques for Anomaly Detection in Communication Networks. *IEEE Access* **2022**, *10*, 91006–91017. [CrossRef]
12. Oleiwi, H.W.; Al-Raweshidy, H. Cooperative SWIPT THz-NOMA/6G Performance Analysis. *Electronics* **2022**, *11*, 873. [CrossRef]
13. Oleiwi, H.W.; Saeed, N.; Al-Raweshidy, H.S. A Cooperative SWIPT-Hybrid-NOMA Pairing Scheme Considering SIC Imperfection for THz Communications. In Proceedings of the 2022 IEEE 4th Glob Power, Energy Commun Conf GPECOM 2022, Cappadocia, Turkey, 14–17 June 2022; pp. 638–643. [CrossRef]
14. Oleiwi, H.W.; Al-Raweshidy, H. SWIPT-Pairing Mechanism for Channel-Aware Cooperative H-NOMA in 6G Terahertz Communications. *Sensors* **2022**, *22*, 6200. [CrossRef] [PubMed]
15. Oleiwi, H.W.; Saeed, N.; Al-Raweshidy, H. Cooperative SWIPT MIMO-NOMA for Reliable THz 6G Communications. *Network* **2022**, *2*, 257–269. [CrossRef]
16. Zhang, J.; Su, Q.; Tang, B.; Wang, C.; Li, Y. DPSNet: Multitask Learning Using Geometry Reasoning for Scene Depth and Semantics. *IEEE Trans. Neural. Netw. Learn. Syst.* **2021**, 1–12. [CrossRef] [PubMed]
17. Gui, G.; Liu, M.; Tang, F.; Kato, N.; Adachi, F. 6G: Opening New Horizons for Integration of Comfort, Security, and Intelligence. *IEEE Wirel. Commun.* **2020**, *27*, 126–132. [CrossRef]
18. Rajagopal, S.; Kundapur, P.P.; Hareesha, K.S. A Stacking Ensemble for Network Intrusion Detection Using Heterogeneous Datasets. *Secur. Commun. Netw.* **2020**, *2020*, 1–9. [CrossRef]
19. Mhawi, D.N.; Oleiwi, H.W.; Saeed, N.H.; Al-Taie, H.L. An Efficient Information Retrieval System Using Evolutionary Algorithms. *Network* **2022**, *2*, 583–605. [CrossRef]
20. Doaa Nteesha Mhawi, A.K. Information Retrieval Using Modified Genetic Algorithm. *Al. Mansour. J.* **2017**, *27*, 15–35. [CrossRef]
21. Oleiwi, H.W.; Saeed, N.; Al-taie, H.L.; Mhawi, D.N. Evaluation of Differentiated Services Policies in Multihomed Networks Based on an Interface-Selection Mechanism. *Sustainability* **2022**, *14*, 3235. [CrossRef]
22. Ghindawi, I.W.; Kadhm, M.S.; Mhawi, D.N. The Weighted Feature Selection Method. *J. Coll. Educ.* **2018**, *3*, 1–12.

23. Mhawi, D.N.; Aldallal, A.; Hassan, S. Advanced Feature-Selection-Based Hybrid Ensemble Learning Algorithms for Network Intrusion Detection Systems. *Symmetry* **2022**, *14*, 1461. [CrossRef]
24. Oleiwi, H.; Saeed, N.; Al-Taie, H.; Nteesha, D. An Enhanced Interface Selectivity Technique to Improve the QoS for the Multi-homed Node. *Eng. Technol. J.* **2022**, *40*, 101–109. [CrossRef]
25. Mhawi, D.N. Proposed Hybrid Correlation Feature Selection Forest Panalized Attribute Approach to advance IDSs. *Mod. Sci.* **2021**, *7*, 15. [CrossRef]
26. Hota, H.S.; Shrivas, A.K. Decision Tree Techniques Applied on NSL-KDD Data and Its Comparison with Various Feature Selection Techniques. In *Smart Innovation, Systems and Technologies*; Springer: Cham, Switzerland, 2014; Volume 27, pp. 205–212.
27. Khammassi, C.; Krichen, S. A GA-LR wrapper approach for feature selection in network intrusion detection. *Comput. Secur.* **2017**, *70*, 255–277. [CrossRef]
28. Moon, S.-H.; Kim, Y.-H. An improved forecast of precipitation type using correlation-based feature selection and multinomial logistic regression. *Atmos. Res.* **2020**, *240*, 104928. [CrossRef]
29. Loey, M.; Manogaran, G.; Taha, M.H.N.; Khalifa, N.E.M. A hybrid deep transfer learning model with machine learning methods for face mask detection in the era of the COVID-19 pandemic. *Measurement* **2020**, *167*, 108288. [CrossRef]
30. Gaikwad, D.; Thool, R. DAREnsemble: Decision tree and rule learner based ensemble for network intrusion detection system. *Proc. Smart Innov. Syst. Technol.* **2016**, *50*, 185–193. [CrossRef]
31. Pajouh, H.H.; Dastghaibyfard, G.; Hashemi, S. Two-tier network anomaly detection model: A machine learning approach. *J. Intell. Inf. Syst.* **2015**, *48*, 61–74. [CrossRef]
32. Kanakarajan, N.K.; Muniasamy, K. Improving the accuracy of intrusion detection using gar-forest with feature selection. *Proc. Adv. Intell. Syst. Comput.* **2016**, *404*, 539–547.
33. Mittal, M.; de Prado, R.; Kawai, Y.; Nakajima, S.; Muñoz-Expósito, J. Machine Learning Techniques for Energy Efficiency and Anomaly Detection in Hybrid Wireless Sensor Networks. *Energies* **2021**, *14*, 3125. [CrossRef]
34. Jaw, E.; Wang, X. Feature Selection and Ensemble-Based Intrusion Detection System: An Efficient and Comprehensive Approach. *Symmetry* **2021**, *13*, 1764. [CrossRef]
35. Gupta, N.; Jindal, V.; Bedi, P. CSE-IDS: Using cost-sensitive deep learning and ensemble algorithms to handle class imbalance in network-based intrusion detection systems. *Comput. Secur.* **2022**, *112*, 102499. [CrossRef]

electronics

MDPI

Article

Generalized Code-Abiding Countermeasure

Pierre-Antoine Tissot *, Lilian Bossuet and Vincent Grosso

CNRS Laboratoire Hubert Curien UMR 5516, 42000 Saint-Etienne, France
* Correspondence: pierre.antoine.tissot@univ-st-etienne.fr

Abstract: The widely used countermeasures against fault attacks are based on spatial, temporal, or information redundancy. This type of solution is very efficient, but it can be very expensive in terms of implementation cost. Thus, trying to propose a secure and efficient countermeasure for a lightweight cipher is a hard challenge, as the goal of a lightweight cipher is to be the lightest possible. This paper considers information redundancy based on parity bit code, with code-abiding transformations of the operations. This error detection code, with the code-abiding notion added, is very efficient against single fault injection and has a small overcost. The solution is tested on the LED lightweight cipher to measure its overhead. Moreover, a bitslice version of the cipher is used with the parity bit code applied to be robust against all the single-word fault injections. The challenge is to adapt the cipher functions in a way in which the parity bit is always considered, but without considering a heavy implementation. The advantage of our solution is that this countermeasure leads to a 100% fault coverage, with a reasonable overhead.

Keywords: fault attack; error detection; code abiding; overcost; bitslice cipher

Citation: Tissot, P.-A.; Bossuet, L.; Grosso, V. Generalized Code-Abiding Countermeasure. *Electronics* **2023**, *12*, 976. https://doi.org/10.3390/electronics12040976

Academic Editors: Tao Huang, Shihao Yan, Guanglin Zhang, Li Sun, Tsz Hon Yuen, YoHan Park and Changhoon Lee

Received: 27 January 2023
Revised: 13 February 2023
Accepted: 14 February 2023
Published: 15 February 2023

1. Introduction

Cryptographic implementations are prone to physical attacks. Physical attacks take advantage of physical properties of a device while running a cryptographic algorithm to break the security. Most popular physical attacks are fault attacks [1] (taking advantage of the circuit's tend to perturbations) and side-channel attacks [2] (taking advantage of the circuit's leakage). This work focuses on fault attacks. The principle of fault attacks is to use means, such as laser injection or clock glitching, in order to inject faults during an encryption and to extract information by analyzing the circuit's behavior after the injection.

To counter fault attacks, various countermeasures have been developed, using mainly redundancy [3–7]. Redundancy allows one to create multiple information sources, and these multiple sources are compared at the end of the computation to detect fault injection. Redundancy can be applied at three different levels: temporal, spacial, and informational. Temporal redundancy is based on the multiple encryptions of a plaintext by the same physical cipher (same circuit) and on the comparison between the resulting ciphertexts. Spacial redundancy is based on the multiple encryptions of a plaintext by different physical ciphers (different circuits). Moreover, additional information can be added to the plaintext to create an information redundancy. This information is data-dependant and is used to detect if a fault is present. In all the cases, the potential leakages of the cipher are more numerous, and the side-channel attacks (SCA) are thus more efficient [8]. Therefore, when designing a countermeasure against fault attacks, the designer should then take into account vulnerabilities that the countermeasure considered able with regard to including for the side channel adversary. The objective in that case is to make the overcost of the countermeasure as small as possible, especially when the countermeasure is implemented on a lightweight cipher.

1.1. Related Work

Simon et al. [9] presented a solution of error detection that hardly increased the SCA vulnerability. However, this solution is restrictive for any designer that would prefer to apply code-abiding to an existing cipher, and more particularly, work-oriented block ciphers. Our goal is then to generalize the code abiding method to any existing or new word-oriented block cipher.

Bertoni et al. [10] used the parity bit code to detect the fault injected on AES. Bertoni et al. described a modification of the algorithm with the addition of the parity bit matrix, when our objective is to use a bitslice version of an algorithm to add a bitwise countermeasure to a word-oriented cipher. Then, the S-Box used in [10] with half of its entries set to 00..001 is efficient in a software way, but in a hardware implementation, a big number of logical gates would be used. Moreover, the protected cipher would be robust against 1-bit fault injection, but our solution would prevent any 1-word fault injection.

Lac et al. [11] used an internal redundancy countermeasure: every data block is duplicated k times and surrounded by n reference blocks. The k copies allow us to detect up to k fault injections by comparing the results. Moreover, the reference blocks would detect a fault, even if it affects all the copies of the data block. Indeed, reference blocks are known pairs of plaintext/ciphertext, and a check is done of the cipher reference blocks to detect an injected fault. The blocks are randomly distributed in the register. Thus, for each data block, we have $k + n$ blocks of overhead. In our paper, the solution adds 1 bit for each data block. The overhead is then much lighter in our solution.

1.2. Contributions

Our first contribution is an exhibition of a fault injection realized in precise conditions during a computation of a Friet operation [9] that results into an undetected error. The conditions are presented, with two countermeasures that can be applied to allow the detection of the error.

The second contribution is the application of the code-abiding method to an existing cipher with an example on the lightweight LED cipher [12]. The countermeasure is designed to obtain the smallest overcost possible. The secured solution presented in our work should be 25% more expensive than the original LED implementation, as only one parity bit is added for each nibble. Our work thus focuses on the cost optimization of the countermeasure, in terms of the number of gates and memory space needed, as well as power consumption.

This work should allow implementations to be robust against a single injection fault with an optimization of the overcost brought by the countermeasure.

2. Background

In this section, we briefly introduce notions on coding theory that are useful for the countermeasure presented. We also recall the operation of LED block cipher [12] on which we apply our countermeasure as a proof of efficiency of our method.

2.1. Error Detection

The solutions presented in this paper use the code-abiding concept introduced in [9]. This solution is based on computation over data encoded with error detection. In the following, we give the goal and the principle of error detection, which is a set of techniques that makes it possible to detect errors during the transmission of information.

Definition 1 (Error detection code). *Let E be a set and $\mathcal{C} \subset E$. We denote $\bar{\mathcal{C}} = E \backslash \{\mathcal{C}\}$. $\mathcal{C}$ is an error detection code if and only if:*

- $\forall x \in \mathcal{C}, \forall y \in \bar{\mathcal{C}}, x + y \in \bar{\mathcal{C}}$
- $\forall x \in \mathcal{C}, \forall z \in \mathcal{S} \subsetneq \mathcal{C}, x + z \in \mathcal{C}$

In this case, $+$ is the addition operator, according to the set E.

An error detection code allows one to divide a set into two different subsets with a minimal Hamming distance between a word of a subset and a word from the other.

Definition 2 (Parity bit). *Let x be a n-bit word. We denote x_i as the i-th bit of x, then we have $x = x_{n-1}||x_{n-2}||...||x_1||x_0$, where $||$ is the concatenation operator. The parity bit x_p is the sum of all the x_i (using XOR operator): $x_p = x_{n-1} \oplus x_{n-2} \oplus ... \oplus x_1 \oplus x_0$. We use the even parity in our case, so the XOR of all the bits (including the parity bit) is 0. Its purpose is to detect an odd number of fault in the output.*

Example 1 (Parity bit). *Let $x = 011010$. The parity bit $x_p = 0 \oplus 1 \oplus 1 \oplus 0 \oplus 1 \oplus 0 = 1$.*

The parity bit method is the error detection scheme that is used during this work. The two subsets are composed by the words with an even parity for the first one and the words with an odd parity for the second one.

Definition 3 (Check function). *The* `CheckFunction` *applied to a word verifies its parity characteristic. The function returns a Boolean with* `TRUE` *when an even parity is verified and* `FALSE` *when odd parity is verified.*

Example 2 (Check function).

- `CheckFunction`$(011011) =$ `TRUE`
- `CheckFunction`$(110010) =$ `FALSE`

2.2. Code Abiding

We now want to implement the error detection scheme into an encryption algorithm. Then, we need to use functions that keep the parity characteristic of the words. With this intention, we use the code abiding notion. Code abiding was introduced in [9]. In this work, the idea was to build permutation over the space E. In order to detect fault, the permutation built must respect separation of the space. In other words, the permutation over E can be seen as two permutations, one over C, the other over $\complement$. The separation between the two spaces should allows detection of every single fault injection.

Definition 4 (Code abiding function). *f is a C code abiding function if and only if:*

- $\forall x \in C, f(x) \in C$
- $\forall x \notin C, f(x) \notin C$

The algorithm has to be composed by code abiding functions to keep the parity property of the words and to propagate the error injected until the detection of the fault.

2.3. LED Cipher

The LED Cipher, presented in [12], is a lightweight block cipher. Its purpose is to offer a very small silicon footprint in comparison with other block ciphers, as well as to be secure against related-key attacks by using AES-like security proofs.

This cipher is a 64-bit block cipher using mostly 64-bit keys and 128-bit keys. However, any length between 64 bits and 128 bits can be used if the length is divisible by four. In this sense, 80-bit keys are also often used. In our work, we focus on the 64-bit key length. However, the results presented are valid for any key length and are not limited to the LED cipher. Indeed, the code abiding solution can be added on any block cipher.

A 64-bit state St is conceptually divided into sixteen 4-bit nibbles ($St = st_0||st_1||\ldots||st_{15}$) and arranged in a square array, as described in Matrix *state*.

$$state = \begin{pmatrix} st_0 & st_1 & st_2 & st_3 \\ st_4 & st_5 & st_6 & st_7 \\ st_8 & st_9 & st_{10} & st_{11} \\ st_{12} & st_{13} & st_{14} & st_{15} \end{pmatrix}$$

Using the same process, the key K is divided into subkeys k_i (Matrix K).

$$K = \begin{pmatrix} k_0 & k_1 & k_2 & k_3 \\ k_4 & k_5 & k_6 & k_7 \\ k_8 & k_9 & k_{10} & k_{11} \\ k_{12} & k_{13} & k_{14} & k_{15} \end{pmatrix}$$

The cipher process is the combination of two operations: **AddRoundKey** and **step** (see Figure 1). The **step** operation is computed s times while the **AddRoundKey** is computed $s+1$ times. This value depends on the key length: $s = 8$ for a 64-bit key and $s = 12$ for a 128-bit key.

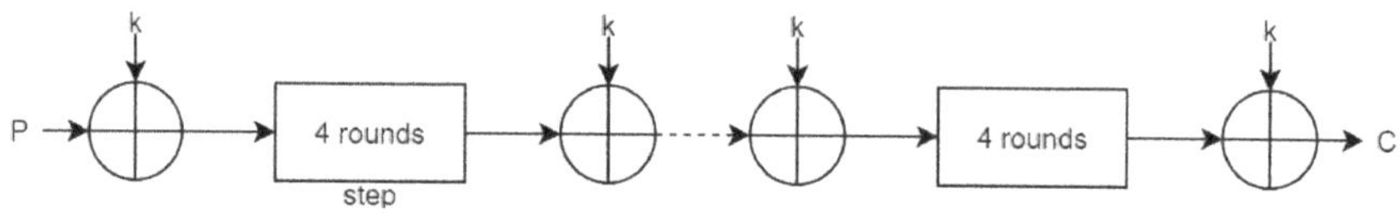

Figure 1. Representation of the LED encryption.

The **step** operation is composed by four rounds themselves composed by four operations, `AddConstants`, `SubCells`, `ShiftRows`, and `MixColumnsSerial` (See Figure 2), while the **AddRoundKey** operation is the combination of the state and the subkeys using XOR.

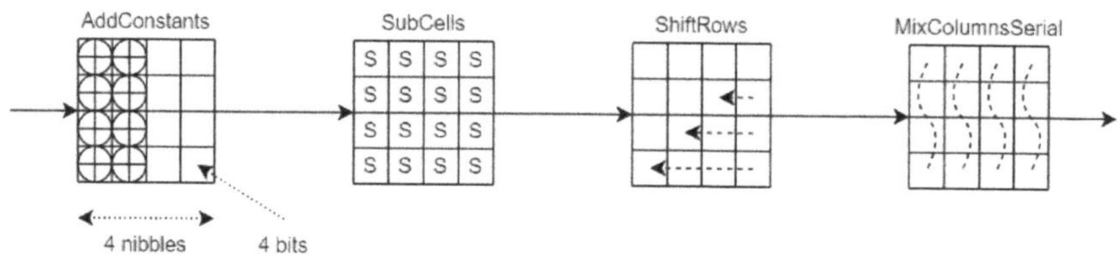

Figure 2. A round of LED composed by the functions `AddConstants`, `SubCells`, `ShiftRows` and `MixColumnsSerial`.

`AddConstants`. Six bits, $rc_5, rc_4, rc_3, rc_2, rc_1$, and rc_0 (initialized to zero), are shifted to the left ($rc_5 = rc_4; \ldots; rc_1 = rc_0$) and $rc_0 = rc_5 \oplus rc_4 \oplus 1$. Those computations are done each round before using the constant. Moreover, the key size (written in its bit form $ks_7||ks_6||ks_5||ks_4||ks_3||ks_2||ks_1||ks_0$) is used to create the constant. Then, the values are combined into a round constant (see Matrix *constant*), and this constant is added (using bitwise exclusive or) to the state.

$$constant = \begin{pmatrix} 0 \oplus (ks_7||ks_6||ks_5||ks_4) & (rc_5||rc_4||rc_3) & 0 & 0 \\ 1 \oplus (ks_7||ks_6||ks_5||ks_4) & (rc_5||rc_4||rc_3) & 0 & 0 \\ 2 \oplus (ks_3||ks_2||ks_1||ks_0) & (rc_5||rc_4||rc_3) & 0 & 0 \\ 3 \oplus (ks_3||ks_2||ks_1||ks_0) & (rc_5||rc_4||rc_3) & 0 & 0 \end{pmatrix}$$

`SubCells`. The actual state is substituted by the new state using the `PRESENT` S-box presented in Table 1. This function adds some confusion and non-linearity during the process.

Table 1. PRESENT S-box.

x	0	1	2	3	4	5	6	7	8	9	A	B	C	D	E	F
S_x	C	5	6	B	9	0	A	D	3	E	F	8	4	7	1	2

`ShiftRows`. The rows of the the state are rotated: row i is rotated i positions.

`MixColumnsSerial`. The state array is post-multiplied by the matrix M (see Matrix M). For the sake of efficiency, we use the matrix A (see Matrix A) with $A^4 = M$ and post-multiply it four times to the state.

$$A^4 = \begin{pmatrix} 0 & 1 & 0 & 0 \\ 0 & 0 & 1 & 0 \\ 0 & 0 & 0 & 1 \\ 4 & 1 & 2 & 2 \end{pmatrix}^4 = M = \begin{pmatrix} 4 & 1 & 2 & 2 \\ 8 & 6 & 5 & 6 \\ B & E & A & 9 \\ 2 & 2 & F & B \end{pmatrix}$$

After the s **step** and $s+1$ **AddRoundKey**, the state becomes the ciphertext.

Another approach of the cipher is its bitslice version [13], and this approach brings some important properties for this work. Let us suppose that the machine used has 64-bit length registers. Then, the 64-bit state is stored in a single register. The bitslice transformation of the cipher stores the state in 64 registers, each containing 1 bit of data. This approach allows us to have a bit-oriented cipher, rather than a word-oriented one. This is very important for the implementation of the parity scheme.

Another advantage of the bitslice version is the parallel encryptions. In the same conditions as the previous point, instead of using 64 registers containing only 1 bit of useful data, we can encrypt n plaintexts in parallel and then store 64 n-bit useful data. As the bitslice version is bitwise, the cipher cannot interfere between the different states. It is this method that induces detection of any 1-word fault injection. Indeed, as every machine word is seen as the concatenation of a single bit of n states and as any 1-bit injection in a state would be detected, then up to 1-word fault injection could be detected here.

The state transformation is presented in Figure 3 within a blue register of the machine and within a red state of the cipher.

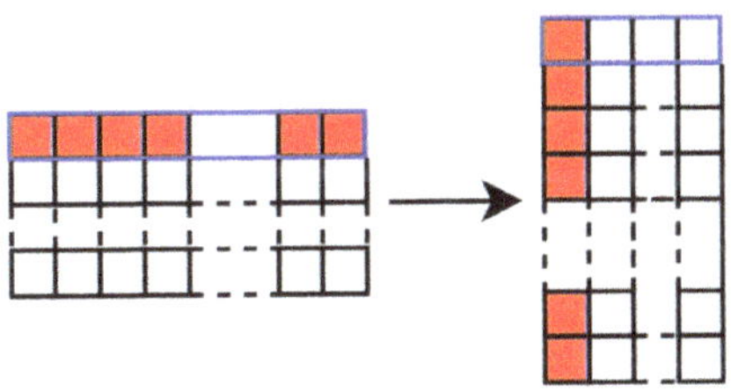

Figure 3. Bitslice transformation of the state

3. Error Compensation Issue

In this section, we exhibit a potential fault attack against Friet [9]. Indeed, in particular scenarios, we show that a fault injected can create several errors that can be compensated during the parity check, so the error is not detected. The scenario has a small probability of success, depending on the attack model and some requirements about implementation characteristics. We then present a countermeasure to prevent this kind of attack in a strong model where the adversary can inject the fault of his choice at the position and time one chooses.

3.1. Issue Example

We bring out the vulnerabilities with an example, and we next generalize it. We assume that the attacker can add a fault to one value during the computation.

Let $\epsilon = 2^{128} - 1$ as Friet manipulates 128-bit data. For the sake of simplicity, we call a, b, c, and d the inputs of the μ_2 operation. For the same reason, a', b', c', and d' are the outputs of the operation. During these operations, the parity equation followed is $d = a \oplus b \oplus c$. This parity equation is checked to ensure that no fault is injected. We inject the additive fault ϵ into the word c during the μ_2 operation, and, after that, the rotated word c is added to a and to b before the addition. The fault injection is illustrated in Figure 4, with the red line showing the modification.

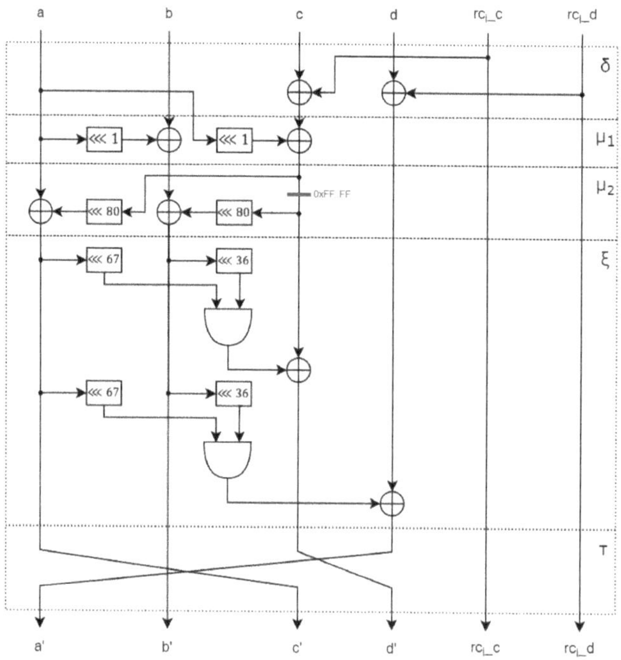

Figure 4. Round of FRIET-P.

When such a fault is injected, the outputs of two branches are modified: the second b and the third c. We denote b' and c' as the second and third words of the output of the faulty μ_2 operation. Then, we have two equations:

$$b' = b \oplus ((c \oplus \texttt{0xFF..FF}) \lll 80) = b \oplus (c \lll 80) \oplus (\texttt{0xFF..FF} \lll 80) = b \oplus (c \lll 80) \oplus \texttt{0xFF..FF}$$
$$c' = c \oplus \texttt{0xFF..FF}.$$

At the output of the faulty μ_2, we have $(a', b', c', d') = (a, b \oplus (c \lll 80) \oplus \texttt{0xFF..FF}, c \oplus \texttt{0xFF..FF}, d)$. Then, the check subroutine does not detect the injected fault, since the fault on the two branches cancel out when applying a XOR operation on the values.

In the previous example, we saw that the fault 0xFF..FF is not detected because it remains unchanged with the shift by 80 bits (that is the shift of μ_2). However, this is not the only fault that is not detected with this shift. Indeed, all the faults that remain unchanged with the 80-bit shift have this property. With the Algorithm 1 we can identify all the valid fault that are undetected. Indeed, we begin with the word $i = 1$, and we shift this word by 80 bits, and we test when the word come back to the initial value 1. The cycle length found is 8, and the number of cycles is then $\frac{word\_length}{cycle\_length} = \frac{128}{8} = 16$. That means that every fault composed by 8 same 16-bit concatenated words is not detected. Thus, we have $2^{16} - 1$ undetected faults (the value 0x00..00 is not a fault) over 2^{128} different faults. In terms of probability, we have a probability around 2^{-112} to have an undetected fault. As this probability is very tiny, then with some random faults it is difficult to identify such a weakness. Moreover, the undetected fault is a 128-bit injection, and when the registers are strictly smaller than 128-bit long, then the necessary fault would affect two registers so two faults would be needed. This constraint places the error outside of the study.

Algorithm 1 Find the length of a cycle.

Require: Size of the shift (here 80)
Ensure: Length of a cycle (how many shifts to recover the former value)
  $i \leftarrow 1$
  $t_{cycle} \leftarrow 1$
  while $(i \lll 80)\%128 \neq 1$ **do**
    $t_{cycle} \leftarrow t_{cycle} + 1$
    $i \leftarrow (i \lll 80)\%128$
  return t_{cycle}

The same analysis can be done for μ_1, and it is easy to see that the only undetected fault is the all 1 fault. For the χ, the bitwise and between two branches make the fault non-detection probabilistic in function of the data in the second branch.

3.2. Countermeasures

We assume that the registers are wide enough to ensure that the undetected faults are still in the limits of the study. An obvious solution to this issue is to increase the cycle length to limit the number of undetected faults. With a shift of 1 bit, the cycle length is maximum with a value of 128. Indeed, only the words composed by 128 same 1-bit words are undetected. However, the fault `0xFF..FF` remains undetected (`0x00..00` is still not a fault), and we have to modify the former operation to implement our solution.

Another solution must be found to detect all the faults without changing the cryptographic primitives of the cipher. We copy every variable used more than once and check if the copies are equal. A Boolean `flag` is used to express the error detection (`flag` obtains the value 0 when an error is detected). With this principle, any fault injected during an operation only affects one copy and is detected before using the copies. The following Algorithm 2 presents the copies and the checks on the operation μ_2 of the FRIET-P round and shows the overcost of this solution in comparison with the classical FRIET-P presented in Algorithm 3.

The overcost of the countermeasure lies on the three copies of the value c and the comparison of these three copies. This solution is used in the rest of the paper to avoid undetected fault injection.

Algorithm 2 Protected μ_2 operation of the FRIET-P round.

Require: Four 128-bit words a, b, c and d
Ensure: Four 128-bit words a', b', c' and d' computed by the protected μ_2 operation
  $c_0 \leftarrow c$
  $c_1 \leftarrow c$
  $c_2 \leftarrow c$
  `flag` $\leftarrow$ `flag` & $(c_0 == c_1)$ & $(c_0 == c_2)$
  $a' \leftarrow a \oplus (c_0 \lll 80)$
  $b' \leftarrow b \oplus (c_1 \lll 80)$
  $c' \leftarrow c_2$
  $d' \leftarrow d$
  return(a', b', c', d')

Algorithm 3 Classical μ_2 operation of the FRIET-P round.

Require: Four 128-bit words a, b, c and d
Ensure: Four 128-bit words a', b', c' and d' computed by the original μ_2 operation
  $a' \leftarrow a \oplus (c \lll 80)$
  $b' \leftarrow b \oplus (c \lll 80)$
  $c' \leftarrow c$
  $d' \leftarrow d$
  return(a', b', c', d')

4. Code Abiding on LED

In this section, we present a generic method to apply code-abiding countermeasures to word-oriented block ciphers and illustrate this technique on LED cipher [12]. Word-oriented ciphers are often implemented with tables (S-boxes for the substitution layer and multiplicative tables for the diffusion layer) and with XOR operation on words in the same column. We then need to consider error detection codes at word, column, and state levels.

The basic code is defined at word level and is simply extended to the column and state level. Indeed, the columns and the state are only a concatenation of the words. Thus, if we have a code $\mathcal{C}$ of parameters $[n, k, d]$ at word level, the concatenation of l word is a code $\mathcal{C}'$ of parameters $[ln, lk, d]$ at column level.

The principle of code abiding protection is to apply permutation on different codes. Thus, we have two cases if we apply always permutation to the full state, and then, either we are in the code and stay in the code or we are not in the code and stay outside the code. Since we target only one fault injection, we can at most change one set, and, due to our construction, any single fault injection forces us to change from a word of the code to a word outside the code. The last property is obtained thanks to bitslice representation and check of non-modification when we use the same variable in a different place. (Note that we can hope for security, and fault detection, for multiple random fault with high probability, thanks to parallelism we use). We next present, in more details, the adaptation made for each operation.

4.1. State Modification

Let S be the 64-bit state of the unprotected LED cipher. In order to detect fault, we need to add a redundant part. In our case, we select the 5-bit parity check code. Thus, we need to add a parity bit for each 4-bit nibble of the state. Indeed, if we denote S_i the ith bit of S, we have: $S_{64+i} = S_{4\times i} \oplus S_{4\times i+1} \oplus S_{4\times i+2} \oplus S_{4\times i+3}$, where $S_{4\times i}, S_{4\times i+1}, S_{4\times i+2}, S_{4\times i+3}$ are the bits of the nibble i. Eventually, we have a 80-bit state composed of 64 data bits and 16 parity bits.

In the Section 2, we presented the bitslice version of LED. This is the version that is used in this work, and we thus have to add the parity property by adding 16 registers. The Figure 5 illustrates this step within blue for a register of the machine, in red the data bits of a state, and in pink the parity bits of the red state. The code abiding notion is more bit-oriented than word-oriented, and then the bitslice approach allows us to use the code abiding notion on a classical word-oriented cipher.

These transformation functions are summarized in the following Algorithms 4 and 5.

Algorithm 4 State S transformed into its bitslice version.

Require: State S
Ensure: State S in its bitslice version
  for j in range(64) **do**
    $Sj_i \leftarrow Si_j$
  return S

Algorithm 5 Parity bits added to the bitslice state S.

Require: Bitslice state S
Ensure: State S with parity bits
  for i in range(16) **do**
    $S_{64+i} \leftarrow S_{4\times i} \oplus S_{4\times i+1} \oplus S_{4\times i+2} \oplus S_{4\times i+3}$
  return S

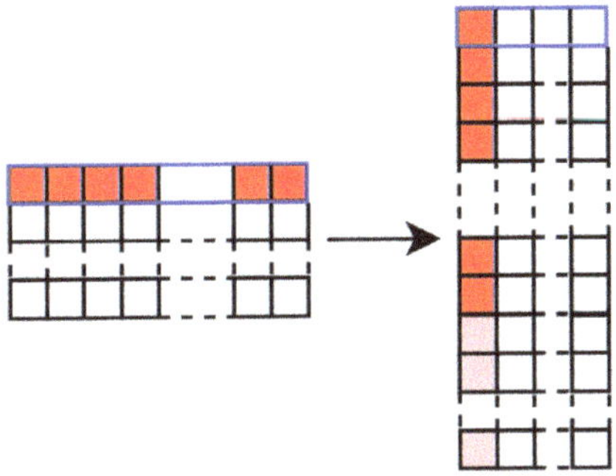

Figure 5. State after Bitslice and Parity transformations.

4.2. *Key and Constant*

We assume that the key and the constant are stored in an encoded manner.

We copy the key and the constant at the beginning of the computation and use the copy for the all computation at the end, and we check that the copy used stayed unchanged. Thus, any charge in the key during the encryption is detected. Since the attacker can only inject one fault, modification of the key is detected. An adversary that modifies the key may inject fault at each key addition. However, by using copy and checking at the end and thanks to the absence of key schedule in LED-64, the attacker cannot use this method for multiple fault injection.

The only method should be to modify the stored key. However, LED is known for resistance against related key attacks and, thus, no exploitable information can be obtained by the attacker.

If a fault is injected on the key, the XOR operation with the state propagates the error on the state until the parity check of the state.

4.3. *AddConstant*

We calculate the constant presented in Section 2. This constant is a 64-bit value that we transform into 80 n-bits values (bitslice + parity transformations) that are computed to the state using XOR operation. If n encryptions are performed in parallel, the constant must fit the n-bit length of the registers, and then the 80 bits have to be duplicated n times. This is illustrated in Algorithm 6 (0xFF..FF is composed by $\frac{n}{4}$ F).

Algorithm 6 Constant c bitsliced and duplicated.

Require: 64-bit constant c
Ensure: 80 n-bit (with duplication of the bit) constants c_i with parity and bitslice transformations.
  for i in range(64) **do**
    $c_i \leftarrow ((c \ggg (63 - i)) \ \& \ 1) \times \texttt{0xFF..FF}$
  return c_i

If a fault is injected on the constant, the error is propagated on the state until the check parity function.

4.4. ShiftRows

This function is the same operation as the former `ShiftRows` operation. The parity bits are shifted among the nibble from which they have been computed. The bit S_{64+i} is shifted $\frac{i}{4}$ bits to the left. This is illustrated in Figure 6.

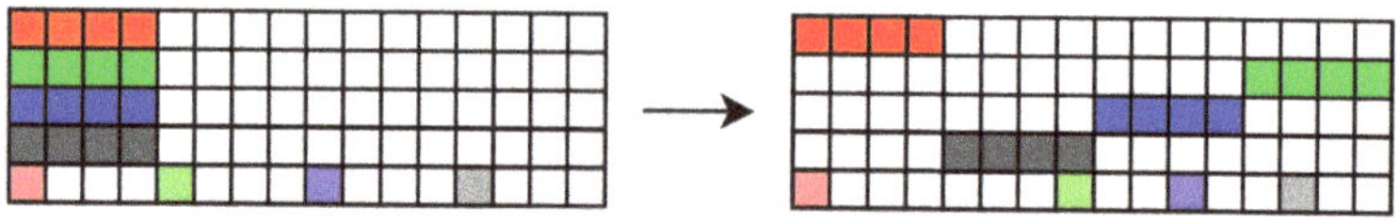

Figure 6. ShiftRows on the bitslice parity state.

During this operation, a fault can be injected on the state and stays on it until its detection. Moreover, as the state is only shifted, its value remains the same, then a fault injected before the operation is propagated on the output.

4.5. SubCells

The substitute operation brings confusion and non-linearity to encryption. It is then a critical function of the block cipher construction. The extension from the code $\mathcal{C}$ to the code $\mathcal{C}'$ requires us to represent the 4-bit S-box by a 5-bit S-box, and this projection brings a choice of the 5-bit S-box.

We present, in Section 5, a way to construct the protected S-box. Here, we present the results on the `PRESENT` S-box represented by the Table 3. However, we use an alternative form of the S-box composed only by logical gates, the algebraic normal form. This form gives five equations, where x_i is the ith bit of the input and y_i the ith bit of the output ($x_0..x_3$ are the data bits and x_4 is the parity bit).

$$
\begin{aligned}
y_0 &= x_3x_2x_1 \oplus x_3x_2x_0 \oplus x_3x_1x_0 \oplus x_3 \oplus x_2x_1 \oplus x_2 \oplus x_0 \oplus 1 \\
y_1 &= x_3x_2x_0 \oplus x_3x_2 \oplus x_3x_1x_0 \oplus x_3x_0 \oplus x_2x_0 \oplus x_1 \oplus x_0 \oplus 1 \\
y_2 &= x_3x_2x_1 \oplus x_3x_2x_0 \oplus x_3x_1x_0 \oplus x_2x_0 \oplus x_2 \oplus x_1x_0 \oplus x_0 \\
y_3 &= x_4x_3x_2x_1x_0 \oplus x_4x_3x_1x_0 \oplus x_3x_2x_1x_0 \oplus x_3x_1x_0 \oplus x_3 \oplus x_2x_1 \oplus x_1 \oplus x_0 \\
y_4 &= x_4x_3x_2x_1x_0 \oplus x_4x_3x_1x_0 \oplus x_4 \oplus x_3x_2x_1x_0 \oplus x_3x_2x_0 \oplus x_3x_2 \oplus x_3x_0 \oplus x_3 \oplus x_2 \oplus x_1x_0 \oplus x_1 \oplus x_0
\end{aligned}
$$

This function is presented in Algorithm 7. We can denote the copies of the values used more than once to avoid error compensation presented in Section 3, and as the bit x_4 is used only in the last two equations, this bit is copied only twice. Indeed, the output $S_{4\times i+m}$ only lies on the values $x_{m..}$, and then a fault is injected on a copy only affecting one output, and the parity characteristic allows the error detection.

If a fault is injected before or during the function, and the separation of the codes in the S-box representation keeps the word out of the code $\mathcal{C}$, and the fault is propagated into the space.

4.6. MixColumnsSerial

This operation is composed by the four post-multiplications with the matrix A (see Section 2). The state is decomposed into four columns of four 5-bit nibbles each. These nibbles are the 4-bit data and the parity bit associated. In our operation, only a multiplication by two is used (a multiplication by four is just two multiplications by two). The Algorithm 8 show the multiplication by two operation. This operation is a shift of the bits and a XOR with the LSB of the data word on the second bit of the nibble. The computation on the parity bit is thus only a XOR with this LSB.

The Algorithm 9 presents the state divided into columns and the multiplication with the matrix A. Same as in the `SubCells` function, we create a copy of each element used more than once to avoid a future error compensation.

With the same observations than the previous operations, if a fault is injected before or during the `MixColumnsSerial` operation, this error is propagated through the operation on the state.

All the LED functions are converted into code abiding functions to keep the parity characteristic of the state and to allow the fault injection detection. The next section focuses on the 5-bit representation of a 4-bit S-box.

Algorithm 7 `SubCells` function.

Require: State $S, i,$ `flag`
Ensure: State S after the `SubCells` operation and the flag detection `flag`
  for j in range(5) **do**
    $x_{j0} \leftarrow S_{4\times i+0}$
    $x_{j1} \leftarrow S_{4\times i+1}$
    $x_{j2} \leftarrow S_{4\times i+2}$
    $x_{j3} \leftarrow S_{4\times i+3}$
    if $j > 2$ **then**
      $x_{j4} \leftarrow S_{4+i}$
  for j in range(5) **do**
    $\texttt{flag} \leftarrow \texttt{flag}\ \&\ (x_{0j} == x_{1j})\ \&\ (x_{0j} == x_{2j})\ \&\ (x_{0j} == x_{3j})\ \&\ (x_{0j} == x_{4j})$
  $S_{4\times i+0} \leftarrow x_{03}x_{02}x_{01} \oplus x_{03}x_{02}x_{00} \oplus x_{03}x_{01}x_{00} \oplus x_{03} \oplus x_{02}x_{01} \oplus x_{02} \oplus x_{00} \oplus 1$
  $S_{4\times i+1} \leftarrow x_{13}x_{12}x_{10} \oplus x_{13}x_{12} \oplus x_{13}x_{11}x_{10} \oplus x_{13}x_{10} \oplus x_{12}x_{10} \oplus x_{11} \oplus x_{10} \oplus 1$
  $S_{4\times i+2} \leftarrow x_{23}x_{22}x_{21} \oplus x_{23}x_{22}x_{20} \oplus x_{23}x_{21}x_{20} \oplus x_{22}x_{20} \oplus x_{22} \oplus x_{21}x_{20} \oplus x_{20}$
  $S_{\times i+3} \leftarrow x_{34}x_{33}x_{32}x_{31}x_{30} \oplus x_{34}x_{33}x_{31}x_{30} \oplus x_{33}x_{32}x_{31}x_{30} \oplus x_{33}x_{31}x_{30} \oplus x_{33} \oplus x_{32}x_{31} \oplus x_{31} \oplus x_{30}$
  $S_{64+i} \leftarrow x_{44}x_{43}x_{42}x_{41}x_{40} \oplus x_{44}x_{43}x_{41}x_{40} \oplus x_{44} \oplus x_{43}x_{42}x_{41}x_{40} \oplus x_{43}x_{42}x_{40} \oplus x_{43}x_{42} \oplus x_{43}x_{40} \oplus x_{43} \oplus x_{42} \oplus x_{41}x_{40} \oplus x_{41} \oplus x_{40}$

Algorithm 8 Multiplication by 2.

Require: Nibble *nibble*
Ensure: Nibble $nibble \times 2$
  function `mc2`(*nibble*)
    $nib_{30} \leftarrow nibble[3]$
    $nib_{31} \leftarrow nibble[3]$
    $nib_{32} \leftarrow nibble[3]$

    $\texttt{flag} \leftarrow \texttt{flag}\ \&\ (nib_{30} == nib_{31})\ \&\ (nib_{30} == nib_{32})$

    $nibble[0], nibble[1], nibble[2], nibble[3], nibble[4] \leftarrow nib_{30}, nibble[0] \oplus nib_{31}, nibble[1], nibble[2], nibble[4] \oplus nib_{32}$

Algorithm 9 `MixColumnsSerial` function.

Require: Column *col* composed by five 4-bit nibbles
Ensure: Column *col* composed by five 4-bit nibbles after post-multiply with the matrix M
  function `MixSingleColumn`(*col*)
    $nibble[0] \leftarrow [col[0], col[1], col[2], col[3], col[16]]$
    $nibble[1] \leftarrow [col[4], col[5], col[6], col[7], col[17]]$
    $nibble[2] \leftarrow [col[8], col[9], col[10], col[11], col[18]]$
    $nibble[3] \leftarrow [col[12], col[13], col[14], col[15], col[19]]$
    for i in range(4) **do**
      $nibble[0], nibble[1], nibble[2], nibble[3] \leftarrow nibble[1], nibble[2], nibble[3], \texttt{mc2}(\texttt{mc2}(nibble[0])) \oplus nibble[1] \oplus \texttt{mc2}(nibble[2]) \oplus \texttt{mc2}(nibble[3])$

Require: State *RS*
Ensure: State *RS* after the `MixColumnsSerial` operation

Algorithm 9 *Cont.*

function MixColumnsSerial(RS)
$col0 = [RS[0], RS[1], RS[2], RS[3], RS[16], RS[17], RS[18], RS[19], RS[32], RS[33], RS[34], RS[35], RS[48], RS[49], RS[50], RS[51], RS[64], RS[68], RS[72], RS[76]]$
$col1 = [RS[4], RS[5], RS[6], RS[7], RS[20], RS[21], RS[22], RS[23], RS[36], RS[37], RS[38], RS[39], RS[52], RS[53], RS[54], RS[55], RS[65], RS[69], RS[73], RS[77]]$
$col2 = [RS[8], RS[9], RS[10], RS[11], RS[24], RS[25], RS[26], RS[27], RS[40], RS[41], RS[42], RS[43], RS[56], RS[57], RS[58], RS[59], RS[66], RS[70], RS[74], RS[78]]$
$col3 = [RS[12], RS[13], RS[14], RS[15], RS[28], RS[29], RS[30], RS[31], RS[44], RS[45], RS[46], RS[47], RS[60], RS[61], RS[62], RS[63], RS[67], RS[71], RS[75], RS[79]]$

MixSingleColumn($col0$)
MixSingleColumn($col1$)
MixSingleColumn($col2$)
MixSingleColumn($col3$)

5. 5-Bit Representation of a 4-Bit S-Box

In the protected version of LED, the SubCells function uses a 5-bit representation of the PRESENT S-box. This section presents how to create a 5-bit representation from a 4-bit permutation and which representation is the best in terms of cost optimization.

In the last section, the SubCells function requires a representation on 5 bits of the 4-bit PRESENT. The former 4-bit S-box must remain the same with the parity bit added at the end of the words. Indeed, the 5-bit representation is already half filled with the words with an even parity (see Table 2). Then, we have 16^{16} candidates to represent a 4-bit S-box. We must find a way to compare one candidate from another.

Only the S-boxes that correspond to permutations are considered (each output has one and only one related input). Indeed, the parity code used is the 5-bit parity code $\mathcal{C} = [5,4,2]$, but, as we want to consider this code at the state level, the resulting code $\mathcal{C}' = [80,64,2]$ is selected. $\mathcal{C}'$ is only a concatenation of 16 codes $\mathcal{C}$. This concatenation brings the constrains of the permutations on the S-boxes.

Table 2. 5-bit S-box derived from PRESENT to fill.

x	00	01	02	03	04	05	06	07	08	09	0A	0B	0C	0D	0E	0F
S'_x	18	...	...	0A	...	0C	17	...	...	12	00	...	14	...	...	1B
x	10	11	12	13	14	15	16	17	18	19	1A	1B	1C	1D	1E	1F
S'_x	...	06	1D	...	1E	...	...	11	09	...	...	0F	...	03	05	...

5.1. Score Function

To compare the candidates, a score to the S-boxes must be attributed and the best score among the candidates is selected. In this work, a focus on the implementation cost is realized. Then, the score of a candidate is the number of logic gates needed to construct the S-box. The algebraic normal form (ANF) of the S-box is used to count the number of AND and XOR gates. With the score function presented in Algorithm 10, the best representation on 5 bits is the S-box with the lowest number of logical gates. The next subsection is the application of the score function to every representation on 5 bits of a 4-bit S-box.

Algorithm 10 Score of a S-box S.

Require: S-box S
Ensure: Score of S (number of logical gates in the ANF)
function score(S)
$anf \leftarrow$ ANF(S)
return count(&) + count($\wedge$)

5.2. Exhaustive List

To fill the 5-bit S-box, a candidate must be selected among all the 16! permutations. The obvious way to choose the best S-box is to score every candidate and to keep the one with the lowest score. This process is summarized in Algorithm 11 and is the most precise way to find the lowest score. Indeed, we would have the score of each function and then select the best one according to the criteria of implementation cost. However, it requires us to browse all of the 16! permutations, and this can be a very long task. A new solution based on the construction of the 5-bit representation can be as efficient and very easier to achieve.

Algorithm 11 Selection of the S-box with the lowest number of gates.

Require: List of all the 5-bit permutations derived from a 4-bit S-box $\texttt{Permutation}_{\texttt{LIST}}$
Ensure: Permutation with the lowest score and its score
  function `score_selection`($\texttt{Permutation}_{\texttt{LIST}}$)
    `low_score` $\leftarrow$ 1000
    for $S \in \texttt{Permutation}_{\texttt{LIST}}$ **do**
      $s \leftarrow \texttt{score}(S)$
      if $s <$ `low_score` **then**
        `low_score` $\leftarrow s$
        `selected` $\leftarrow S$
    return `low_score, selected`

5.3. Construction

A new selection method is introduced with a construction approach instead of an exhaustive approach. In this paragraph, an *even* word denotes a word that verifies the parity characteristic, and an *odd* word is one which does not. Every even word is only 1 bit away from an odd word. The LSB is used to separate an even from an odd word (`0x18` and `0x19` are only 1 bit away from each other, and this bit is the LSB). Each even input is substituted by an even output, and each odd input is substituted by an odd output. The 5-bit S-box is constructed with the following rule: an odd input is substituted by the odd word 1-bit away from the even output linked to the even input 1-bit away from the odd input. Indeed, each even pair of input/output have a 1-bit away odd pair of input/output. This construction is explained in the Algorithm 12. With this method, the representation of PRESENT is shown in Table 3 and consists of 62 logical gates. Several S-Boxes (found with an exhaustive search) with good cryptographic properties were tested, and none has an ANF constructed with less than 94 logical gates (the biggest one was created with 124 logical gates). We now have to test the robustness of the protected cipher.

Algorithm 12 Construction of a code abiding 5-bit representation from a 4-bit S-box.

Require: S-box S half-filled
Ensure: S-box S full-filled
  for i in range(32) **do**
    if i is *odd* **then**
      $S[i] \leftarrow S[i \oplus 1] \oplus 1$

Table 3. 5-bit code abiding representation constructed from `PRESENT`.

x	00	01	02	03	04	05	06	07	08	09	0A	0B	0C	0D	0E	0F
S'_x	18	19	0B	0A	0D	0C	17	16	13	12	00	01	14	15	1A	1B
x	10	11	12	13	14	15	16	17	18	19	1A	1B	1C	1D	1E	1F
S'_x	07	06	1D	1C	1E	1F	10	11	09	08	0E	0F	02	03	05	04

6. Experimental Results

This section presents the various tests done on the protected LED to determine its robustness against fault injection.

6.1. Robustness

To test the robustness of the protected LED cipher, three scenarios are tested. The detection of a fault injected simply sets a variable `flag` to 0. During the tests, the fault injection is simulated, so there is no case where a fault does not create an error.

Scenario 1: A bit of the state is toggled at a random place of the state and at a random moment of the encryption. This bit-flip induces a change on the parity characteristic of the nibble where it belongs. With the code abiding properties of the functions used during the encryption, the error persists until the parity check function and thus is always detected.
Scenario 2: A bit of the key or of the constant of the `AddConstant` function is toggled at a random place and a random round of the encryption. As the XOR operation is a code abiding operation, the fault is transmitted from the constant to the state and persists until the parity check. The error is thus always detected.
Scenario 3: A fault is injected on data used more than once during a function at a random place and a random round of the encryption. The copies done before the use of the data are then not equal, and the test sets the `flag` to 0. The fault is thus always detected.

In all the scenarios, the fault is always detected, and then the code abiding solution is robust against 1-bit fault attack.

In all the scenarios, 1,000,000 faults have been injected, and the countermeasure (combining code abiding property and copies of the elements used more than once) always leads to a fault detection. The code abiding solution is then robust against 1-word fault attack. The results are presented in Table 4.

Table 4. Robustness results of the secured implementation.

Fault Injections	Scenario 1	Scenario 2	Scenario 3
1,000,000	100% detected	100% detected	100% detected

6.2. Overcost of the Countermeasure

Adding the parity scheme to the LED cipher has a cost. Indeed, we convert an encryption algorithm working on 4-bit words to an encryption algorithm working on 5-bit words. Thus, the new round functions have a bigger price than the former ones. Moreover, our n states are 80-bit long instead of 64-bit long (we encrypt n plaintexts in parallel, and in the tests, we fix $n = 64$). Thus, our implementation takes a bigger place in the memory and one secure encryption takes longer than an unprotected encryption. We differentiate several implementations: the *classical* implementation refers to the soft implementation using lookup tables; the *bitslice* is the bitslice version of LED without any protection; the *code abiding* implementation is the addition of the parity bit during the encryption; and the *code abiding + copies* implementation combines the code abiding properties with copies of values used more than once. The cost can be summarized in the Table 5. The compiler used was the GNU GCC Compiler without any optimization. The CPU used is the Intel Core i5 CPU. The results are presented as a ratio to have a better understanding of the overcost of countermeasures from one implementation to another. The results a must be put into perspective as the classical implementation encrypts only one plaintext at the time when the other implementations can encrypt up to 64 plaintexts at the same time (on a 64-bit length machine). The overcost of the code abiding countermeasure is then better than expected. Indeed, in terms of time overcost, a 25% rise was expected (25% more bits are computed) when an only 12% is measured. However, with the copies countermeasure, an overcost of 79% is reached.

Table 5. Implementation results and cost comparison of the encryptions.

	Ratio *Classical*	Ratio *Bitslice*	Ratio *Code Abiding*
classical	1	-	-
bitslice	1.83	1	-
code abiding	2.04	1.12	1.00
CA + copies	3.28	1.79	1.6

Moreover, another comparison on each round function allows us to precisely understand where the countermeasure has the biggest impact (see Table 6). The heaviest functions from the classical implementation to the other ones are clearly the `subCells`, as the function does not use any lookup table and the `addConstant` as the constant used must be transformed into a bitslice and parity constant. However, as mentioned before, it is more interesting to compare the bitslice versions as they encrypt the same number of plaintexts and are based on the same principles. With these comparisons, the biggest overcost is the `mixColumns` function with all the copies brought.

Table 6. Implementation results and cost comparison of the round functions.

Ratio Classical	*Bitslice*	*CA*	*CA + Copies*
`addConstant`	6.6	8.6	8.6
`subCells`	6.0	7.2	9.6
`ShiftRows`	0.8	0.9	0.9
`mixColumns`	1.9	2.1	3.6
Ratio Bitslice	*CA*	*CA + Copies*	
`addConstant`	1.3	1.3	
`subCells`	1.2	1.6	
`ShiftRows`	1.2	1.2	
`mixColumns`	1.1	1.9	
Ratio CA	*CA + copies*		
`addConstant`	1.0		
`subCells`	1.3		
`ShiftRows`	1.0		
`mixColumns`	1.7		

7. Conclusions

The principle used in this work to prevent fault injections is to detect them using an error detecting code, the parity bit code. This code relies on a redundancy of the information contained in a word. The parity bit code used is the 5-bit parity code, with 4 data bits and 1 parity bit. This method allows us to detect a 1-bit fault injection on a value during an operation.

This work lightens an issue induced by an error compensation. Indeed, depending on the operation performed, an error injected on a value can be propagated into several computed outputs and with the parity bit code, and this error may compensate with its multiple occurrences. The first step is then to present the conditions on the fault and on the operation to reach the compensation, and then to propose a countermeasure to this error compensation that lies on copying the values used more than once and check for equality of the copies.

In addition to this first measure, a method is presented to apply code-abiding notion to word-oriented ciphers. An example on the LED cipher shows the transformations of the state and the round functions to include the parity bit code to the operations. A protected version of the existing LED cipher is then created. Its robustness against 1-bit fault injection is tested, and the results validate its security. Moreover, with the bitslice method, the robustness reaches 1-word fault injection detection.

The next step is to extend this method to a generic one to include code abiding to new cryptographic primitives. A critical operation is the S-box used, and the projection of this S-box into a larger space to add the parity bit brings many candidates. A way to differentiate them is to give them a score based on their implementation cost and select the cheapest S-box.

Eventually, future works could focus on applying the code-abiding method to a larger cipher, such as AES, rather than lightweight ciphers, as well as to evaluate the overcost of the countermeasure compared to other error detecting solutions. Moreover, 1-bit error detection has its limitations [14], and a work on multiple faults detection and correction would be interesting.

Author Contributions: Conceptualization, P.-A.T. and V.G.; methodology, P.-A.T.; software, P.-A.T.; validation, P.-A.T.; investigation, P.-A.T.; writing—original draft preparation, P.-A.T.; writing—review and editing, P.-A.T., L.B. and V.G.; visualization, P.-A.T.; supervision, L.B. and V.G.; project administration, L.B.; funding acquisition, L.B. All authors have read and agreed to the published version of the manuscript.

Funding: Part of this was support by the French Agence Nationale de la Recherche under the grant ANR-22-CE39-0008 (project PROPHY).

Institutional Review Board Statement: Not applicable.

Data Availability Statement: Not applicable.

Conflicts of Interest: The authors declare no conflict of interest.

Abbreviations

The following abbreviations are used in this manuscript:

CA	Code Abiding
CA + Copies	Code Abiding with copies included

References

1. Biham, E.; Shamir, A. Differential Fault Analysis of Secret Key Cryptosystems. In *Lecture Notes in Computer Science, Proceedings of the Advances in Cryptology—CRYPTO'97, 17th Annual International Cryptology Conference, Santa Barbara, CA, USA, 17–21 August 1997*; Kaliski, B.S.K., Jr. Ed.; Springer: Berlin/Heidelberg, Germany, 1997; Volume 1294, pp. 513–525. [CrossRef]
2. Kocher, P.C.; Jaffe, J.; Jun, B. Differential Power Analysis. In *Lecture Notes in Computer Science, Proceedings of the Advances in Cryptology—CRYPTO'99, 19th Annual International Cryptology Conference, Santa Barbara, CA, USA, 15–19 August 1999*; Wiener, M.J., Ed.; Springer: Berlin/Heidelberg, Germany, 1999; Volume 1666, pp. 388–397. [CrossRef]
3. Baksi, A.; Bhasin, S.; Breier, J.; Chattopadhyay, A.; Kumar, V.B.Y. Feeding Three Birds With One Scone: A Generic Duplication Based Countermeasure To Fault Attacks. In Proceedings of the Design, Automation & Test in Europe Conference & Exhibition, DATE 2021, Grenoble, France, 1–5 February 2021; pp. 561–564. [CrossRef]
4. Breier, J.; Hou, X.; Liu, Y. On Evaluating Fault Resilient Encoding Schemes in Software. *IEEE Trans. Dependable Secur. Comput.* **2021**, *18*, 1065–1079. [CrossRef]
5. Kiaei, P.; Mercadier, D.; Dagand, P.; Heydemann, K.; Schaumont, P. Custom Instruction Support for Modular Defense Against Side-Channel and Fault Attacks. In *Lecture Notes in Computer Science, Proceedings of the Constructive Side-Channel Analysis and Secure Design—11th International Workshop, COSADE 2020, Lugano, Switzerland, 1–3 April 2020*; Revised Selected Papers; Bertoni, G.M., Regazzoni, F., Eds.; Springer: Berlin/Heidelberg, Germany, 2020; Volume 12244, pp. 221–253. [CrossRef]
6. Lee, S.; Jho, N.; Kim, M. Table Redundancy Method for Protecting Against Fault Attacks. *IEEE Access* **2021**, *9*, 92214–92223. [CrossRef]
7. Patrick, C.; Yuce, B.; Ghalaty, N.F.; Schaumont, P. Lightweight Fault Attack Resistance in Software Using Intra-instruction Redundancy. In *Lecture Notes in Computer Science, Proceedings of the Selected Areas in Cryptography-SAC 2016—23rd International Conference, St. John's, NL, Canada, 10–12 August 2016*; Revised Selected Papers; Avanzi, R., Heys, H.M., Eds.; Springer: Berlin/Heidelberg, Germany, 2016; Voume 10532, pp. 231–244. [CrossRef]
8. Regazzoni, F.; Eisenbarth, T.; Breveglieri, L.; Ienne, P.; Koren, I. Can Knowledge Regarding the Presence of Countermeasures Against Fault Attacks Simplify Power Attacks on Cryptographic Devices? In *IEEE Computer Society, Proceedings of the 23rd IEEE International Symposium on Defect and Fault-Tolerance in VLSI Systems (DFT 2008), Boston, MA, USA, 1–3 October 2008*; Bolchini, C., Kim, Y., Gizopoulos, D., Tehranipoor, M., Eds.; 2008; pp. 202–210. [CrossRef]

9. Simon, T.; Batina, L.; Daemen, J.; Grosso, V.; Massolino, P.M.C.; Papagiannopoulos, K.; Regazzoni, F.; Samwel, N. Friet: An Authenticated Encryption Scheme with Built-in Fault Detection. In *Lecture Notes in Computer Science, Proceedings of the Advances in Cryptology-EUROCRYPT 2020—39th Annual International Conference on the Theory and Applications of Cryptographic Techniques, Zagreb, Croatia, 10–14 May 2020*; Part I; Canteaut, A., Ishai, Y., Eds.; Springer: Berlin/Heidelberg, Germany, 2020; Volume 12105, pp. 581–611. [CrossRef]
10. Bertoni, G.; Breveglieri, L.; Koren, I.; Maistri, P.; Piuri, V. A Parity Code Based Fault Detection for an Implementation of the Advanced Encryption Standard. In Proceedings of the 17th IEEE International Symposium on Defect and Fault-Tolerance in VLSI Systems (DFT 2002), Vancouver, BC, Canada, 6–8 November 2002; pp. 51–59. [CrossRef]
11. Lac, B.; Canteaut, A.; Fournier, J.J.A.; Sirdey, R. Thwarting Fault Attacks using the Internal Redundancy Countermeasure (IRC). *IACR Cryptol. ePrint Arch.* **2017**, *910*, 1–26.
12. Guo, J.; Peyrin, T.; Poschmann, A.; Robshaw, M.J.B. The LED Block Cipher. In *Lecture Notes in Computer Science, Proceedings of the Cryptographic Hardware and Embedded Systems-CHES 2011—13th International Workshop, Nara, Japan, 28 September–1 October 2011*; Preneel, B., Takagi, T., Eds.; Springer: Berlin/Heidelberg, Germany, 2011; Volume 6917, pp. 326–341. [CrossRef]
13. Bao, Z.; Luo, P.; Lin, D. Bitsliced Implementations of the PRINCE, LED and RECTANGLE Block Ciphers on AVR 8-Bit Microcontrollers. In *Lecture Notes in Computer Science, Proceedings of the Information and Communications Security—17th International Conference, ICICS 2015, Beijing, China, 9–11December 2015*; Revised Selected Papers; Qing, S., Okamoto, E., Kim, K., Liu, D., Eds.; Springer: Berlin/Heidelberg, Germany, 2015; Volme 9543, pp. 18–36. [CrossRef]
14. Colombier, B.; Grandamme, P.; Vernay, J.; Chanavat, É.; Bossuet, L.; de Laulanié, L.; Chassagne, B. Multi-Spot Laser Fault Injection Setup: New Possibilities for Fault Injection Attacks. In *Lecture Notes in Computer Science, Proceedings of the Smart Card Research and Advanced Applications—20th International Conference, CARDIS 2021, Lübeck, Germany, 11–12 November 2021*; Revised Selected Papers; Grosso, V., Pöppelmann, T., Eds.; Springer: Berlin/Heidelberg, Germany, 2021; Volume 13173, pp. 151–166. [CrossRef]

electronics

MDPI

Communication

Cooperative Jamming with AF Relay in Power Monitoring and Communication Systems for Mining

Wei Meng [1], Yidong Gu [1], Jianjun Bao [1], Li Gan [2], Tao Huang [3] and Zhengmin Kong [2,*]

1 Tiandi (Changzhou) Automation Co., Ltd., CCTEG Changzhou Research Institute, Changzhou 213015, China
2 School of Electrical Engineering and Automation, Wuhan University, Wuhan 430072, China
3 College of Science and Engineering, James Cook University, Smithfield, QLD 4878, Australia
* Correspondence: zmkong@whu.edu.cn

Abstract: In underground mines, physical layer security (PLS) technology is a promising method for the effective and secure communication to monitor the mining process. Therefore, in this paper, we investigate the PLS of an amplify-and-forward relay-aided system in power monitoring and communication systems for mining, with the consideration of multiple eavesdroppers. Explicitly, we propose a PLS scheme of cooperative jamming and precoding for a full-duplex system considering imperfect channel state information. To maximize the secrecy rate of the communications, an effective block coordinate descent algorithm is used to design the precoding and jamming matrix at both the source and the relay. Furthermore, the effectiveness and convergence of the proposed scheme with high channel state information uncertainty have been proven.

Keywords: physical layer security; multiple eavesdroppers; full-duplex; underground mine; amplify-and-forward relay

Citation: Meng, W.; Gu, Y.; Bao, J.; Gan, L.; Huang, T.; Kong, Z. Cooperative Jamming with AF Relay in Power Monitoring and Communication Systems for Mining. *Electronics* **2023**, *12*, 1057. https://doi.org/10.3390/electronics12041057

Academic Editor: Cheng-Chi Lee

Received: 21 January 2023
Revised: 10 February 2023
Accepted: 16 February 2023
Published: 20 February 2023

1. Introduction

Underground mining promotes the economy's growth, but the dust and poisonous gases formed during mining make it a dangerous and complex operation. Therefore, a reliable communication system is needed to monitor the mining process and communicate with external management offices to ensure the safety and maximum production of the underground mine. Wireless communication technology is applied to realize information exchange in underground mines due to its simple construction.

However, due to the complex structure of underground mines, there exists significant attenuation of radio wave transmission in wireless communications [1]. To solve these problems, relay-aided wireless communications have been studied to improve the reliability and have also been used to enhance the coverage of a broader range of networks. According to the forwarding protocol adopted by the relay, cooperation relay can be divided into amplify-and-forward (AF) and decode-and-forward (DF) relay [2]. AF is the simplest protocol, and it processes the received signals linearly and then forwards them to the destination [3]. Offering a reasonable trade-off between actual implementation costs and benefits, AF is considered the most promising solution [4].

To guarantee the communication rate in wireless communications, full-duplex (FD) relays are studied in Refs. [5,6]. FD technology allows radios to receive and transmit simultaneously on the same frequency band, which can improve spectrum efficiency [7]. Furthermore, in addition to doubling the spectral efficiency of the physical layer, FD can help to solve the throughput losses due to congestion and large point-to-point delays in existing wireless networks.

In addition, due to the openness and sharing of wireless media, any wireless device connected to the communication system can access messages exchanged through the connection, making wireless channels easy to be eavesdropped on and inject with malicious information [8]. Worse still, relay-aided wireless networks may suffer severe security risks from malicious users since they may eavesdrop on the messages from both the source and the relay. Physical layer security (PLS) can effectively protect the privacy among the transmitter and the legitimate receivers [9]. Shannon conducted pioneering research on secret communications and established the concept of perfect secrecy [10]. Unlike Shannon, Wyner proposed a degraded wiretap channel model in Ref. [11]. After the degraded wiretap channel, the fading wiretap channels and multiple-input-multiple-output wiretap channels have been investigated in Refs. [12,13] and Refs. [14,15], respectively.

The work in Ref. [5] investigated a FD communication system, and the transmission block is divided into an energy harvesting phase and an information transmission phase. Different from Ref. [5], in Ref. [6], an FD is designed to capture energy from the source while forwarding information to the legitimate receivers. With the presence of passive colluding wireless eavesdroppers, Ref. [16] studied the effective secrecy throughput to the physical layer security of in-home and broadband PLC systems. In Ref. [17], the authors investigated the optimal trunk position of FD relay systems with DF and the minimal outage probability criterion considered.

Above all, to the best of our knowledge, the existing contributions fail to ensure secure communications in the challenging FD relay-aided wireless communications scenario in the face of multiple eavesdroppers and imperfect channel state information (CSI). Therefore, in this paper, we propose a PLS scheme of cooperative jamming and precoding for FD-DF relay-assisted wireless communications system considering imperfect CSI, which combines cooperative precoding for legitimate users to improve the quality of legitimate channels and cooperative jamming for illegal users to reduce the quality of eavesdropping channels. Considering the imperfect CSI and multiple eavesdroppers, we use an effective BCD algorithm to design the precoding and jamming matrix at both the source and the relay, in which maximizing the secrecy rate of the FD-AF relay-assisted wireless communications system is emphasized.

This paper is organized as follows. Section 2 describes the system model. The secrecy rate optimization problem is proposed and transformed into a solvable form in Section 3, which also gives the algorithm. Section 4 characterizes the numerical results in different scenarios. Finally, the conclusion is presented in Section 5.

Notation: The $\mathbf{W}^T, \mathbf{W}^H, \text{vec}(\mathbf{W}), \|\mathbf{W}\|$and $\text{tr}(\mathbf{W})$ denote the transpose, conjugate transpose, vectorization, Frobenius norm, and trace of the matrix, respectively. $\otimes$ denote the Kronecker product, and $\mathbf{W}^K$ represents the $\mathbf{W}\mathbf{W}^H$ along with $\log|\mathbf{E}+\mathbf{CD}| = \log|\mathbf{E}+\mathbf{DC}|$. $\mathbf{E}$ is the identity matrix.

2. System Model

Consider a MIMO wireless system, as shown in Figure 1, where a source, a relay, a user, and two eavesdroppers have N_S, N_R, N_D, and N_E channels, respectively. We assume that there is no direct link between the source and the user for the long-distance path loss. For simplicity, the eavesdroppers represent all the eavesdroppers eavesdropping the same legitimate in the same time phase. More specifically, in the first time phase, eavesdroppers eavesdrop E_1 message from the source, and in the second time phase, eavesdropper eavesdrop E_2 message from the relay.

In wireless communications system, messages are transmitted through MIMO wireless communications channels. We describe each path between two nodes by CSI $\mathbf{H}_{ij,k}$ as the matrix of channel coefficients, where $i \in \{S, R\}$, $j \in \{R, D, E_1, E_2\}$, and $k = 1,2$ denote the transmitter, receiver, and transmission time phases, respectively. It is worth noting that $\mathbf{H}_{RR,1}$ refers to the self-interference matrix because of self-interference and in the process of transmission $\mathbf{H}_{ij,k}$ stays constant because of the short transmission time.

In this paper, the uncertainty of CSI is taken into consideration, i.e., the CSI of the wireless communications system cannot be perfectly known at the source or the relay due to factors such as the limited capacity of the feedback channel. As a result, the deterministic uncertainty model [18] is introduced to characterize the imperfect CSI, as follows:

$$\mathbf{H}_{ij,k} \in \mathcal{H}_{ij,k} = \left\{ \mathbf{H}_{ij,k} | \mathbf{H}_{ij,k} = \overline{\mathbf{H}}_{ij,k} + \Delta_{ij,k}, \left\| \Delta_{ij,k} \right\| \leq \delta_{ij,k} \right\}, \tag{1}$$

where $\Delta_{ij,k}$ denotes the channel uncertainty as the degree of deviation from the mean CSI $\overline{\mathbf{H}}_{ij,k}$.

In Figure 1, during the first time phase, the source sends confidential signals to the relay while E_1 eavesdrops on the signals from the source. To interrupt E_1, the relay emits jamming signals to E_1. More specifically, the message transmitted by the source is secret data symbol $\mathbf{S} \in \mathcal{CN}(\mathbf{0}, 1)$ precoded by the precoding vector $\mathbf{L} \in \mathbb{C}^{N_S \times 1}$. Then, we can formulate the progress at the source as follows:

$$\mathbf{X}_S = \mathbf{LS}, \tag{2}$$

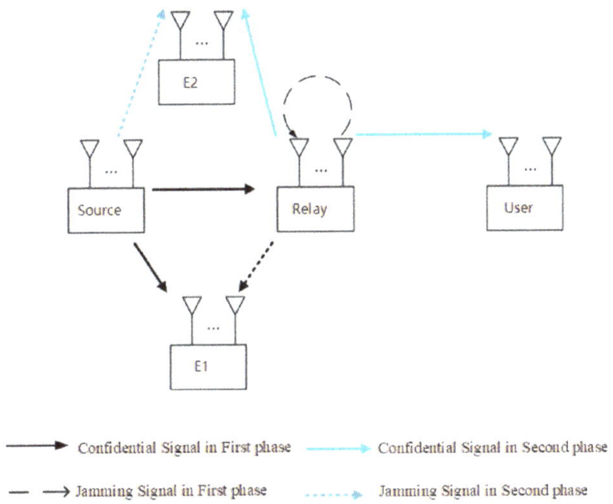

Figure 1. Wireless communication system model with a single relay.

Next, we formulate the messages emitted by the relay. Note that the relay in this time phase only emits jamming to disrupt E_1 so the messages can be formulated as follows :

$$\mathbf{X}_R = \mathbf{J_1 Z}_1, \tag{3}$$

where we utilize the jamming precoding vector $\mathbf{J_1} \in \mathbb{C}^{N_R \times 1}$ and jamming symbol $\mathbf{Z}_1 \in \mathcal{CN}(\mathbf{0}, 1)$.

Considering the self-interference of the relay, we can formulate the messages received by the relay:

$$\mathbf{Y}_{R1} = \mathbf{H}_{SR,1}\mathbf{LS} + \mathbf{H}_I \mathbf{J_1 Z}_1 + \mathbf{n}_{R1}, \tag{4}$$

where $\mathbf{n}_{R1}$ is Additive White Gaussian Noise (AWGN) at the relay and $\mathbf{H}_I$ is the self-interference matrix. Meanwhile, E_1 eavesdrops on both of the messages from the source and the relay, so the messages eavesdropped by E_1 can be expressed as

$$\mathbf{Y}_{E1} = \mathbf{H}_{SE,1}\mathbf{LS} + \mathbf{H}_{RE,1}\mathbf{J_1 Z}_1 + \mathbf{n}_{E1}, \tag{5}$$

where $\mathbf{n}_{E1}$ is AWGN at E_1.

In the second time phase, the source emits the jamming signals $\mathbf{X}_{S2}$ to E_2 where $\mathbf{J_2}$,$\mathbf{Z}_2 \in \mathcal{CN}(\mathbf{0}, 1)$ represent the jamming precoding vector and jamming symbol, respectively.

$$\mathbf{X}_{S2} = \mathbf{J_2}\mathbf{Z}_2, \tag{6}$$

Then, the relay amplifies the messages it received in the first time phase and forwards them to the user,

$$\mathbf{X}_{R2} = \mathbf{G}\mathbf{Y}_{R1} = \mathbf{G}(\mathbf{H}_{SR,1}\mathbf{L}\mathbf{S} + \mathbf{H}_I\mathbf{J_1}\mathbf{Z}_1 + \mathbf{n}_{R1}), \tag{7}$$

$$\mathbf{Y}_D = \mathbf{H}_{RD,2}\mathbf{X}_{R2} + \mathbf{n}_D = \mathbf{H}_{RD,2}\mathbf{G}(\mathbf{H}_{SR,1}\mathbf{L}\mathbf{S} + \mathbf{H}_I\mathbf{J_1}\mathbf{Z}_1 + \mathbf{n}_{R1}) + \mathbf{n}_D, \tag{8}$$

where $\mathbf{G} \in \mathbb{C}^{N_R \times N_R}$ is the amplifying matrix at the relay and $\mathbf{n}_D$ is AWGN at the users, and $\mathbf{X}_{R2}$, $\mathbf{Y}_D$ represent the messages transmitted by the relay and received by the users, respectively.

E_2 receive both the signals from the relay and the jamming signals from the source, i.e.,

$$\mathbf{Y}_{E2} = \mathbf{H}_{SE,2}\mathbf{J_2}\mathbf{Z}_2 + \mathbf{H}_{RE,2}\mathbf{G}(\mathbf{H}_{SR,1}\mathbf{L}\mathbf{S} + \mathbf{H}_I\mathbf{J_1}\mathbf{Z}_1 + \mathbf{n}_{R1}) + \mathbf{n}_{E2}, \tag{9}$$

where $\mathbf{n}_{E2}$ is AWGN at E_2.

Above all, to formulate the problem in a mathematical form, we calculate the signal-noise ratio (SNR) at the users, E_1 and E_2, respectively.

$$\mathbf{\Gamma}_D = (\mathbf{H}_{RD,2}\mathbf{G}\mathbf{H}_{SR,1}\mathbf{L})^K\mathbf{Q}_D^{-1}, \tag{10}$$

where $\mathbf{Q}_D = (\mathbf{H}_{RD,2}\mathbf{G}\mathbf{H}_{RR,1}\mathbf{J_1})^K + \sigma_R^2(\mathbf{H}_{RD,2}\mathbf{G})^K + \sigma_D^2\mathbf{E}$.

$$\mathbf{\Gamma}_{E1} = (\mathbf{H}_{SE,1}\mathbf{L})^K\mathbf{Q}_{E1}^{-1}, \tag{11}$$

where $\mathbf{Q}_{E1} = (\mathbf{H}_{RE,1}\mathbf{J_1})^K + \sigma_E^2\mathbf{E}$.

$$\mathbf{\Gamma}_{E2} = (\mathbf{H}_{RE,2}\mathbf{G}\mathbf{H}_{R1}\mathbf{L})^K\mathbf{Q}_{E2}^{-1}, \tag{12}$$

where $\mathbf{Q}_{E2} = (\mathbf{H}_{SE,2}\mathbf{J_2})^K + (\mathbf{H}_{RE,2}\mathbf{G}\mathbf{H}_I\mathbf{J_1})^K + \sigma_R^2(\mathbf{H}_{RE,2}\mathbf{G})^K + \sigma_E^2\mathbf{E}$ and σ_i is the noise amplitude of the corresponding AWGN $\mathbf{n}_i$.

Then, we can arrive at the achievable secrecy rate of the legitimate users [11]:

$$R_D = \log|\mathbf{E} + \mathbf{\Gamma}_D|, \tag{13}$$

In the non-colluding strategy, each eavesdropper processes messages individually. Therefore, the achievable secrecy rate of the non-colluding [11] eavesdroppers is

$$R_E = \max\{\log|\mathbf{E} + \mathbf{\Gamma}_{E1}|, \log|\mathbf{E} + \mathbf{\Gamma}_{E2}|\} \tag{14}$$

Finally, we can gain the achievable secrecy rate of the wireless communications system,

$$R_S = R_D - R_E \tag{15}$$

3. Optimization Problem Transformation

In this part, the goal is to maximize the secrecy rate of the communication system. Then according to the system model, we can formulate the optimization problem of the secrecy rate with the transmit power constraint as follows.

$$\max_{\mathbf{L},\mathbf{J_1},\mathbf{J_2},\mathbf{G}} \min_{\mathbf{H}_{ij,k}\in\mathcal{H}_{ij,k}} R_S \tag{16a}$$

$$\text{s.t. } \|\mathbf{L}\|^2 \leqslant P_S,\ \|\mathbf{J_1}\|^2 \leqslant P_S, \|\mathbf{J_2}\|^2 \leqslant P_R, \tag{16b}$$

$$tr((\mathbf{GH}_{SR,1}\mathbf{L})^K + (\mathbf{GH}_I\mathbf{J_1})^K + \sigma_R^2\mathbf{G}^K) \leqslant P_R \quad \forall \mathbf{H}_{ij,k} \in \mathcal{H}_{ij,k} \tag{16c}$$

However, due to the non-convexity of the optimization problem, it is difficult to solve. To deal with the high non-convexity of the function $-\log|\cdot|$, the objective function in (16a) is transformed into an equivalent counterpart through the WMMSE algorithm, which can be solved by the BCD method. The following introduces the WMMSE algorithm.

Lemma 1. *Define the mean-square error (MSE) matrix*

$$\widehat{\mathbf{N}} \triangleq (\mathbf{TH}\text{-}\mathbf{E})^K + \mathbf{TRT}^H \tag{17}$$

where $\mathbf{R} \succ \mathbf{0}$*. Then we have*

$$-\log|\mathbf{N}| = \max_{\mathbf{K}\succ 0} \log|\mathbf{K}| - \mathrm{tr}(\mathbf{KN}) + \mathrm{tr}(\mathbf{E}) \tag{18}$$

$$\log\left|\mathbf{I} + \mathbf{R}^{-1}\mathbf{H}^K\right| = \max_{\mathbf{K}\succ 0,\mathbf{T}} \log|\mathbf{K}| - \mathrm{tr}(\mathbf{K}\widehat{\mathbf{N}}) + \mathrm{tr}(\mathbf{E}) \tag{19}$$

To reformulate the parts of $-\log|\cdot|$ in the objective function, we apply Lemma 1 on (13) and introduce the MSE matrix $\mathbf{N}_i$ and auxiliary matrices $\mathbf{K}_i, \mathbf{T}_i$,which have been defined in (17) and (19). So, the achievable secrecy rate of the legitimate can be reorganized as

$$\begin{aligned} R_D &= \log|\mathbf{E} + \mathbf{\Gamma}_D| = \log\left|\mathbf{E} + (\mathbf{H}_{RD,2}\mathbf{GH}_{SR,1}\mathbf{L})^K\mathbf{Q}_D^{-1}\right| \\ &= \max_{\mathbf{K}_D\succ 0,\mathbf{D}_D} \log|\mathbf{K}_D| - \mathrm{tr}(\mathbf{K}_D\mathbf{N}_D) + \mathrm{tr}(\mathbf{E}) \end{aligned} \tag{20}$$

where

$$\mathbf{N}_D = (\mathbf{T}_D\mathbf{H}_{RD,2}\mathbf{GH}_{SR,1}\mathbf{L} - \mathbf{E})^K + \mathbf{T}_D\mathbf{Q}_D\mathbf{T}_D^H \tag{21}$$

Applying Lemma 1 on (14), the achievable rates of E_1 and E_2 can be transformed as (22) and (23).

$$\begin{aligned} &-\log|\mathbf{E} + \mathbf{\Gamma}_1| = \log|\mathbf{Q}_{E1}| - \log\left|(\mathbf{H}_{SE,1}\mathbf{L})^K + \mathbf{Q}_{E1}\right| \\ &= \underbrace{\log\left|\mathbf{E} + \sigma_E^{-2}(\mathbf{H}_{RE,1}\mathbf{J_1})^K\right|}_{C_{E11}} \underbrace{- \log\left|\mathbf{E} + \sigma_E^{-2}\left((\mathbf{H}_{SE,1}\mathbf{L})^K + (\mathbf{H}_{RE,1}\mathbf{J_1})^K\right)\right|}_{C_{E12}} \end{aligned} \tag{22}$$

$$
\begin{aligned}
&-\log|\mathbf{E}+\boldsymbol{\Gamma}_2| = \log|\mathbf{Q}_{E2}| - \log\left|(\mathbf{H}_{RE,2}\mathbf{G}\mathbf{H}_I\mathbf{L})^K + \mathbf{Q}_{E2}\right| \\
&= \underbrace{\log\left|\mathbf{E}+\sigma_E^{-2}\left((\mathbf{H}_{SE,2}\mathbf{J_2})^K + (\mathbf{H}_{RE,2}\mathbf{G}\mathbf{H}_I\mathbf{J_1})^K + \sigma_R^2(\mathbf{H}_{RE,2}\mathbf{G})^K\right)\right|}_{C_{E21}} + \\
&\underbrace{-\log\left|\mathbf{E}+\sigma_E^{-2}\left((\mathbf{H}_{RE,2}\mathbf{G}\mathbf{H}_I\mathbf{L})^K + (\mathbf{H}_{SE,2}\mathbf{J_2})^K + (\mathbf{H}_{RE,2}\mathbf{G}\mathbf{H}_I\mathbf{J_1})^K + \sigma_R^2(\mathbf{H}_{RE,2}\mathbf{G})^K\right)\right|}_{C_{E22}}
\end{aligned} \tag{23}
$$

Then, the auxiliary variables C_{E11}, C_{E12}, C_{E21} and C_{E22} can be rewritten according to Lemma 1 as

$$C_{E11} = \max_{\mathbf{K}_{E11}\succ 0,\mathbf{T}_{E1}} \log|\mathbf{K}_{E11}| - \mathrm{tr}(\mathbf{K}_{E11}\mathbf{N}_{E11}) + \mathrm{tr}(\mathbf{E}) \tag{24}$$

$$C_{E12} = \max_{\mathbf{K}_{E12}\succ 0} \log|\mathbf{K}_{E12}| - \mathrm{tr}(\mathbf{K}_{E12}\mathbf{N}_{E12}) + \mathrm{tr}(\mathbf{E}) \tag{25}$$

$$C_{E21} = \max_{\mathbf{K}_{E21}\succ 0,\mathbf{T}_{E2}} \log|\mathbf{K}_{E21}| - \mathrm{tr}(\mathbf{K}_{E21}\mathbf{N}_{E21}) + \mathrm{tr}(\mathbf{E}) \tag{26}$$

$$C_{E22} = \max_{\mathbf{K}_{E22}\succ 0} \log|\mathbf{K}_{E22}| - \mathrm{tr}(\mathbf{K}_{E22}\mathbf{N}_{E22}) + \mathrm{tr}(\mathbf{E}) \tag{27}$$

where

$$
\begin{aligned}
\mathbf{N}_{E11} &= (\mathbf{D}\mathbf{T}_{E1}\mathbf{H}_{RE,1}\mathbf{J_1} - \mathbf{E})^K + \sigma_E^2\mathbf{T}_{E1}^K \\
\mathbf{N}_{E12} &= \sigma_E^{-2}\left((\mathbf{H}_{SE,1}\mathbf{L})^K + (\mathbf{H}_{RE,1}\mathbf{J_1})^K\right) + \mathbf{E}
\end{aligned}
$$

$$
\begin{aligned}
\mathbf{N}_{E21} = (&\mathbf{T}_{E21}\mathbf{H}_{SE,2}\mathbf{J_2}\mathbf{X} + \mathbf{T}_{E22}\mathbf{H}_{RE,2}\mathbf{G}\mathbf{H}_I\mathbf{V}\mathbf{X} + \\
&\sigma_R\mathbf{T}_{E23}\mathbf{H}_{RE,2}\mathbf{G} - \mathbf{E})^K + \sigma_E^2\left(\mathbf{T}_{E21}^K + \mathbf{T}_{E22}^K + \mathbf{T}_{E23}^K\right)
\end{aligned}
$$

$$\mathbf{N}_{E22} = \sigma_E^{-2}((\mathbf{H}_{RE,2}\mathbf{G}\mathbf{H}_I\mathbf{L})^K + (\mathbf{H}_{SE,2}\mathbf{J_2})^K + (\mathbf{H}_{RE,2}\mathbf{G}\mathbf{H}_I\mathbf{J_1})^K + \sigma_R^2(\mathbf{H}_{RE,2}\mathbf{G})^K) + \mathbf{E}$$

and note the decomposition $\mathbf{T}_{E2} = \begin{bmatrix} \mathbf{T}_{E21} & \mathbf{T}_{E22} & \mathbf{T}_{E23} \end{bmatrix}$ and $\mathbf{X} = \begin{bmatrix} 1 & \mathbf{0} \end{bmatrix} \in \mathbb{C}^{1\times Nr}$.

After substituting (24)–(27) into (16a), the secrecy rate of the system is equivalently rewritten as

$$\max_{\mathbf{L},\mathbf{J_1},\mathbf{J_2},\mathbf{G},\mathbf{K}_i\succ 0,\mathbf{T}_i} \ \min_{\mathbf{H}_{ij,k}\in\mathcal{H}_{ij,k}} f(\mathbf{L},\mathbf{J_1},\mathbf{J_2},\mathbf{G},\mathbf{S_i},\mathbf{D_i}) \tag{28}$$

$$\text{s.t. (16c)} \tag{29}$$

$$
\begin{aligned}
f \triangleq &\log|\mathbf{K}_D| - \mathrm{tr}(\mathbf{K}_D\mathbf{N}_D) + \min\{\log|\mathbf{K}_{E11}| - \mathrm{tr}(\mathbf{K}_{E11}\mathbf{N}_{E11}) + \log|\mathbf{K}_{E12}| - \mathrm{tr}(\mathbf{K}_{E12}\mathbf{N}_{E12}), \\
&\log|\mathbf{K}_{E21}| - \mathrm{tr}(\mathbf{K}_{E21}\mathbf{N}_{E21}) + \log|\mathbf{K}_{E22}| - \mathrm{tr}(\mathbf{K}_{E22}\mathbf{N}_{E22})\}
\end{aligned} \tag{30}
$$

where the function $f(\mathbf{L},\mathbf{J_1},\mathbf{J_2},\mathbf{G},\mathbf{K_i},\mathbf{T_i})$ is defined in (30).

To solve the proposed problem and constraint (16c), the slack variables β_i ($i \in \{T, E11, E12, E21, E22, P\}$) are introduced to transform (28) into an optimization problem.

$$\mathrm{tr}(\mathbf{K}_i\mathbf{N}_i) \le \beta_i, \forall \mathbf{H}_{ij,k} \in \mathcal{H}_{ij,k} \tag{31}$$

We can further rewrite the problem (28) as

$$\max_{\mathbf{L},\mathbf{J_1},\mathbf{J_2},\mathbf{G},\mathbf{K}_i \succ 0,\mathbf{T}_i} g(\mathbf{L},\mathbf{J_1},\mathbf{J_2},\mathbf{G},\mathbf{S_i},\mathbf{D_i}) \tag{32}$$

$$\text{s.t. (16c), (31)} \tag{33}$$

$$g \overset{\Delta}{=} \log|\mathbf{K}_D| - \beta_D + \min\{\log|\mathbf{K}_{E11}| - \beta_{E11} + \log|\mathbf{K}_{E12}| - \beta_{E12}, \\ \log|\mathbf{K}_{E21}| - \beta_{E21} + \log|\mathbf{K}_{E22}| - \beta_{E22}\} \tag{34}$$

where $g(\mathbf{L},\mathbf{J_1},\mathbf{J_2},\mathbf{G},\mathbf{K_i},\mathbf{T_i})$ is defined in (34), respectively. However, the semi-infinite inequalities (31) are non-convex and need further transformation. In the next step, (31) is transformed into a convex form. In fact, all the inequalities $\mathrm{tr}(\mathbf{K}_i\mathbf{N}_i) \le \beta_i$ can be transformed into a convex form in a similar way. Such as, when $i = D$, the semi-definite constraint $\mathrm{tr}(\mathbf{K}_D\mathbf{N}_D)$ can be rewritten as

$$\mathrm{tr}(\mathbf{K}_D\mathbf{N}_D) = \left\| \underbrace{\begin{bmatrix} \mathrm{vec}(\mathbf{F}_D(\mathbf{T}_D\mathbf{H}_{RD,2}\mathbf{G}\mathbf{H}_{SR,1}\mathbf{L} - \mathbf{E})) \\ \mathrm{vec}(\mathbf{F}_D\mathbf{T}_D\mathbf{H}_{RD,2}\mathbf{G}\mathbf{H}_{RR,1}\mathbf{J}_1) \\ \mathrm{vec}(\sigma_R\mathbf{F}_D\mathbf{T}_D\mathbf{H}_{RD,2}\mathbf{G}) \\ \mathrm{vec}(\sigma_D\mathbf{F}_D\mathbf{T}_D) \end{bmatrix}}_{\phi_D} \right\|^2 \tag{35}$$

by applying $\mathbf{T}_D = \mathbf{F}_D^H\mathbf{F}_D$ and the equality $\mathrm{tr}(\mathbf{W}^K) = \|\mathrm{vec}(\mathbf{W})\|^2$.

Then we need to extract the uncertain CSI from (35).

$$\phi_D = \bar{\phi}_D + \underbrace{\sum_j \mathbf{\Omega}_{Dj}\mathrm{vec}(\mathbf{\Delta}_j)}_{\mathbf{\Delta}_D} + \underbrace{\sum_k \alpha_k \mathrm{vec}(\mathbf{\Delta}_{k1})\mathrm{vec}^H(\mathbf{\Delta}_{k2})}_{\tilde{\mathbf{\Delta}}_D} \tag{36}$$

where the identity $\mathrm{vec}(\mathbf{ABC}) = (\mathbf{C}^T \otimes \mathbf{A})\mathrm{vec}(\mathbf{B})$ is applied and $j \in \{RR,1; RD,2; SR,1\}$. Note that $k1, k2$ denote the coupling parts of CSI in ϕ_D in the $\tilde{\mathbf{\Delta}}_D$ part. In fact, the uncertainty of the CSI is small enough to make its quadratic forms negligible. As a result, the ϕ_D can be represented as its asymptotic form as

$$\phi_D = \bar{\phi}_D + \underbrace{\sum_j \mathbf{\Omega}_{Dj}\mathrm{vec}(\mathbf{\Delta}_j)}_{\mathbf{\Delta}_D} \tag{37}$$

where

$$\bar{\phi}_D = \begin{bmatrix} \mathrm{vec}(\mathbf{F}_D(\mathbf{T}_D\bar{\mathbf{H}}_{RD,2}\mathbf{G}\bar{\mathbf{H}}_{SR,1}\mathbf{L} - \mathbf{E})) \\ \mathrm{vec}(\mathbf{F}_D\mathbf{T}_D\bar{\mathbf{H}}_{RD,2}\mathbf{G}\bar{\mathbf{H}}_{I}\mathbf{J}_1) \\ \mathrm{vec}(\sigma_R\mathbf{F}_D\mathbf{T}_D\bar{\mathbf{H}}_{RD,2}\mathbf{G}) \\ \mathrm{vec}(\sigma_D\mathbf{F}_D\mathbf{T}_D) \end{bmatrix} \tag{38}$$

$$\mathbf{\Omega}_{DSR,1} = \begin{bmatrix} \mathbf{L}^T \otimes \mathbf{F}_D\mathbf{T}_D\bar{\mathbf{H}}_{RD,2}\mathbf{G} \\ \mathbf{0} \\ \mathbf{0} \\ \mathbf{0} \end{bmatrix} \tag{39}$$

$$\mathbf{\Omega}_{DSD,2} = \begin{bmatrix} (\mathbf{G}\bar{\mathbf{H}}_{SR,1}\mathbf{L})^T \otimes \mathbf{F}_D\mathbf{T}_D \\ (\mathbf{G}\bar{\mathbf{H}}_{RR,1}\mathbf{J}_1)^T \otimes \mathbf{F}_D\mathbf{T}_D \\ \sigma_R\mathbf{G}^T \otimes \mathbf{F}_D\mathbf{T}_D \\ \mathbf{0} \end{bmatrix} \tag{40}$$

$$\mathbf{\Omega}_{DRR,1} = \begin{bmatrix} \mathbf{0} \\ \mathbf{J_1}^T \otimes \mathbf{F}_D\mathbf{D}_D\bar{\mathbf{H}}_{RD,2}\mathbf{G} \\ \mathbf{0} \\ \mathbf{0} \end{bmatrix} \tag{41}$$

Then, we exploit the Schur complement lemma to recast the constraint (31) as a matrix inequality by substituting (35) and (37).

$$\begin{bmatrix} \beta_D & \bar{\phi}_D^H \\ \bar{\phi}_D & \mathbf{E} \end{bmatrix} \succ - \begin{bmatrix} 0 & \mathbf{\Delta}_D^H \\ \mathbf{\Delta}_D & \mathbf{0} \end{bmatrix} \tag{42}$$

To eliminate the $\mathbf{\Delta}_D$, the sign-definiteness lemma is applied.

Lemma 2. *Defined matrix* $\mathbf{U}$ *and* $\{\mathbf{P}_i, \mathbf{Q}_i\}$, $i \in \{1, 2, \ldots, N\}$ *with* $\mathbf{U} = \mathbf{U}^H$, *the semi-infinite Linear Matrix Inequality (LMI) of the form*

$$\mathbf{U} \succ \sum_i^N \left(\mathbf{P}_i^H \mathbf{Y}_i \mathbf{Q}_i + \mathbf{Q}_i^H \mathbf{Y}_i^H \mathbf{P}_i \right), \ \|\mathbf{Y}_i\| \le \delta_i \tag{43}$$

Holds if and only if there exist nonnegative real numbers $\lambda_1, \lambda_2, \ldots, \lambda_N$ such that

$$\begin{bmatrix} \mathbf{U} - \sum_{i=1}^N \lambda_i \mathbf{Q}_i^H \mathbf{Q}_i & -\delta_1 \mathbf{P}_1^H & \cdots & -\delta_N \mathbf{P}_N^H \\ -\delta_1 \mathbf{P}_1 & \delta_1 \mathbf{E} & \cdots & \mathbf{0} \\ \vdots & \vdots & \ddots & \vdots \\ -\delta_N \mathbf{P}_N & \mathbf{0} & \cdots & \delta_N \mathbf{E} \end{bmatrix} \succ \mathbf{0} \tag{44}$$

Appropriately choose the parameters below

$$\mathbf{U}_D = \begin{bmatrix} \beta_D & \bar{\phi}_D^H \\ \bar{\phi}_D & \mathbf{I} \end{bmatrix} \tag{45}$$

$$\mathbf{Q}_{D1} = \mathbf{Q}_{D2} = \mathbf{Q}_{D3} = [-\mathbf{1}\mathbf{0}] \tag{46}$$

$$\mathbf{P}_{D1} = \begin{bmatrix} \mathbf{0} & \mathbf{\Omega}_{DSR,1}^H \end{bmatrix} \tag{47}$$

$$\mathbf{P}_{D2} = \begin{bmatrix} \mathbf{0} & \mathbf{\Omega}_{DSD,2}^H \end{bmatrix} \tag{48}$$

$$\mathbf{P}_{D3} = \begin{bmatrix} \mathbf{0} & \mathbf{\Omega}_{DRR,1}^H \end{bmatrix} \tag{49}$$

Apply Lemma 2 to transform (42) as

$$\begin{bmatrix} \begin{bmatrix} \beta_D - \lambda_{D1} - \lambda_{D2} - \lambda_{D3} & \bar{\phi}_D^H \\ \bar{\phi}_D & \mathbf{E} \end{bmatrix} & \mathbf{\Theta}_D^H \\ \mathbf{\Theta}_D & \mathrm{diag}(\lambda_{D1}\mathbf{E}, \lambda_{D2}\mathbf{I}, \lambda_{D3}\mathbf{I}) \end{bmatrix} \succ 0 \tag{50}$$

where $\mathbf{\Theta}_D = -[\delta_{DSR,1}\mathbf{P}_{D1}^T, \delta_{DSD,2}\mathbf{P}_{D2}^T, \delta_{DRR,1}\mathbf{P}_{D3}^T,]^T$. Similarly, the other constraint $\mathrm{tr}(\mathbf{K}_i\mathbf{N}_i) \leq \beta_i$ is written as follows.

$$\begin{bmatrix} \begin{bmatrix} \beta_i - \sum\limits_{k=l}^{j} \lambda_k & \bar{\phi}_i^H \\ \bar{\phi}_i & \mathbf{E} \end{bmatrix} & \mathbf{\Theta}_i^H \\ \mathbf{\Theta}_i & [\lambda_l \mathbf{E}, \cdots, \lambda_j \mathbf{E}] \end{bmatrix} \succ 0 \tag{51}$$

By assembling all the components, the problem can now be written as

$$\max_{\mathbf{L},\mathbf{J_1},\mathbf{J_2},\mathbf{G},\mathbf{F}_i \succ 0,\mathbf{T}_i,\lambda_i \geq 0,\beta_i} h(\mathbf{L}, \mathbf{J_1}, \mathbf{J_2}, \mathbf{G}, \mathbf{F_i}, \mathbf{T_i}, \lambda_i, \beta_i) \tag{52}$$

$$\text{s.t. (16c), (50), (51)} \tag{53}$$

$$h \triangleq 2\log|\mathbf{F}_D| - \beta_D + \min\{2\log|\mathbf{F}_{E11}| - \beta_{E11} + 2\log|\mathbf{F}_{E12}| - \beta_{E12}, \\ 2\log|\mathbf{F}_{E21}| - \beta_{E21} + 2\log|\mathbf{F}_{E22}| - \beta_{E22}) \tag{54}$$

where the function $h(\mathbf{L}, \mathbf{J_1}, \mathbf{J_2}, \mathbf{G}, \mathbf{F_i}, \mathbf{T_i}, \lambda_i, \beta_i)$ is defined in (54). The problem (52) remains non-convex. However, it becomes a convex optimization problem when fixing some of the optimization variables. In other words, after proper manipulations, its sub-problems can become convex, which are readily solvable. Therefore, a BCD algorithm is employed to solve the nonconvex problem (52), which is summarized in Algorithm 1.

Algorithm 1 AN-BF scheme to solve the optimization problem

input l= 0, precoding vector$\mathbf{L}$=$\mathbf{L}^{(0)}$; jamming precoding vector$\mathbf{J_1}$=$\mathbf{J_1}^{(0)}$,$\mathbf{J_2}$=$\mathbf{J_2}^{(0)}$; $\mathbf{F}_i = \mathbf{F}_i^{(0)}, \mathbf{G} = \mathbf{G}^{(0)}$;

repeat

1: Begin BCD to deal with the (52) with $\mathbf{L}$=$\mathbf{L}^{(l-1)}$,$\mathbf{J_1}$=$\mathbf{J_1}^{(l-1)}$,$\mathbf{J_2}$=$\mathbf{J_2}^{(l-1)}$; $\mathbf{F}_i = \mathbf{F}_i^{(l-1)}, \mathbf{G} = \mathbf{G}^{(l-1)}$, and gain the $\mathbf{D}_i^{(l)}$;

2: Solve (52) with $\mathbf{L}$=$\mathbf{L}^{(l-1)}$,$\mathbf{J_1}$=$\mathbf{J_1}^{(l-1)}$,$\mathbf{J_2}$=$\mathbf{J_2}^{(l-1)}$; $\mathbf{D}_i = \mathbf{D}_i^{(l)}, \mathbf{G} = \mathbf{G}^{(l-1)}$, and gain the $\mathbf{F}_i^{(l)}$;

3: Solve (52) to attain $\mathbf{J_1}^{(l)}$, $\mathbf{J_2}^{(l)}$and$\mathbf{L}^{(l)}$ with $\mathbf{D}_i = \mathbf{D}_i^{(l)}, \mathbf{G} = \mathbf{G}^{(l-1)}, \mathbf{F}_i = \mathbf{F}_i^{(l)}$;

4: Solve (52) to gain $\mathbf{G}^{(l)}$ with $\mathbf{L}$=$\mathbf{L}^{(l)}$,$\mathbf{J_1}$=$\mathbf{J_1}^{(l)}$,$\mathbf{J_2}$=$\mathbf{J_2}^{(l)}$, $\mathbf{F}_i = \mathbf{F}_i^{(l)}, \mathbf{D} = \mathbf{D}^{(l)}$;

until $\left|y^{(l)} - y^{(l-1)}\right| \leq \varepsilon$.

4. Results

In this section, numerical simulations are provided to evaluate the performance of the proposed scheme in terms of the average secrecy rate. In this part, we consider a wireless communications system with $N_S = N_R = N_D = N_E = N = 2$. Besides, for simplicity, the CSI uncertainty bound $\delta_{ij,k}$ is represented as the corresponding determinant of mean CSI multiplied by one certain coefficient, or $\delta_{ij,k} = \mu\left\|\overline{\mathbf{H}}_{ij,k}\right\|$.

Figure 2 portrays the average secrecy rate versus numbers of iterations with $P_S = P_R = P = 10$ dB. By the proposed scheme, the average secrecy rate always converges within about 40 iterations. It indicates that the CSI uncertainty has a destructive effect on the secrecy rate and the BCD algorithm converges faster with larger uncertainty. Additionally, the proposed scheme achieves a better average secrecy rate with more ports of legitimate users and fewer ports of eavesdroppers, which is especially obvious in small uncertainty scenarios. It can be explained that the number of ports suggests the ability to receive or intercept the information.

Figure 3 shows the impact of a different transmit power of the proposed scheme. It can be observed that the average secrecy rate increases with the increase of transmitting power. In addition, it is observed that the security rate does not improve significantly when the transmitted power is more than 10 dB under the condition of more ports of eavesdroppers and greater CSI uncertainty. It can be explained that the increase in transmitting power increases the capacity of not only legitimate users but also eavesdroppers, resulting in a slight change in the security rate.

We compare the proposed schemes with a similar one without jamming by presenting the numerical results in Figure 4. Our proposed scheme achieves better performance in terms of the average secrecy rate, especially with lower uncertainty and higher transmit power. Therefore, to some extent, jamming can disturb the interception of eavesdroppers even with higher uncertainty.

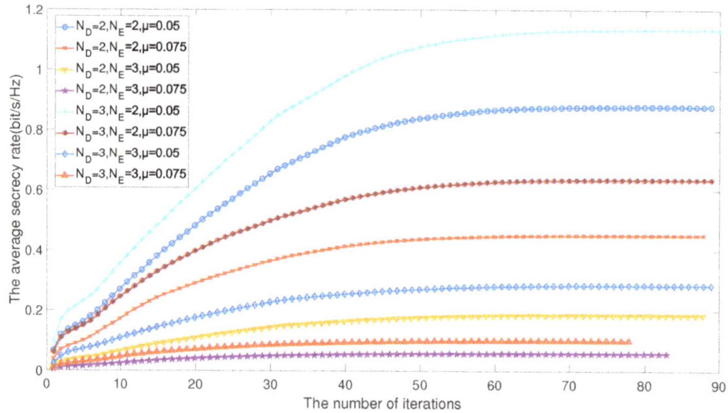

Figure 2. Average secrecy rate versus the number of iterations, a comparison of different ports number and CSI uncertainty.

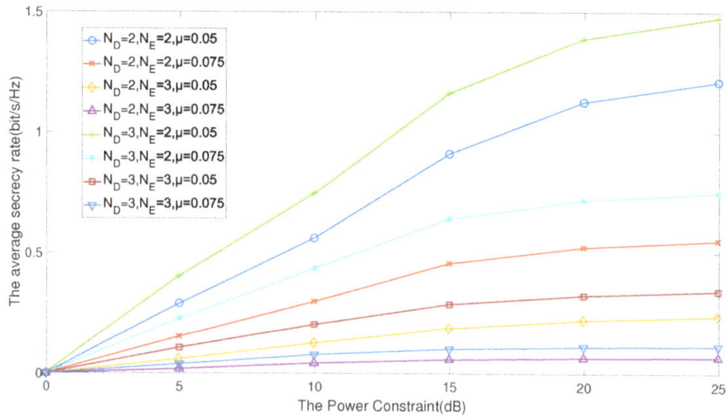

Figure 3. Average secrecy rate versus power constraint, a comparison of different antenna numbers and CSI uncertainty.

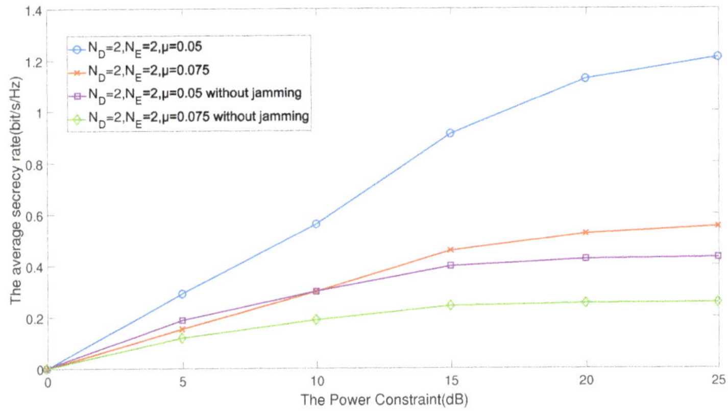

Figure 4. Average secrecy rate versus power constraint comparison of different schemes.

5. Discussion

In this paper, the precoding jamming scheme has been proposed to enhance the security of AF relay-aided power monitoring and communication systems, where the CSI uncertainty and colluding eavesdroppers are considered. Such a system can be used in an underground mining process to guarantee the communication with management offices to ensure the safety. The scheme combined cooperative precoding for users and cooperative jamming for eavesdroppers. Numerical results have shown that the proposed scheme outperforms the scheme without jamming. Furthermore, the effectiveness of the proposed scheme with high CSI uncertainty has been proven.

Author Contributions: Methodology, W.M.; writing—original draft, Y.G. and L.G.; formal analysis, J.B.; validation, T.H. and Z.K.; writing—review and editing, T.H. and Z.K. All authors have read and agreed to the published version of the manuscript.

Funding: This research was supported by the National Natural Science Foundation of China under Grant 62173256.

Data Availability Statement: Not applicable.

Conflicts of Interest: The authors declare no conflict of interest.

References

1. Wang, J.; Al-Kinani, A.; Zhang, W.; Wang, C.-X.; Zhou, L. A general channel model for visible light communications in underground mines. *China Commun.* **2018**, *15*, 95–105. [CrossRef]
2. Lv, Y.; He, Z.; Rong, Y. Two-Way AF MIMO Multi-Relay System Design Using MMSE-DFE Techniques. *IEEE Trans. Wirel. Commun.* **2020**, *20*, 389–405. [CrossRef]
3. Lv, Y.; He, Z.; Rong, Y. Multiuser Multi-Hop AF MIMO Relay System Design Based on MMSE-DFE Receiver. *IEEE Access* **2019**, *7*, 42518–42535. [CrossRef]
4. Sanguinetti, L.; D'Amico, A.A.; Rong, Y. A Tutorial on the Optimization of Amplify-and-Forward MIMO Relay Systems. *IEEE J. Sel. Areas Commun.* **2012**, *30*, 1331–1346. [CrossRef]
5. Zhong, C.; Suraweera, H.A.; Zheng, G.; Krikidis, I.; Zhang, Z. Wireless Information and Power Transfer With Full Duplex Relaying. *IEEE Trans. Commun.* **2014**, *62*, 3447–3461. [CrossRef]
6. Zeng, Y.; Zhang, R. Full-Duplex Wireless-Powered Relay With Self-Energy Recycling. *IEEE Wirel. Commun. Lett.* **2015**, *4*, 201–204. [CrossRef]
7. Song, X.; Ni, Y.; Han, X.; Xu, S. Optimal Power Splitting of Full Duplex Wireless Powered Communication Networks with Two-Way Relay. In Proceedings of the 2018 3rd Interntional Conference on Mechanical, Control and Computer Engineering (ICMCCE), Huhhot, China, 14–16 September 2018; pp. 374–378. [CrossRef]
8. Xiong, T.; Lou, W.; Zhang, J.; Tan, H. MIO: Enhancing Wireless Communications Security Through Physical Layer Multiple Inter-Symbol Obfuscation. *IEEE Trans. Inf. Forensics Secur.* **2015**, *10*, 1678–1691. [CrossRef]

9. Li, R.; Cui, J.; Huang, T.; Yang, L.; Yan, S. Optimal Pulse-Position Modulation Order and Transmit Power in Covert Communications. *IEEE Trans. Veh. Technol.* **2022**, *71*, 5570–5575. [CrossRef]
10. Shannon, C.E. Communication Theory of Secrecy Systems*. *Bell Syst. Tech. J.* **1949**, *28*, 656–715. [CrossRef]
11. Wyner, A.D. The Wire-Tap Channel. *Bell Syst. Tech. J.* **1975**, *54*, 1355–1387. [CrossRef]
12. Parada, P.; Blahut, R. Secrecy capacity of SIMO and slow fading channels. *Int. Symp. Inf. Theory* **2005**, 2152–2155. [CrossRef]
13. Liang, Y.; Poor, H.V.; Shamai, S. Secure Communication Over Fading Channels. *IEEE Trans. Inf. Theory* **2008**, *54*, 2470–2492. [CrossRef]
14. Mukherjee, A.; Swindlehurst, A.L. Robust Beamforming for Security in MIMO Wiretap Channels With Imperfect CSI. *IEEE Trans. Signal Process.* **2010**, *59*, 351–361. [CrossRef]
15. Shlezinger, N.; Zahavi, D.; Murin, Y.; Dabora, R. The Secrecy Capacity of Gaussian MIMO Channels With Finite Memory. *IEEE Trans. Inf. Theory* **2017**, *63*, 1874–1897. [CrossRef]
16. Camponogara, A.; Souza, R.D.; Ribeiro, M.V. The Effective Secrecy Throughput of a Broadband Power Line Communication System Under the Presence of Colluding Wireless Eavesdroppers. *IEEE Access* **2022**, *10*, 85019–85029. [CrossRef]
17. Yu, B.; Yang, L.; Cheng, X.; Cao, R. Relay Location Optimization for Full-Duplex Decode-and-Forward Relaying. In Proceedings of the MILCOM 2013–2013 IEEE Military Communications Conference 2013, San Diego, CA, USA, 18–20 November 2013; pp. 13–18. [CrossRef]
18. Kong, Z.; Yang, S.; Wang, D.; Hanzo, L. Robust Beamforming and Jamming for Enhancing the Physical Layer Security of Full Duplex Radios. *IEEE Trans. Inf. Forensics Secur.* **2019**, *14*, 3151–3159. [CrossRef]

electronics

Article

Design of Low Probability Detection Signal with Application to Physical Layer Security

Lintao Li [1,*], Jiayi Lv [2], Xin Ma [1], Yue Han [3] and Jiaqi Feng [1]

1 School of Computer and Communication Engineering, University of Science and Technology Beijing, Beijing 100083, China
2 School of Information and Electronics, Beijing Institute of Technology, Beijing 100081, China
3 School of Cyberspace Science and Technology, Beijing Institute of Technology, Beijing 100081, China
* Correspondence: lilintao@ustb.edu.cn

Abstract: In this work, we mainly focus on low probability detection (LPD) and low probability interception (LPI) wireless communication in cyber-physical systems. An LPD signal waveform based on multi-carrier modulation and an under-sampling method for signal detection is introduced. The application of the proposed LPD signal for physical layer security is discussed in a typical wireless-tap channel model, which consists of a transmitter (Alice), an intended receiver (Bob), and an eavesdropper (Eve). Since the under-sampling method at Bob's end depends very sensitively on accurate sampling clock and channel state information (CSI), which can hardly be obtained by Eve, the security transmission is initialized as Bob transmits a pilot for Alice to perform channel sounding and clock synchronization by invoking the channel reciprocal principle. Then, Alice sends a multi-carrier information-bearing signal constructed according to Bob's actual sampling clock and the CSI between the two. Consequently, Bob can coherently combine the sub-band signals after sampling, while Eve can only obtain a destructive combination. Finally, we derived the closed-form expressions of detection probability at Bob's and Eve's ends when the energy detector is employed. Simulation results show that the bit error rate (BER) at Alice's end is gradually decreased with the increase in the signal-to-noise ratio (SNR) in both the AWGN and fading channels. Meanwhile, the BER at Eve's end is always unacceptably high no matter how the SNR changes.

Keywords: communication system security; physical layer; wireless communication; precoding; wire-tap channel

Citation: Li, L.; Lv, J.; Ma, X.; Han, Y.; Feng, J. Design of Low Probability Detection Signal with Application to Physical Layer Security. *Electronics* **2023**, *12*, 1075. https://doi.org/10.3390/electronics12051075

Academic Editor: Christos J. Bouras

Received: 31 January 2023
Revised: 14 February 2023
Accepted: 17 February 2023
Published: 21 February 2023

1. Introduction

Cyber-physical systems (CPSs) are networked systems that integrate computation, communication, and control elements. The principal goal of CPSs is to monitor and (if necessary) change the behavior of a physical process to ensure that it functions correctly, reliably, and efficiently. Nowadays, it has been applied in various domains, such as smart grids, health management, vehicular management, and military applications [1]. As CPSs advance rapidly in the degree of informatization and intelligence, their security issues have attracted both scholarly and industrial attention. Security issues of CPSs cover various aspects, including sensing security, computing security, communication security, and control security [2,3]. For the CPSs that are networked in nature, information sharing and interactions should be built on secure and reliable links among various terminals. As a result, communication security is crucial to CPSs [4,5]. Due to the broadcast nature of radio propagation, secure wireless transmission is a challenge. Malicious attacks on communication systems in CPS are classified as passive attacks and active attacks. Passive attacks are those where the attacker listens to network traffic in order to gain access to sensitive information. Yulong Zou studies the intercept behavior of an industrial wireless sensor network, and propose an optimal sensor scheduling scheme aiming at maximizing the secrecy capacity of wireless transmissions from sensors to the sink [6]. In this paper, we develop a practical countermeasure for passive attacks and propose a physical layer security communication scheme for CPS applications.

A well-designed secure wireless link should have LPD and LPI properties with respect to illegal users [7,8]. The concept of perfect secrecy was first introduced in Shannon's fundamental paper [9]. He also proposed that security of communication could be guaranteed only when the transmitter and receiver have a certain degree of cooperation, and perfect secrecy could be achieved if a one-time pad protocol were employed. Traditional encryption techniques are based on the complexity of mathematical tasks, such as the computation of discrete logarithms in large finite fields. With the rapid development of computer hardware and computing technologies, such as distributed computing and cloud computing, the security of traditional encryption techniques has become questionable [10]. Quantum communication can provide almost perfect security through the use of quantum laws to detect any possible information leak [11]. However, its application to wireless and mobile communications is confined because the line of sight for the transmission of optical quantum is not always available, particularly in urban areas crowded by large buildings. The classical spread spectrum communication systems have good LPD, and LPI characteristics and are widely used. However, the random and noise-like properties of pseudo-noise spreading sequences are usually deterministic and periodical in actual systems. With the rapid development of blind signal detection techniques [12], the spreading sequences may be cracked by illegal users. Then, the traditional spread spectrum techniques are also not as secure as expected.

Physical layer security is to develop a secure transmission that exploits the physical properties of transceivers without relying on source encryption [13]. Wyner introduced the concept of secrecy capacity over wire-tap channels [14]. In Wyner's model, the wire-tap channel is a degraded version of the main channel; thus, the eavesdropper can only receive a noisy version of the signal received at the intended receiver. Wyner's work was extended to single input multiple output (SIMO) systems in the presence of one eavesdropper [15]. Hero proposed an information theoretical framework to investigate information security in wireless multiple-input multiple-output links [16]. Another important line of research is the design of a practical system to achieve near-optimal physical layer security performance [17]. Zheng proposed a low-complexity polar-coded cooperative jamming scheme for the general two-way wire-tap channel, without any constraint on channel symmetry or degradation [18–22]. The research mentioned above is unexceptionally confined to the information-theoretic perspective, which only focuses on the LPI performance. Therefore, the main contribution of our work is to design an LPD signal waveform and investigate its application in physical layer security.

Motivated by achieving an LPD signal waveform, we previously proposed an under-sampling spectrum-sparse signal based on active aliasing [23]. In this work, we extend our earlier work to a more practical scenario. Application of the LPD signal for physical layer security is investigated, and a typical wire-tap channel model with three users, namely, the transmitter (Alice), the intended receiver (Bob), and the eavesdropper (Eve), is considered. Since the under-sampling method may be effective only when the sub-band signals are accurately aligned after the sampling process, Alice can shift the central frequencies of the transmitted sub-band signals according to the clock offset between Alice and Bob, to make sure that Bob can collect the signal power on all sub-carriers coherently. Furthermore, a precoding technique based on CSI can be employed to maximize Bob's SNR at the sampling stage. The sampling clock frequency offset and CSI between Alice and Bob are treated as security keys which can be determined at Alice's end according to the reciprocal principle. Meanwhile, Alice and Eve do not have a negotiation of compensation for the sampling rate and CSI; Eve can only use incoherent demodulation techniques. Finally, the LPD and LPI performance of the proposed scheme is evaluated by the detection probability of the received signal and BER, respectively.

The rest of this paper is organized as follows. Section 2 presents the construction of the LPD signal waveform, the principle of the signal detection method. Section 3 presents the application of the designed LPD signal for physical layer security. A practical secure transmission scheme based on channel reciprocity is proposed. Section 4 analyses the LPD performance of the designed signal in the Wire-tap channel. Section 5 investigates the signal and information security performance in terms of detection probability and BER at both Bob's and Eve's ends by simulations. Finally, the conclusions are drawn in Section 6.

2. LPD Signal Design and Detection Method

2.1. LPD Signal Waveform Design

The basic strategy of LPD signal design is to reduce the level of radio frequency energy; the DSSS signal is an example. In this section, an LPD signal waveform based on multi-carrier modulation is designed. The differences between our design and the traditional multi-carrier modulation method lie in the following aspects: signal structure and receiving method. In our design, signals modulated by the sub-carriers are the same, and the sub carriers should be equally spaced. Furthermore, under-sampling method based on active aliasing is employed for signal detection. The designed LPD signal can be expressed as

$$x(t) = \sum_{k=L}^{L+N-1} s(t) \cdot \alpha_k \exp(k \cdot j\omega_c t) \tag{1}$$

where the scaled factor α_k satisfied power constraint as $\sum_{k=1}^{N} |\alpha_k|^2 = 1$, N implies the total number of sub-carriers. Thus, the mean power of signal $x(t)$ is equivalent to that of signal $s(t)$, and L is the number of null subcarriers from the zero frequency to the first signal carrier. $s(t)$ is the original modulated signal with bandwidth ω_B. The carrier spacing can be given by $\omega_c = D \cdot \omega_B$, where D is the ratio between the carrier spacing and baseband width of signal $s(t)$. The parameter D should be no less than 2, or aliasing may occur between adjacent channels. Moreover, artificial noise can be added over the gaps among useful sub-band signals to enhance the covertness of the transmitted signal. In such a case, D should be determined cautiously to avoid aliasing between artificial noise and useful signals.

The comparison diagram of the spectrum structure between the modulated signal $s(t)$ and the proposed LPD signal $x(t)$ is shown in Figure 1. The bandwidth of $x(t)$ is N times of $s(t)$ while the power is consistent. As a result, the power spectrum density of signal $x(t)$ will be significantly reduced, which may be even lower than the background noise provided if N is large enough. Furthermore, $x(t)$ also performs sparsity in the frequency domain when $\omega_c \gg \omega_B$. These two characteristics are similar to that of direct sequence spread spectrum and frequency hopping signals.

2.2. Principle of Under-Sampling Method for Intended Receiver

As previously mentioned, the proposed LPD signal has a low power spectrum and is sparse in the frequency domain. Therefore, detecting the LPD signal at the intended receiver becomes a problem. In this part, the under-sampling method based on active aliasing is presented. The sampling rate is determined by the subcarrier spacing, and the sampling and combination for the proposed LPD signal can be finished simultaneously. For simplicity, the principle of the sampling process is explained in frequency domain. The complex-valued signal at the receiver can be given by

$$r(t) = \sum_{k=L}^{L+N-1} h_k s(t) \cdot \alpha_k \exp(k \cdot j\omega_c t) + w(t) \tag{2}$$

where h_k is the channel coefficient over the kth sub-channel determined by the channel environment [24]. For the additive white Gaussian noise (AWGN) channel, channel coefficient h_k is considered to be 1 for all k. For the fading channel, the channel coefficients can be given by $h_k = |h_k| \exp(j\varphi_k)$, which means the signal transmitted over the kth sub-channel is scaled by the attenuation factor $|h_k|$ and phase-shifted by φ_k. In this work, $|h_k|$ is subjected to Rayleigh distribution and φ_k is subjected to uniform distribution. $w(t)$ is the independent complex additive noise with power spectrum density N_0. The sampling process can be modeled as a pulse modulation process, and the sampling pulse is a periodic ideal pulse sequence given by $p(t) = \sum_{n=-\infty}^{+\infty} \delta(t - nT_s)$, where $\delta(t)$ is the unit impulse function. The sampling frequency can be calculated by $f_s = 1/T_s$.

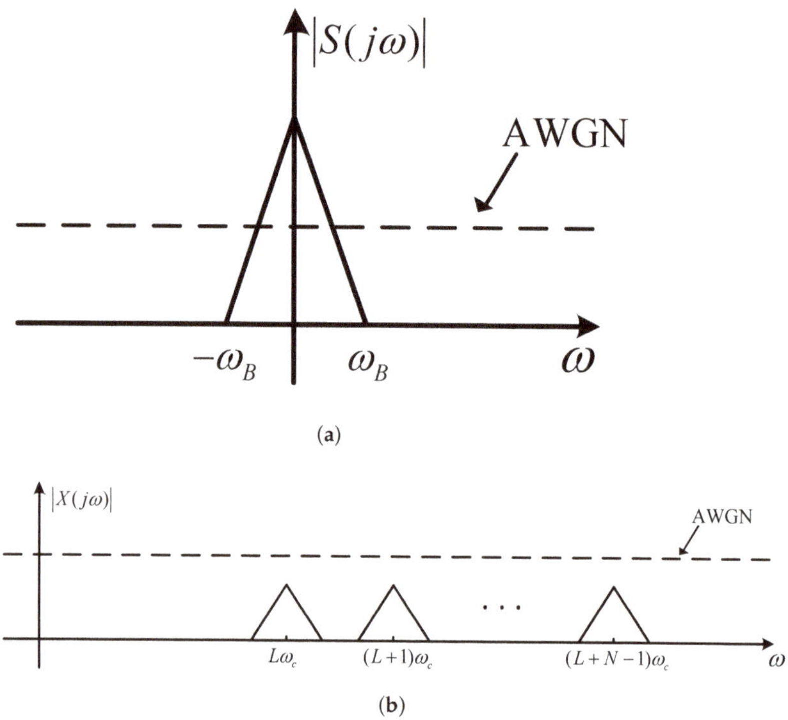

Figure 1. Comparison diagram of the spectrum between $s(t)$ and $x(t)$. (**a**) Spectrum of modulated signal $s(t)$. (**b**) Spectrum of the proposed LPD signal $x(t)$.

The frequency representation of the proposed sampling process is illustrated in Figure 2. As shown in Figure 2a, signal $X(j\omega)$ consists of N sub-band signals $S_k(j\omega)$ with sub-band spacing ω_c, and the total bandwidth of $X(j\omega)$ is $N\omega_c$. The frequency domain representation of the sampling function is illustrated in Figure 2b. The spectrum of the sampled signal can be represented as a convolution of $X(j\omega)$ and $P(j\omega)$. For each sub-band signal $S_k(j\omega)$, a replica of $S_k(j\omega)$ remains at each integer multiple of ω_s. If the sampling rate is chosen as $f_s = f_c$, replicas of sub-band signals $S_k(j\omega)$ may be aligned and added coherently, as shown in Figure 2c. The mean power of the sampled signal increases by N times. Consequently, $S(j\omega)$ can be recovered from the sampled signal with an ideal low-pass filter. Otherwise, these sub-band signals would not be aligned as shown in Figure 2d if $f_s \neq f_c$, aliasing between adjacent sub-band signals can hardly be eliminated.

2.3. Practical Receiver Design

As illustrated in the last section, the feasibility of the under-sampling method has been proved. However, the proposed LPD signal waveform does not exhibit a constant envelope. The sampling phase plays an important role in the sampling process. Here, a practical receiver-based on a multiphase clock [25,26] is presented. The block diagram of the receiver is shown in Figure 3. At the front end of the receiver, the pass band of the analog band pass filter (BPF) is $[(L-1/2)\omega_c, (L+N-1/2)\omega_c]$, and the bandwidth of the pass band is $N\omega_c$. Then, frequency contents out of the pass band will be filtered out by the analog BPF. The SNR of signal $y(t)$ can be given by $SNR_y = P_s/NN_0\omega_c$, where P_s denotes the average transmit power. Thereafter, the multiphase clock, which can produce several sampling clocks with the same frequency but different phases, are employed. They can be modeled as $p_m(t) = \sum_{n=-\infty}^{+\infty} \delta(t - nT_s - m\Delta T_s)$ where $m = 0, 1, 2, \cdots, M-1$ and $\Delta = 1/M$. Thus, a total of M sampled signals can be obtained. Comparing the mean power of these sampled signals, we can select the sampled signal with the maximum mean power as input for the LPF. The LPF is considered to be an ideal LPF with cut-off frequency ω_B. It's important to note that artificial noise (if it exists) will be filtered out by the LPF.

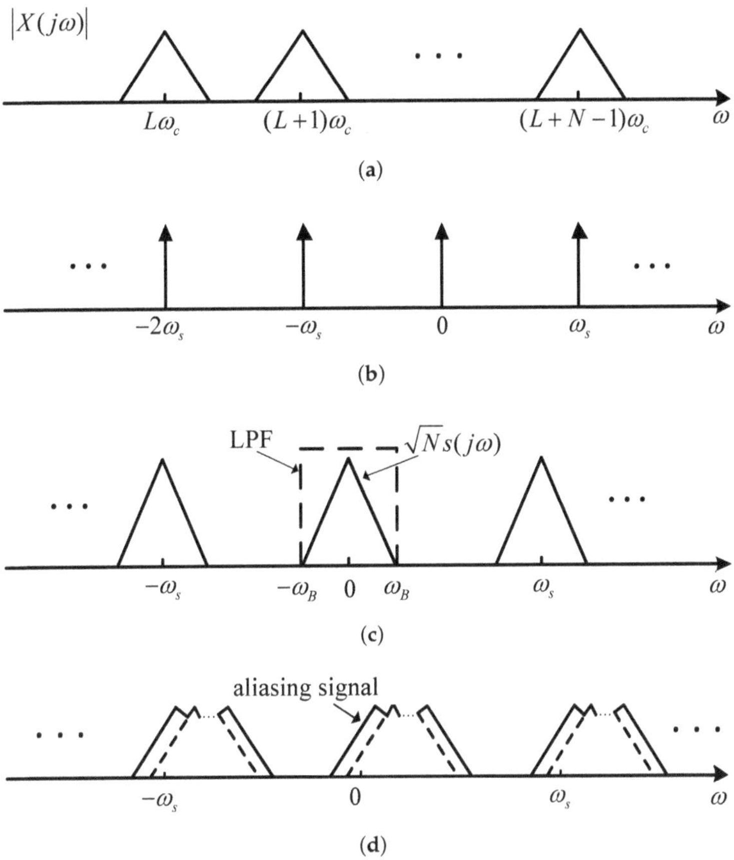

Figure 2. Frequency domain representation of proposed sampling method. (**a**) Spectrum of $x(t)$. (**b**) Spectrum of sampling function. (**c**) Spectrum of sampled signal with $f_s = f_c$. (**d**) Spectrum of sampled signal with $f_s \neq f_c$.

For the noise component, $w(t)$ can be written by summation of N sub-band noise elements as $w(t) = \sum_{k=0}^{N-1} w_k(t) \exp[(k+L) \cdot j\omega_c t]$, where $w_k(t)$ is an independent zero-mean band-limited AWGN with bandwidth ω_c and power spectrum density N_0. After sampling, these sub-band noises are added incoherently, and the power spectrum density becomes NN_0. Following that, the SNR of signal $\tilde{s}[n]$ can be given by

$$SNR_d = \frac{E\left[\tilde{s}^2[n]\right]}{NN_0\omega_B} = \frac{\left|\sum_{k=L}^{L+N-1} \sqrt{1/N} \cdot \exp(j2\Delta k\pi)\right|^2 P_s}{NN_0\omega_B} \tag{3}$$

and then the receiving gain can be achieved as

$$\eta = \frac{SNR_d}{SNR_y} = D \cdot \left| \sum_{k=L}^{L+N-1} \sqrt{1/N} \exp(j2\Delta k\pi) \right|^2 \quad (4)$$

As a result, the maximum receiving gain becomes ND if the sampling phase is synchronized perfectly when $\Delta = 0$. Followed by the LPF, signal $\tilde{s}[n]$ can be demodulated in traditional ways.

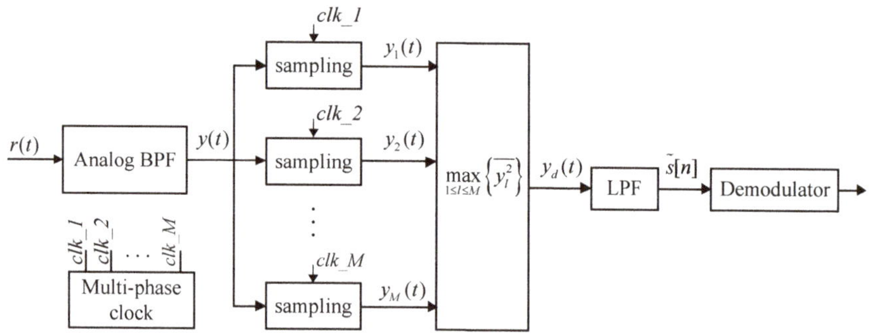

Figure 3. Block diagram of receiver.

2.4. Complexity Analysis of Receiver

In this section, the complexity of the proposed receiver is investigated. For a spectrum sparse signal with bandwidth NDf_B, the wideband bandpass filter is used for signal extraction. Different from the traditional receiver, the multiphase clock should be employed to obtain L-sampled copies. The sampled signal, which has the highest power, is chosen for processing in the following steps. Therefore, a total of M analog to digital converters(ADCs) is needed. Assuming the power of the sampled signal is calculated over Q samples, the selection combining step consumes QM times multiplier, $(Q-1)M$ times add operation, and $log_2(L)$ times comparison operation.

There are also two other possible architectures of receivers for the designed LPD signal. The first receiver architecture uses parallel demodulators for each subcarrier and post-detection combining to recover the signal $s(t)$. Each demodulator needs a narrow band filter and ADC. The complexity and power consumption of the receiver will grow in direct proportion to the subcarrier number N. The second receiver architecture uses direct baseband sampling or radio frequency bandpass sampling method. The sampling rate should be at least twice the bandwidth of the LPD signal as $2NDf_B$ that performance requirements for ADCs will be ultra high. As mentioned above, the proposed under-sampling detection method has lower implementation complexity and hardware requirements.

3. Design of Physical Layer Security Communication System Using Proposed LPD Signal

The analyses in Section 2 show that the designed LPD signal can be exactly detected if and only if the sampling rate is synchronized perfectly. Otherwise, a different sampling rate may lead to a destructive combination after sampling. Then, the intrinsic sampling clock offset between the transmitter and the receiver can be used for secure transmission. In this section, the application of the designed LPD signal for physical layer security is discussed.

3.1. Wireless-Tap Channel Model

In this work, the typical wire-tap channel models consisting of Alice, Bob, and Eve are considered. The secure transmission model is shown in Figure 4, and details of the transmission protocol are presented as follows:

1. Bob sends a transmission request signal to Alice, followed by pilot signals in the same frequency that Alice is going to use for secure data transmission.
2. Alice estimates the sampling clock offset and CSI via the pilot signals.
3. According to the sampling clock offset between Alice and Bob, Alice shifts the central frequencies of sub-band signals to ensure that they are aligned at Bob's side after sampling.
4. As the CSI of the channel from Bob to Alice has become known, the CSI from Alice to Bob can also be informed according to the channel reciprocal principle. Then, a precoding scheme is employed to enhance both capacity and security.
5. Alice securely transmits the modified LPD signal to Bob.

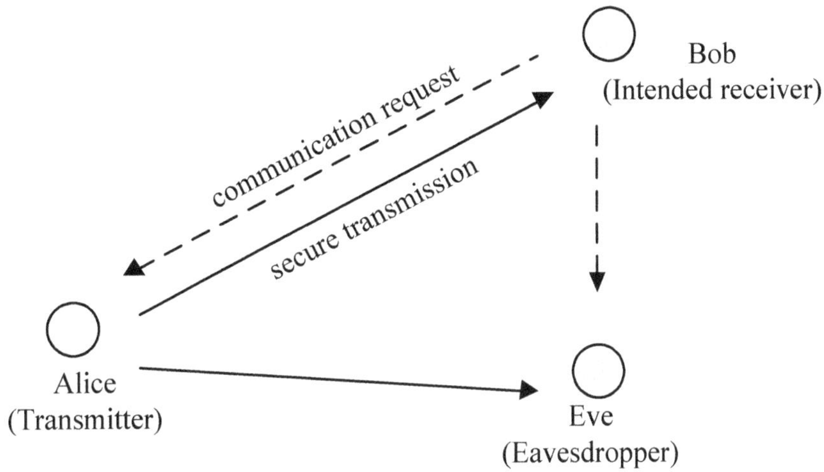

Figure 4. Secure communication system model.

Details of the sampling clock compensation and precoding scheme for security enhancement are presented in this section. These two methods exploit the physical properties of the sampling clock and channel characteristics between Alice and Bob, respectively. For simplification, the sampling phase offset is considered to be $\Delta = 0$ in what follows unless stated otherwise.

3.2. Sampling Clock Offset Compensation

The sampling clock offset for the same frequency ω_c between Alice and Bob is defined as $\kappa_\omega = \omega_{B,c} - \omega_{A,c}$, where $\omega_{A,c}$ and $\omega_{B,c}$ indicate the actual clock frequency of Alice and Bob, respectively. The sampling clock offset can be estimated nearly perfectly only if the SNR is sufficiently high or the number of pilot symbols is sufficiently large. According to the secure transmission protocol, κ_ω can be estimated in step 2. Then, the LPD signal is modified as

$$x_s(t) = \sum_{k=L}^{L+N-1} s(t) \cdot \alpha_k \exp[k \cdot j(\omega_{A,c} + \kappa_\omega)t] \tag{5}$$

The central frequencies of sub-band signals are shifted according to κ_ω, which is considered to be a shared key between Alice and Bob. As the sampling clock offset has been compensated at the transmitter, the sampling clock synchronization between Alice and Bob would be realized. When Bob sampled the received signal with sampling clock $\omega_{B,c}$, sub-band signals in the transmitted signal would be aligned naturally, as shown in Figure 3. Then, modulated signal $s(t)$ may be recovered. Taking the weighted factor α_k and the channel coefficients h_k into account, the SNR of sampled signal $\widetilde{s}[n]$ at Bob is given by

$$SNR_{B,d} = \frac{\left|\sum_{k=L}^{L+N-1} \alpha_k h_k \exp(j2\Delta k\pi)\right|^2 \cdot P_s}{NN_0\omega_B} \quad (6)$$

For Eve, the sampling frequency can hardly be the same as that of Bob. Although the receiving method is known to Eve, sub-band signals cannot always be aligned at the baseband after sampling. The spectrum of the sampled signal at Eve would be the same as that in Figure 3. The sampled signal is a summation of sub-band signals with different carrier frequency offsets, which can hardly be eliminated. Such a sampled signal cannot be used for demodulation, and interception by Eve cannot be realized.

3.3. Precoding Scheme for Fading Channel

According to the proposed secure transmission protocol, the channel fading coefficients from Bob to Alice can be estimated by Alice. Then, the channel coefficient from Alice to Bob can also be known based on the channel reciprocal principle. Therefore, a precoding scheme that exploits the channel characteristics could be employed to improve the receiving gain at Bob. Meanwhile, the precoding scheme can also optimize the power allocation of sub-band signals. As stated, the SNR of the sampled signal at Bob is given by

$$SNR_{B,d} = \frac{\left|\sum_{k=L}^{L+N-1} \alpha_k h_k\right|^2 \cdot P_s}{NN_0\omega_B} \quad (7)$$

when $\Delta = 0$. Weighted factor α_k and channel fading coefficient h_k can be written as $1 \times N$ vectors by $\boldsymbol{\alpha} = [\alpha_L, \alpha_{L+1}, \cdots, \alpha_{L+N-1}]$ and $\mathbf{H} = [h_L, h_{L+1}, \cdots, h_{L+N-1}]$. The Cauchy–Schwarz inequality [27] states that for all vectors $\boldsymbol{\alpha}$ and $\mathbf{H}$ of an inner product space, the following equation holds true:

$$| < \boldsymbol{\alpha}, \mathbf{H} > |^2 \leq ||\boldsymbol{\alpha}||^2 \cdot ||\mathbf{H}||^2 \quad (8)$$

where $< \boldsymbol{\alpha}, \mathbf{H} >$ denotes the inner product of vectors $\boldsymbol{\alpha}$ and $\mathbf{H}$, and the notion $||\cdot||$ denotes the Euclidean norm. Moreover, the equality holds only when $\boldsymbol{\alpha}$ and $\mathbf{H}$ are linearly dependent. It can be written by $\boldsymbol{\alpha} = \lambda \mathbf{H}^*$, where λ is a nonzero constant. The superscript $*$ denotes a conjugate operation. In addition, the weighted factor is constrained by $\sum_{k=L}^{L+N-1} |\alpha_k|^2 = 1$. In order to make the equality in Equation (8) hold true, the weighted factor α_k should be given by $\alpha_k = h_k^*/||\mathbf{H}||$. Then, the SNR of the sampled signal at Bob can be given by $SNR_{B,d} = ||\mathbf{H}||^2 \cdot P_s/(NN_0\omega_B)$. We can conclude that the optimal power allocation strategy for the frequency selective fading channel is to make the SNR over each sub-band identical.

Receiving gain η versus sampling phase offset Δ is shown in Figure 5. We assumed that channel coefficients for sub-band signals are independent, identically distributed, and subject to Rayleigh distribution. Let $\hat{h}_k$ denote the estimates of h_k that can be written by $\hat{h}_k = h_k + h_k^e$. Two different scenarios are explored: (1) perfect CSI, the channel coefficients are perfectly known as $h_k^e = 0$; (2) imperfect CSI, h_k^e is supposed to be a Gaussian random variable with zero mean, and the estimation error is defined by $\rho = E\{|h_k^e|^2/|h_k|^2\}$.

The simulation results show that the sampling phase offset plays an important role in the proposed scheme. It reveals that the accuracy requirement for sampling phase offset is higher with the increase in sub-carrier number N. In addition, an estimation error of CSI may result in inaccurate precoding on Alice's side. It may lead to a performance loss of receiving gain on Bob's side, but will not influence the effectiveness of the under-sampling method.

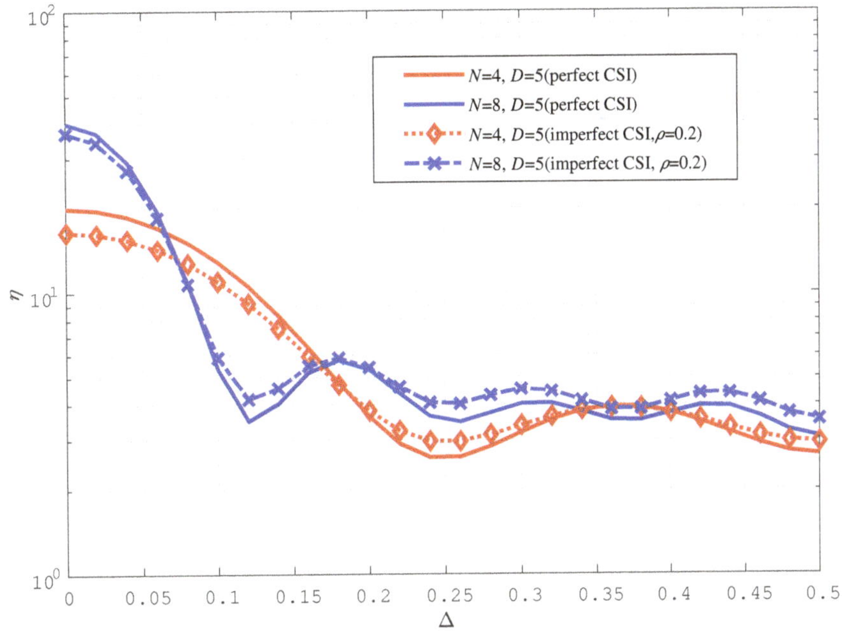

Figure 5. Receiving gain η versus sampling phase offset Δ over fading channel.

4. Performance Evaluation

In this section, we will investigate the signal security performance in terms of probability of detection at Bob's and Eve's ends. It is assumed that Bob uses the proposed under-sampling method, while Eve can only use the energy detection method because of the sampling clock offset and channel differences.

4.1. Energy Detection Method

The signal detection problem can be modeled as a binary hypothetical testing problem with hypotheses $\mathcal{H}_0$ and $\mathcal{H}_1$ defined as

$$\begin{cases} \mathcal{H}_0 : r = w \\ \mathcal{H}_1 : r = x + w \end{cases} \tag{9}$$

where $\mathcal{H}_0$ represents the null hypothesis, and $\mathcal{H}_1$ represents the alternative hypothesis that a useful signal exists. The energy of the received signal is calculated in a bandwidth of W Hz over a period of T_{int}. Users are to detect whether $\mathcal{H}_0$ or $\mathcal{H}_1$ is true based on the test statistic V.

The performance of the ED method is always evaluated by two probabilities, P_d and P_{fa}. P_d implies the probability of detection that $\mathcal{H}_1$ is accepted when $\mathcal{H}_1$ is true, while P_{fa} is the false alarm probability that $\mathcal{H}_1$ is assumed when $\mathcal{H}_0$ is true. The probability density function of normalized decision statistic $Y = 2V/N_0$ has a central chi-square distribution with $v = 2T_{\text{int}}W$ degrees of freedom when $\mathcal{H}_0$ is true. It can be written by

$$P_{\mathcal{H}_0}(Y) = \frac{1}{2^{v/2}\Gamma(v/2)} y^{(v-2)/2} e^{-Y/2} \tag{10}$$

where $\Gamma(u)$ is Gamma function defined by $\Gamma(u) = \int_0^\infty t^{u-1}\exp(-t)dt$.

Meanwhile, the decision statistic obeys a non-central chi-square distribution with v degrees of freedom and non-central parameter $\lambda = 2E/N_0$ when $\mathcal{H}_1$ is true. The E implies the signal energy in the time period T_{int}. The PDF can be written by

$$P_{\mathcal{H}_1}(Y) = \frac{1}{2}\left(\frac{Y}{\lambda}\right)^{(v-2)/4} e^{-(Y+\lambda)/2} I_{(v-2)/2}\left(\sqrt{Y\lambda}\right) \tag{11}$$

where $I_n(u)$ is the Bessel function of the first kind of order n. Therefore, the performance of ED can be described by

$$P_{fa} = \int_{2V_T/N_0}^{\infty} P_{\mathcal{H}_0}(Y)dY \tag{12}$$

and

$$P_d = \int_{2V_T/N_0}^{\infty} P_{\mathcal{H}_1}(Y)dY \tag{13}$$

where V_T denotes the decision threshold.

According to the central limit theorem, $P_{\mathcal{H}_0}(Y)$ and $P_{\mathcal{H}_1}(Y)$ will converge to a Gaussian distribution as v goes to infinity. The approximated PDF can be written by

$$P_{\mathcal{H}_0}(Y) \approx \frac{1}{\sqrt{2\pi}\sigma_w} e^{-\frac{(Y-\mu_w)^2}{2\sigma_w^2}} \tag{14}$$

$$P_{\mathcal{H}_1}(Y) \approx \frac{1}{\sqrt{2\pi}\sigma_{sw}} e^{-\frac{(Y-\mu_{sw})^2}{2\sigma_{sw}^2}} \tag{15}$$

where $\mu_w = 2T_{\text{int}}W$, $\sigma_w^2 = 4T_{\text{int}}W$, $\mu_{sw} = 2T_{\text{int}}W + 2E/N_0$ and $\sigma_{sw}^2 = 4T_{\text{int}}W + 8E/N_0$.

4.2. Detection Performance at Bob's and Eve's Ends

In order to verify the signal security of the designed waveform, detection performance at Bob's and Eve's ends will be analyzed in this section. It is assumed that both Bob and Eve use the ED method. However, the under-sampling method was employed at Bob's end owing to the negotiation with Alice, and the sampling clock offset and CSI can be perfectly known. Moreover, the constant false alarm rate algorithm is applied.

From Equations (14) and (15), we can conclude that

$$\begin{aligned} P_{fa} &= \frac{1}{\sqrt{2\pi}\sigma_w}\int_{\psi}^{\infty} \exp\left(-\frac{(Y-\mu_w)^2}{2\sigma_w^2}\right)dY \\ &= Q\left(\frac{\psi-\mu_w}{\sigma_w}\right) \end{aligned} \tag{16}$$

$$\begin{aligned} P_d &= \frac{1}{\sqrt{2\pi}\sigma_{sw}}\int_{\psi}^{\infty} \exp\left(-\frac{(Y-\mu_{sw})^2}{2\sigma_{sw}^2}\right)dY \\ &= Q\left(\frac{\psi-\mu_{sw}}{\sigma_{sw}}\right) \end{aligned} \tag{17}$$

where $Q(u)$ is Q function defined by $Q(u) = \frac{1}{\sqrt{2\pi}}\int_u^{+\infty} \frac{\exp(x^2)}{2}dx$. Given a predetermined false alarm probability $\hat{P}_{fa}$, the decision threshold can be calculated by

$$\psi^* = \sigma_w Q^{-1}(\hat{P}_{fa}) + \mu_w \tag{18}$$

where $Q^{-1}(u)$ is inverse function of Q(u). Substituting Equation (18) for Equation (17), we can get

$$
\begin{aligned}
P_d &= Q\left(\frac{\psi^* - \mu_{sw}}{\sigma_{sw}}\right) \\
&= Q\left(\frac{\sigma_w Q^{-1}(\hat{P}_{fa}) + \mu_w - \mu_{sw}}{\sigma_{sw}}\right)
\end{aligned} \tag{19}
$$

For the intended user Bob, the time–bandwidth product is approximated as $T_{\text{int}}W = 1$ when the under-sampling method is employed. Furthermore, the ratio of instance symbol energy and power spectrum density is

$$
(E/N_0)_{Bob} = \frac{\|\mathbf{H}\|_2^2 E_s}{NN_0} \tag{20}
$$

where E_s denotes the average symbol energy of $s(t)$. As a result, the decision threshold at Bob's end is $\psi^*_{Bob} = 2Q^{-1}(\hat{P}_{fa}) + 2$, and the detection probability is

$$
\begin{aligned}
P_{d,Bob} &= Q\left(\frac{\psi^*_{Bob} - \mu_{sw}}{\sigma_{sw}}\right) \\
&= Q\left(\frac{Q^{-1}(\hat{P}_{fa}) - (E/N_0)_{Bob}}{\sqrt{1 + 2\cdot(E/N_0)_{Bob}}}\right)
\end{aligned} \tag{21}
$$

For the illegal user Eve, the time–bandwidth product is approximated as $T_{\text{int}}W = ND$. Furthermore, the E/N_0 at Eve's end can be written by

$$
(E/N_0)_{Eve} = \frac{\left|\sum_{k=1}^{N} h_k g_k\right|^2}{\|\mathbf{H}\|_2^2} \cdot \frac{E_s}{NN_0} \tag{22}
$$

where g_k is the channel coefficient of the wire-tap channel that is independent of h_k. In AWGN channel, $g_k = 1 (k = 1, 2, \cdots, N)$ and Equation (22) is simplified as $(E/N_0)_{Eve} = E_s/(NN_0)$. Furthermore, the decision threshold at Eve's end can be given by $\psi^* = 2\sqrt{ND}Q^{-1}(\hat{P}_{fa}) + 2ND$ and the detection probability can be written by

$$
\begin{aligned}
P_{d,Eve} &= Q\left(\frac{\psi^* - \mu_{sw}}{\sigma_{sw}}\right) \\
&= Q\left(\frac{\sqrt{ND}Q^{-1}(\hat{P}_{fa}) - (E/N_0)_{Eve}}{\sqrt{ND + (E/N_0)_{Eve}}}\right)
\end{aligned} \tag{23}
$$

5. Simulation Results

In this section, a number of experiments are designed to evaluate both the reliability and security of the proposed secure transmission system. The receiving gain and BER are chosen as indicators to assess the feasibility and security of the proposed physical layer security communication system. The receiving gain, which was defined in section II implies the phenomenon of SNR improvement caused by the under-sampling method on Bob's side. For secure wireless communication systems, it is desired that the BER at Bob's side is decreased rapidly with the increase in received SNR, while the BER at Eve's side is always unacceptably high. To illustrate the robustness of the proposed physical layer security communication system, simulations are conducted over both AWGN and fading channels. In simulations, the signal $s(t)$ is assumed to be a BPSK-modulated signal with a bandwidth of 10MHz, which means $f_B = 10$ MHz. Furthermore, the parameter L is set as $L = 1$. It is noticed that all simulations in this work are implemented using Matlab. The diagram of system model simulations is shown in Figure 6.

5.1. LPD Performance

The objective of LPD property is to guarantee the covertness of the signal waveform, which means Bob can detect the signals transmitted by Alice, while Eve can hardly detect the presence of the transmit signals. In this section, we will investigate the detection performance at Bob's and Eve's ends in both AWGN and fading channels. The detection method is as described in the last section, and the predetermined false alarm rate is $P_{fa} = 1e - 3$.

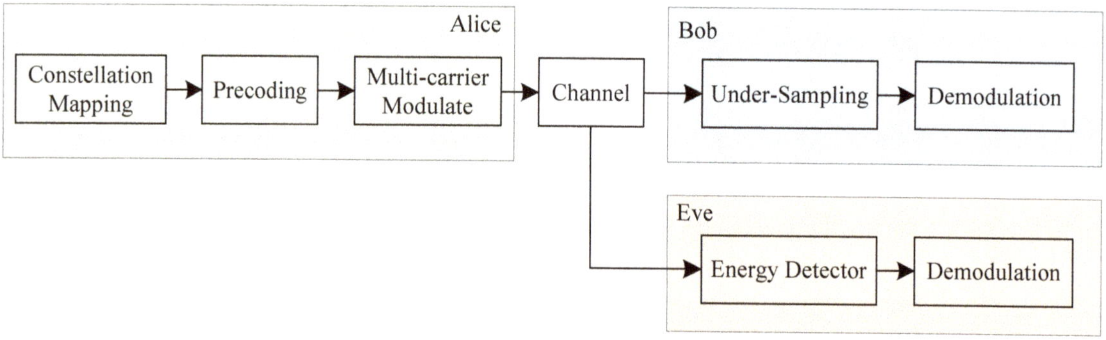

Figure 6. Diagram of the system model in simulations.

Simulation results in Figure 7 show that detection probability at Bob's end is always superior to Eve's when the channel signal-to-noise ratio is less than 10dB over the AWGN channel. There exists a security region depicted by the SNR, in which Bob's detection probability is approaching 1, while that of Eve's is at a low level. For example, when the SNR is in the [−3,4] (dB) interval, the detection probability of Bob is close to 1, while the detection probability of Eve is always lower than 0.1 given $N = 10$ and $D = 4$. In practical applications, Alice can adjust the transmit power so that the received SNR is always in this region, thereby ensuring the covertness of the signal.

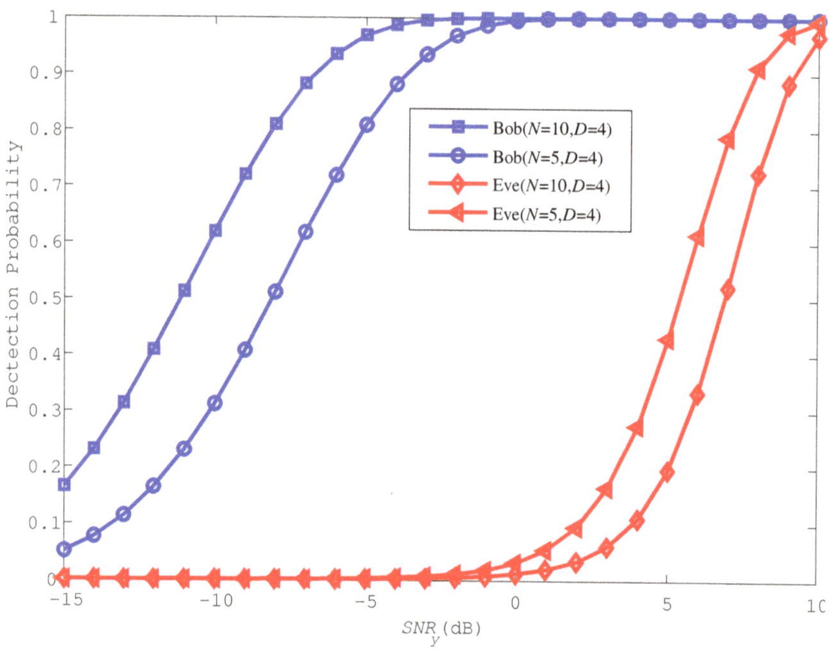

Figure 7. Signal detection performance of Bob and Eve over AWGN channel.

Furthermore, the range of the security region increases with the sub-carrier number N. This means a larger bandwidth may always lead to stronger security in signal covertness.

Such a conclusion is completely consistent with how the larger the spread spectrum ratio is in DSSS, the better the security is in the direct sequence spread spectrum communication system.

Simulation results in Figure 8 show that Bob's detection performance in fading channel is basically the same as that in the AWGN channel, and the precoding scheme is proved to be effective. However, for Eve, the weighted factor α_k and channel coefficients g_k are completely independent, and the SNR at Eve's side is significantly reduced. Therefore, the security region is wider than that in the AWGN channel.

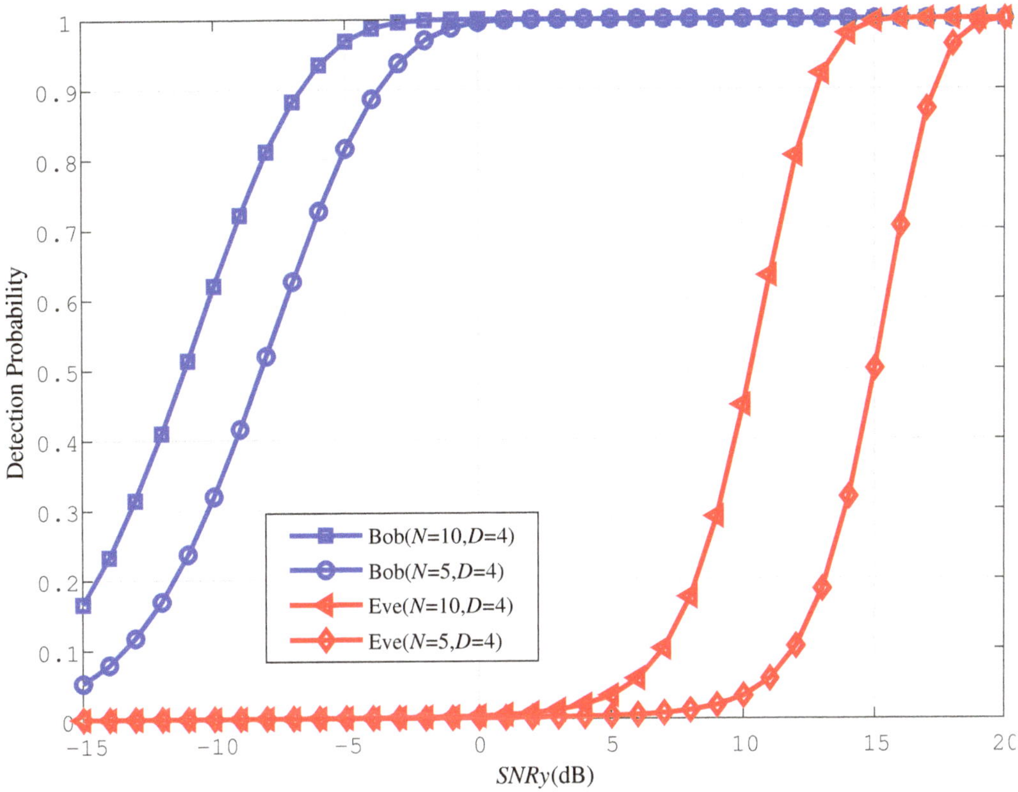

Figure 8. Signal detection performance of Bob and Eve over fading channel.

5.2. Comparison of BER Performance between Bob and Eve

The objective of the proposed physical layer security communication scheme is to simultaneously guarantee the LPD and LPI properties of wireless links. On the one hand, Bob can detect and demodulate the signals transmitted by Alice, while Eve can hardly detect the presence of the transmitted signals. On the other hand, although Eve can detect the transmitted signal, he can hardly extract useful information.

Arguably, BER is an effective and useful measure for both reliability and security. We hope the BER at Bob's side is as low as possible; meanwhile, the BER at Eve's side is (very close to) 0.5, so he essentially cannot recover any information transmitted by Alice. Simulation results demonstrate that the proposed scheme can guarantee that the BER at Eve will always be unacceptably high regardless of the received SNR, while the BER at Bob will be decreased significantly as the received SNR increases.

For the AWGN channel, the security of the proposed communication system is mainly determined by the sampling clock frequency offset between Bob and Eve. According to the communication protocol proposed in Section III, Bob can increase the transmit power or length of pilot signals in order to improve the estimation accuracy. In this way, the sampling clock offset can be estimated nearly perfectly as the SNR of the pilot signal is sufficiently high or the number of pilot symbols is sufficiently large. Meanwhile, the sampling frequency between Alice and Eve can hardly be synchronized because they have no negotiation for sampling frequency synchronization. The BER performance of Bob and Eve is shown in Figure 9. The parameters are set as $D = 5$ and $N = 4$; thus, the sampling frequency is 50 MHz. As the accuracy of the sampling clocks is always at PPM(parts per million) level, we can reasonably assume that the sampling clock offset between Alice and Eve is 1 Hz. The BER versus SNR_y at Bob and Eve are illustrated in Figure 9. The sampling phase offset at Bob is set as $\Delta = 0, 1/8, 1/16$. A significant improvement in BER performance can be achieved when the sampling phase offset decreases. Simulation results show that the BER at Bob decreases rapidly as the SNR increases. Meanwhile, the BER at Eve stays at a high level, and decreases very slowly with the increase in SNR that he can hardly intercept useful information. When some artificial jamming signals are added to the LPD signal, simulation results in Figure 9 show that Bob can still detect and demodulate the LPD signal. The parameter γ in the figure is defined as $\gamma = P_x/P_j$, where P_x denotes the transmit power of useful signals and P_j denotes the transmit power of artificial jamming signals. As a result, the proposed secure communication scheme is proven effective in the AWGN channel.

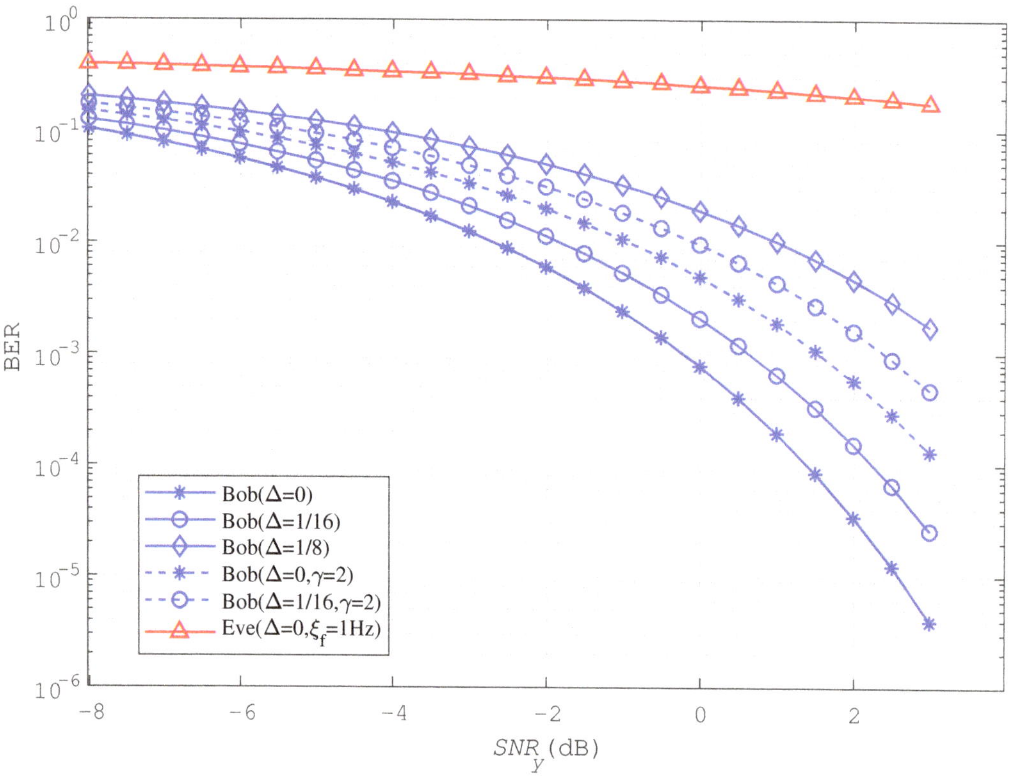

Figure 9. BER performance of Bob and Eve over AWGN channel for $D = 5$ and $N = 4$.

Next, the BER performance of the proposed secure communication scheme over fading channels is shown in Figure 10. For Bob, both perfect CSI and imperfect scenarios are investigated. For the imperfect CSI scenario, the estimation error ρ is assumed to be 0.2 and 0.4. It is not surprising that the BER performance loss is induced by the increase in estimation error under the same channel condition. The results have clearly demonstrated that Bob can detect and demodulate the LPD signal effectively. The BER performance at Eve with different ξ_f is also given in this figure, where ξ_f denotes the sampling frequency offset between Alice and Bob. Simulation results show that the BER at Eve is about 0.5 even $\xi_f = 0$, which means the sampling clock offset between Alice and Eve does not exist. It reveals that Eve can hardly extract useful information only because he has different channel coefficients. It can be seen that Eve will obtain a BER of about 0.5 no matter how the SNR changes. As a result, the proposed secure communication scheme is also proven effective in the fading channel.

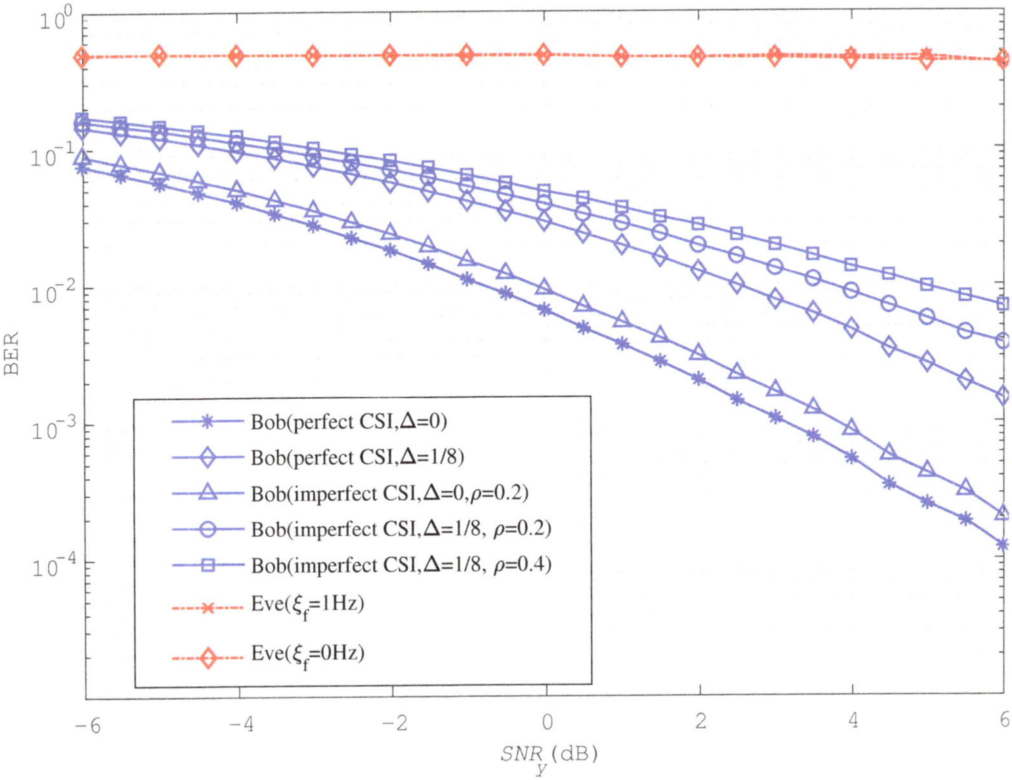

Figure 10. BER performance of Bob and Eve over fading channel for $D = 5$ and $N = 4$.

6. Conclusions

In this work, a physical layer security communication scheme has been proposed for CPS applications. First, a structured LPD signal waveform is designed, and the detection method for the LPD signal is proposed. Analysis shows that the maximum receiving gain is given by ND and decreased with the increase in sapling phase offset. Then, a wireless wire-tap channel is presented, and a secure transmission protocol is proposed. The channel reciprocal principle is applied to achieve the sampling clock offset and CSI between Alice and Bob. Based on such information, the sampling clock compensation method and precoding scheme, which can maximize Bob's SNR at the sampling stage, are proposed. To demonstrate the LPD property, detection probability at both Bob's and Eve's ends are derived with the energy detector model. Simulation results show that there exists a specific SNR interval where Bob's detection probability is approaching 1, while Eve's is well below 0.1. The range is approximately 7 dB and 17 dB in AWGN and fading channel, respectively, when $N = 10$ and $D = 4$. In addition, simulation results in AWNG and fading channel also show that the BER at Bob's end is always decreased with the increase in SNR or the number of sampling phases, while Eve's BER has always been around 0.5 regardless of the SNR. As a result, both the effectiveness and security of the proposed scheme are verified.

Author Contributions: Conceptualization, L.L.; formal analysis, L.L., J.L., and Y.H.; supervision, L.L.; writing—original draft, X.M., Y.H., and J.F.; writing—review and editing, L.L. and J.L. All authors have read and agreed to the published version of the manuscript.

Funding: This work was supported by the Fundamental Research Funds for the Central Universities (FRF-TP-19-052A1) and the Foundation of Beijing Engineering and Technology Center for Convergence Networks and Ubiquitous Services.

Data Availability Statement: Not applicable.

Conflicts of Interest: The authors declare no conflicts of interest.

References

1. Xu, H.; Yu, W.; Griffith, D.; Golmie, N. A Survey on Industrial Internet of Things: A Cyber-Physical Systems Perspective. *IEEE Access* **2018**, *6*, 78238–78259. [CrossRef]
2. Humayed, A.; Lin, J.; Li, F., Luo, B. Cyber-Physical Systems Security—A Survey. *IEEE Internet Things J.* **2017**, *64*, 1802–1831. [CrossRef]
3. Inayat, U.; Zia, M.F.; Mahmood, S.; Berghout, T.; Benbouzid, M. Cybersecurity Enhancement of Smart Grid: Attacks, Methods, and Prospects. *Electronics* **2022**, *11*, 3854. [CrossRef]
4. Cheminod, M.; Durante, L.; Valenzano, A. Review of Security Issues in Industrial Networks. *IEEE Trans. Ind. Inform.* **2013**, *9*, 277–293. [CrossRef]
5. Angueira, P.; Val, I.; Montalbán, J.; Seijo, Ó.; Iradier, E.; Fontaneda, P.S.; Fanari, L.; Arriola, A. A Survey of Physical Layer Techniques for Secure Wireless Communications in Industry. *IEEE Commun. Surv. Tutor.* **2022**, *24*, 810–838. [CrossRef]
6. Zou, Y.; Wang, G. Intercept Behavior Analysis of Industrial Wireless Sensor Networks in the Presence of Eavesdropping Attack. *IEEE Trans. Ind. Inform.* **2016**, *12*, 780–787. [CrossRef]
7. Yan, S.; Zhou, X.; Hu, J.; Hanly, S.V. Low Probability of Detection Communication: Opportunities and Challenges. *IEEE Wirel. Commun.* **2019**, *26*, 19–25. [CrossRef]
8. Shi, C.; Wang, F.; Salous, S.; Zhou, J. Low Probability of Intercept-Based Optimal OFDM Waveform Design Strategy for an Integrated Radar and Communications System. *IEEE Access* **2018**, *6*, 57689–57699. [CrossRef]
9. Shannon, C.E. Communication Theory of Secrecy Systems. *Bell Syst. Tech. J.* **1949**, *28*, 656–715. [CrossRef]
10. Loepp, S.; Wootters, W.K. *Protecting Information From Classical Error Correction to Quantum Cryptography*; Cambridge University Press: Cambridge, UK, 2006.
11. Simon, C. The Foundations of Quantum Information and Feasible Experiments. Ph.D. Dissertation, University Wien, Vienna, Austria, 2000.
12. Zhang, H.G.; Gan, L.; Liao, H.S.; Wei, P.; Li, L.P. Estimating Spreading Waveform of Long-code Direct Sequence Spread Spectrum Signals at a Low Signal-to-noise ratio. *IET Signal Process.* **2012**, *6*, 358–363. [CrossRef]
13. Zhou, X.; Song, L.; Zhang, Y. *Physical Layer Security in Wireless Communications*; CRC Press: Boca Raton, FL, USA, 2013.
14. Wyner, A.D. The Wire-tap Chennel. *Bell Syst. Tech. J.* **1975**, *54*, 1355–1387. [CrossRef]
15. Csiszár, I.; Körner, J. Broadcast Channels With Confidential Messages. *IEEE Trans. Inf. Theory* **1978**, *24*, 339–348. [CrossRef]
16. Hero, A.O. Secure Space-time Communication. *IEEE Trans. Inf. Theory* **2003**, *49*, 3235–3249. [CrossRef]
17. Hong, Y.W.P.; Lan, P.C.; Kuo, C.C.J. Enhancing Physical-Layer Secrecy in Multiantenna Wireless Systems: An Overview of Signal Processing Approaches. *IEEE Signal Process. Mag.* **2013**, *30*, 29–40. [CrossRef]
18. Zhang, W.; Chen, J.; Kuo Y.; Zhou, Y. Artificial-Noise-Aided Optimal Beamforming in Layered Physical Layer Security. *IEEE Commun. Lett.* **2019**, *23*, 72–75. [CrossRef]
19. Sun, C.; Fei, Z.; Li, B.; Wang, X.; Li, N.; Hu, L. Secure Transmission in Downlink Non-orthogonal Multiple Access based on Polar Codes. *China Commun.* **2021**, *18*, 221–235. [CrossRef]
20. Pfeiffer, J.; Fischer, R.F.H. Multilevel Coding for Physical-Layer Security. *IEEE Trans. Commun.* **2022**, *70*, 1999–2009. [CrossRef]
21. Zheng, M.; Tao, M.; Chen, W.; Ling, C. Secure Polar Coding for the Two-Way Wiretap Channel. *IEEE Access* **2018**, *6*, 21731–21744. [CrossRef]
22. Li, J.; Ye, N.; Ma, S.; Bu, X.; An, J. Multi-User Hybrid Beamforming Design for Physical Layer Secured mmWave LOS Communications. *Electronics* **2021**, *10*, 2635 [CrossRef]
23. Li, L.; An, J.; Wang, Z.; Li, X. Under-sampling Spectrum-sparse Signals based on Active Aliasing for Low Probability Detection. *Secur. Commun. Netw.* **2015**, *8*, 4087–4097. [CrossRef]
24. Proakis, J.G.; Salehi, M. *Digital Communications*, 5th ed.; McGraw-Hill: New York, NY, USA, 2007.
25. Dhawan, S.K. Time Measurement With a Multiphase Clock. *IEEE Trans. Nucl. Sci.* **1983**, *30*, 293–296. [CrossRef]

26. Yao, C.; Willson, A.N. A 2.8-3.2GHz Fractional-*N*Digital PLL With ADC-Assisted TDC and Inductively Coupled Fine-Tuning DCO. *IEEE J. -Solid-State Circuits* **2013**, *48*, 698–710. [CrossRef]
27. Gilbert, S. *Linear Algebra and Its Application*, 4th ed.; Cengage Learning: Stamford, CT, USA, 2005.

electronics

MDPI

Article

Secure Modern Wireless Communication Network Based on Blockchain Technology

Radha Raman Chandan [1], Awatef Balobaid [2], Naga Lakshmi Sowjanya Cherukupalli [3], Gururaj H L [4,*], Francesco Flammini [5,*] and Rajesh Natarajan [6]

1 Department of Computer Science, School of Management Sciences (SMS), Varanasi 221001, India
2 Department of Computer Science, Jazan University, Jazan 45142, Saudi Arabia
3 CSE Department, Koneru Lakshmaiah Education Foundation, Green Fields, Vaddeswaram, Andhra Pradesh, Guntur 522302, India
4 Department of Information Technology, Manipal Institute of Technology Bengaluru, Manipal Academy of Higher Education, Manipal 576104, India
5 IDSIA USI-SUPSI, University of Applied Sciences and Arts of Southern Switzerland, 6928 Manno, Switzerland
6 Information Technology Department, University of Technology and Applied Sciences-Shinas, Shinas 324, Oman
* Correspondence: gururaj.hl@manipal.edu (G.H.L.); francesco.flammini@supsi.ch (F.F.)

Abstract: Sixth-generation (6G) wireless networking studies have begun with the global implementation of fifth-generation (5G) wireless systems. It is predicted that multiple heterogeneity applications and facilities may be supported by modern wireless communication networks (MWCNs) with improved effectiveness and protection. Nevertheless, a variety of trust-related problems that are commonly disregarded in network architectures prevent us from achieving this objective. In the current world, MWCN transmits a lot of sensitive information. It is essential to protect MWCN users from harmful attacks and offer them a secure transmission to meet their requirements. A malicious node causes a major attack on reliable data during transmission. Blockchain offers a potential answer for confidentiality and safety as an innovative transformative tool that has emerged in the last few years. Blockchain has been extensively investigated in several domains, including mobile networks and the Internet of Things, as a feasible option for system protection. Therefore, a blockchain-based modal, Transaction Verification Denied conflict with spurious node (TVDCSN) methodology, was presented in this study for wireless communication technologies to detect malicious nodes and prevent attacks. In the suggested mode, malicious nodes will be found and removed from the MWCN and intrusion will be prevented before the sensitive information is transferred to the precise recipient. Detection accuracy, attack prevention, security, network overhead, and computation time are the performance metrics used for evaluation. Various performance measures are used to assess the method's efficacy, and it is compared with more traditional methods.

Keywords: blockchain; wireless communication network; malicious node; security protocol; intrusion detection

Citation: Chandan, R.R.; Balobaid, A.; Cherukupalli, N.L.S.; H L, G.; Flammini, F.; Natarajan, R. Secure Modern Wireless Communication Network Based on Blockchain Technology. *Electronics* **2023**, *12*, 1095. https://doi.org/10.3390/electronics12051095

Academic Editors: Tao Huang, Shihao Yan, Guanglin Zhang, Li Sun, Tsz Hon Yuen, YoHan Park and Changhoon Lee

Received: 24 January 2023
Revised: 19 February 2023
Accepted: 21 February 2023
Published: 22 February 2023

1. Introduction

Over the last several years, the need for contemporary wireless communication networks has increased tremendously. The global deployment of 5G technologies, which has many more capabilities than 4G communications, is approaching. Between 2027 and 2030, the 6G technology, modern wireless communication network architecture with significant AI capability, is anticipated to be introduced into operation. There is an enormous amount of communication as a result of the quick growth of many developing technologies, including artificial intelligence (AI), virtual reality (VR), three-dimensional (3D) media, and the Internet of Everything (IoE). This demonstrates the value of enhancing interaction processes. A civilization with completely autonomous distant administration technologies is what they are moving toward. MWCN systems are gaining popularity in every aspect of life, including business, medicine, transportation, and space exploration [1]. The following list summarizes MWCN's salient features: An ultra-high-density network is needed to support 5 Gigabyte networking deployments, huge connection, and consistent quality. Small-cell networking has been identified as a key component of MWCN systems, specifically the idea of highly dense small channels. Sensor nodes tend to be placed much closer together in tiny channel networks than they are in other types of ad hoc networks. As a consequence of this, there is often a rather high amount of correlation and redundancy in the data that is perceived by several nodes. This system is also anticipated to ensure the effective utilization of cutting-edge encryption and modulating algorithms, as well as a novel waveform architecture. They will need less expensive network hardware, less expensive deployments, and improved power-saving features in both the networking and consumer device sectors. Almost 80% of mobile congestion is produced indoors. This amount of data can be transferred to indoor densely small cells, freeing up costly and important microcell capabilities. Only a few milliseconds or less will separate the beginning and the completion of the transaction [1].

Wireless signals transfer data at the speed of light in the universe using electromagnetic radiation as transport, which significantly aids in the advancement and growth of the community. At the current time, MWCN's data security problems have drawn a lot of attention as depicted in Figure 1. Anybody within the signal-covering region can eavesdrop on or assault the signal at the physiological layer due to the indigenous "genomic" faults of electromagnetic fields that are exposed by the free transmission of wireless communications. However, current security measures are mostly based on the cryptography method utilized in conventional wired communication and are created to a greater extent, making them unable to effectively address security concerns brought on by the accessibility of communication networks [2]. Blockchain innovation has the prospects to substantially improve the safety of physician and Medicare data technologies that cope with data like patient digital wellness data, medical assent, pharmacy supply chains, blockchain-based remote monitoring records, information for investment businesses, and other confidential material related to scientific experiments. The implementation of blockchain technology can increase medical data transfer efficiency, accessibility, security, and accountability. Blockchain technologies, coupled with artificial intelligence (AI) and machine learning, are about to change the medical industry. The distributed design of blockchain technology is being combined with memory innovation to guarantee the confidentiality of the information for the investors utilizing the public ledger method [3]. Various attack types have occurred during communication between nodes, whether it is within transmission range or beyond the spectrum (i.e., an insider threat or an outcast target). As a result, there are security concerns with forwarding, including data gathering, route maintenance, information propagation, etc. [4]. An unauthorized action or behavior that damages the wireless environment is referred to as an intrusion. In other terms, an intrusion is defined as an attack that compromises the privacy, authenticity, or accessibility of data in any way. Safety threats to the MWCN frequently come from both the inner and outside of the network, where legitimate network nodes can become corrupted and occasionally made to behave maliciously. The timely identification, containment, and elimination of rogue nodes inside a network are other crucial security threats. Addressing security-related challenges has drawn a lot of interest and had a significant influence on MWCN's architecture and evolution patterns [5]. Therefore, we suggested using blockchain-based technologies to safeguard MWCN by detecting malicious nodes and preventing attacks in wireless transmission.

This article is organized as follows: Section 1 describes the introduction, Section 2 examines similar works, Section 3 describes the suggested method, Section 4 presents the results and discussion, and Section 5 provides the conclusion.

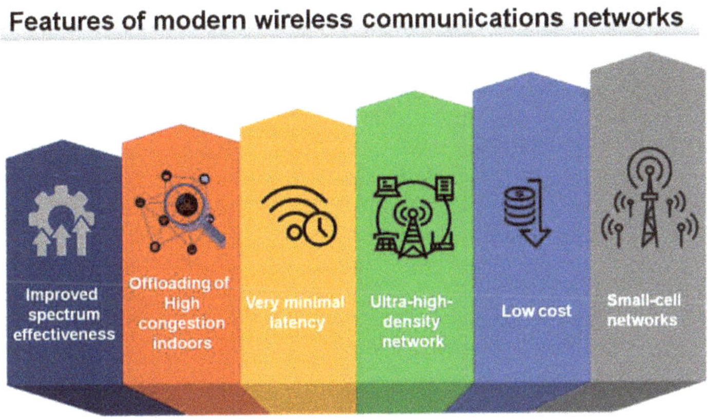

Figure 1. Features of MWCN.

2. Survey of the Literature

The Wireless Multimedia Sensor Network (WNSM) has become more popular among many groups as a result of technical developments in sensors and contemporary gadgets. A network being targeted by many attackers is known as a dispersed assault. Compared to assaults involving a single node, this form of attack greatly worsens network functioning difficulties. An improved machine learning method must be offered to protect the network from the dangers of DoS assaults. An improved Deep Neural Network technique is suggested for WMSN attack detection [6]. A wireless network uses self-organizing modules that are distributed randomly and have a tiny battery capacity to observe the area and allow real activities. Public access is maintained through wireless communication, which encourages a rise in harmful activity inside the network. The majority of network attacks are black hole attacks. In this study, they proposed the Hybrid Deep Learning Prediction (HDLP) framework in the wireless network to maximize battery life and networking reliability [7].

The implementation of fifth-generation (5G) wireless communication technologies was effectively publicized by Non-Orthogonal Multiple Access (NOMA), which is currently regarded as a key innovation in 5G networking. In this study, they created a NOMA model and used a dropping assault to recover a database from the system. Following the use of ML techniques, the retrieved data's detection accuracy for dropping assaults was 95.7%. Additionally, relying on the use of various ML and DL approaches, this study proposes a process for wireless cyber threat identification in 5G technologies [8]. This is the age of smart cognitive radio network innovation, which allows for the effective use of the bandwidth that is now accessible. The goal of cognitive radio innovation should be interference-free frequency availability for consumers. The study addresses various assaults and their causes. The relevance of the authentication system in preventing attacks and ensuring easy frequency use is shown. In this study, the mechanisms and requirements for authentication are examined along with ways to address the safety problems in cognitive networks. The scientific issues surrounding the cognitive radio network's privacy and potential solutions are discussed in this study [9].

Sensors in wireless communication are vulnerable to a variety of security risks. Wireless communication is susceptible to denial-of-service assaults because of these sensors. One of these is a wormhole assault, which alters the network's distribution pathways by using a low-latency connection between two rogue sensor nodes. This assault is harsh because it defies several security techniques and is difficult to detect inside the system. The identification and prevention of wormhole attacks in wireless sensor connectivity is the focus of a thorough assessment of the research in this study [10]. Wireless Sensor Networks (WSNs) are vulnerable to several rogue nodes as a significant information-transmitting technology.

Due to the inadequacy of the current malicious node identification approaches in wireless sensor communications, this research suggested an improved lower energies adaptable clustered hierarch (Enhanced LEACH) routing protocol for harmful node identification based on reputation. A unique method aims to detect rogue nodes in the WSN [11].

The information must be kept secure since it is sent through a wireless channel. The method used in this research helps to ensure that data is safely sent from the origin node to the ground station. This paper proposed a lightweight Bloom filter solution for information transport and packet loss detection in intermediate nodes. They used source data to help them find any malicious packet-discarding nodes and relied on attribution encryption and decryption methods for the model. In this, the data might be discarded while being transmitted [12]. Nodes are vulnerable to several risks as a result of their transparency, among them being deceptive suggestion attacks that provide misleading trust levels that benefit the perpetrator. The artificial bee colony algorithm (ABC) and a fuzzy trust model (FTM-ABC) are utilized in this study to propose a method for identifying malicious nodes. The fuzzy trust model (FTM) is introduced to calculate indirect trust, and the ABC technique is utilized to enhance the trust model to detect false-positive suggestion attacks [13]. In the Malicious Nodes Detection (MND) stage, the Improved Deep Convolutional Neural Network (IDCNN) locates the MN and separates those into the malicious listed box. The Extended K-Means (EKM) method groups the Trusted Nodes (TN) in the energy-efficient stage, and the t-Distribution based Satin Bowerbird Optimization (t-DSBO) method chooses a unique cluster head for every cluster centered on the remaining power of those networks [14].

Malicious assaults (such as wormhole and blackhole assaults) have become a severe problem in wireless transmission in the latest days. Wormhole and blackhole attacks use up more computer resources, network activity, and power. In this study, a brand-new Cross-layer-based Hidden Marko model (C-HMM) is suggested to identify and isolate blackhole and wormhole assaults in wireless ad hoc networks with high efficiency and low transmission costs [14]. The development of wireless communication has only been beneficial to people. In these, data is exchanged between the nodes via the wireless connection at an extremely fast pace. However, one difficulty associated with communication is maintaining confidentiality. They must guarantee that the information packets are transferred privately to the recipient without being accessed by a third party. We provide a technique that uses a node's spatial data attribute to estimate received signal strength (RSS), which is the primary variable for visualizing aggressor nodes in the system and removing the assailant nodes by using clustering methods using a radar grid [15]. This study presents a mechanism for identifying malicious nodes that will make wireless sensor networks much more trustworthy and secure, called density-based spatial clustering of applications with noise (DBSCAN). The major objective of this approach is to design a routing strategy that can detect malicious nodes, has a stronger consistency over time, and has a longer network lifespan. Density-based clustering is a popular and often-used method in many domains. The DBSCAN is a highly popular and effective density-based clustering method that can find clusters of any kind. However, it was unable to identify every node in a network [16]. Numerous drawbacks in the above system, such as low detection accuracy, more energy consumption, and attack prevention, are not effective in wireless communication. Ref. [17] discussed cutting-edge multi-tier authentication techniques that have been presented over the years from 2011 to 2018, their flaws and security concerns, and eventually their solutions for fog computing environments. We compared the various multi-tier authentication solutions based on three criteria: deployment costs, security, and usability. Ref. [18] addressed the multi-stakeholder problem in a fog-enabled cloud. This study proposes a Privacy-Aware Log-preservation Architecture in Fog (PLAF), a comprehensive and automated architecture for proactive forensics in the Internet of Things (IoT). It takes into account the preservation of distributed edge node logs while also being security- and privacy-aware. As previously said, we have created a test bed to implement the specification by combining numerous cutting-edge technologies in one location.

Problem Statement

Modern wireless communication networks (MWCN) serve as a crucial means of information transmission. Because everyone inside a wireless network's service region can seek to penetrate the system, wireless networks have insufficient privacy protection. Destructive cyber-attacks have been recorded regularly at locations with accessible, connected networks, and it has been noted that these locations are most susceptible to a total hack of the smartphone or computer data. They might be attacked by several malicious nodes. It is important to eliminate these MWCN inefficiencies. This research presented a blockchain-based mode, Transaction Verification Denied conflict with spurious node (TVDCSN) technique, in light of the ineffectiveness of the conventional malicious node identification and attack prevention approaches in wireless communication networks.

3. Research Method

Contemporary technologies have advanced technologically, which has increased interest in the MWCN among diverse populations. Although, because of its wide connectivity it faces several security dangers, one of the main problems for network administrators is authenticating communications in MWCN. Each network layer may be the target of several threats. Even though it would be ideal to provide MWCN with enhanced security measures that can identify network intruders and suggest such remedies, we presented the Transaction Verification Denied conflict with a spurious node to provide secure transmission of sensitive information.

3.1. Dataset

Healthcare documents, social media data, and sensor data make up the suggested system's database. Wearable biological and cognitive sensors are used to retrieve the patient's sensory data. People with hyperglycemia and high blood pressure have many variables detected using devices and smart devices. The majority of the signs of diabetes, high blood pressure, and other disorders are covered by the sensed variables. Additional data are also taken out of the person's body. Hospital documents provide information on the therapies that individuals with hypertension and high blood pressure received. They gather patients' health history, which details their health information (including procedures, blood tests, and medication use). This includes the whole patient file in a digital file. This also includes various health information about the patient's condition, including results from testing, responses to questions about one's well-being, and drugs used. A patient's medical state may be evaluated using lab test results from healthcare equipment in the perspective of standards [19].

The content of patients is retrieved from hospital social networking platforms as the initial step of the proposed solution. Nevertheless, further effort is required for this activity, and its success is entirely dependent on the privacy settings of social networking sites.

The application programming interfaces (APIs) of certain social networks are hidden from public view. In a circumstance such as this one, specialized software, such as wrappers, can be utilized to retrieve information (for example, patient posts) [20]. People with diabetes and high blood pressure typically maintain regular contact with their physicians; however, patients with these conditions also require assistance, information, and abilities to personally monitor their healthcare situation. In addition, if patients do not receive useful information from their doctors, social media may be able to perform an important role in satisfying their requirements. As a result, patients can make use of chances provided by social networking platforms such as Facebook and Twitter to acquire sufficient knowledge regarding diabetes and BP and to interact with people who have similar health problems and have had comparable experiences. Patients and medical professionals alike can benefit from the platform that social networks offer for the exchange of information regarding diabetes therapies. To improve patient care and knowledge, we collect data from social media, such as drug reviews and emotional posts made by patients. This allows us to predict the patients' levels of stress and depression, identify the side effects of diabetes medications on diet and lifestyle, and improve patient care.

The data that make up the system that is being suggested include medical records, sensing data, and data from social networking sites. However, due to its inconsistencies, missing information, noise, multiple formats, vast size, and high complexity, real-world big data is notoriously difficult to work with. The results produced by low-quality and noisy data are also of low quality. The phase of preprocessing the data is performed before the processing itself, which both enhances the overall quality of the processing and reduces the amount of time it takes. The pre-analysis of sensor data, preprocessing and filtering of sensor data, preprocessing of medical records, and preparation of sensor data are all components of our system.

3.2. *Transaction Verification Denied Conflict with Spurious Node (TVDCSN)*

Every node in the suggested technique must only utilize the data that is readily accessible to it, without depending on a centrally or localized trustworthy source. This method examines the validity of the WELCOME information rather than constantly verifying it by searching for inconsistencies between the information and the known architecture. This allows for single *MPR* nominations as long as there are no inconsistencies. An *MPR* may be chosen for any two-hop residents for whom it is the only access point, despite any inconsistencies. However, it cannot be proposed as the exclusive *MPR* for two-hop neighbors that are accessible by other routes.

The notations utilized in the technique are as follows:

N denotes the group of all nodes in the network; the victim and attacking nodes are denoted by v, a; S_y is a spurious node that y promotes; the collection of all v's 1-hop neighbors is represented by $HN(v) \subset N$; $HN2\ (v) \subset\ HN\ (v)$ is the collection of all the v's two-hop neighbors; the collection of one-hop nodes of v that designated v as their MPR is known as $MPR\ (v) \subseteq adi\ (v)$; and the collection of one-hop nodes chosen by v to serve as *MPR*s is denoted by $MPR'(v) \subseteq HN\ (v)$.

3.2.1. Conflict Rules

We outline the conditions that should be achieved for a node to recognize the sender of a WELCOME text. Take into account *HN* (v) = b, c, x and *NH2* (v) = d, e. Depending on the protocol, v must choose *MPR* (v) = b, c to encompass *HN2* (v). Assuming that x wants to isolate victim v, y sends a false Welcome text with the following contents: *HN* (x) = v, d, e, S_y. The Laws are:

- If node x broadcasts a WELCOME message with *HN*(y), node v must verify that none of the nodes indicated by x are one of *HN* (v). Nodes b and c must be present in *HN2*(y); therefore, y should choose *MPR*s that would enable it to connect to them. Nevertheless, y may pretend to wish to select v as *MPR* for taking care of a and b;
- If a node y is named in a Welcome text, node v must check to determine if there is a node u *HN*(y) that is (a) not referenced in the recipient's WELCOME text and (b) at least three hops distant from node v. If such criteria are met, a secondary assessment is required: (c) has w been designated as *MPR* to fill in for u by y? Figure 2 shows the finding of conflicts to prevent attacks.
- Accessing the Topology management (*TM*) table might be used to perform assessments (a) and (b). There is a conflict if there isn't an element carrying the *MPR* that x selected and that enables it to go from x to z in just two hops. Keep in mind that if either condition (a) or (b) is not met, conflicts cannot be found. A *TM* text must exist where either y has chosen u or u has chosen x as *MPR* for (c) to be verified. To do this, v must check each u 2 *U*, where *U HN2*(y) is dependent on y's text. Algorithm 1 illustrates the testing of the criterion when the *TM* message's structure is "latest (location), dest (location)";
- A WELCOME text with all *Hn*(v) must be viewed by v as a threat, and necessary action must be taken.

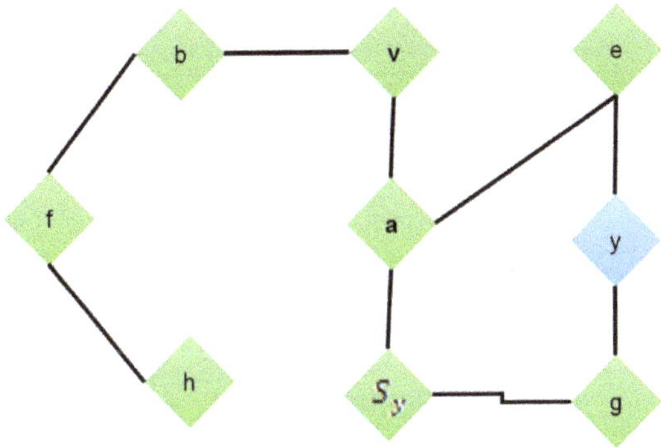

Figure 2. Detecting conflicts.

Algorithm 1: Testing criterion

Testing–criterion $(TM,\ H,\ X,\ V)$
$U \leftarrow \Phi$
For each $r\epsilon TM$ *do*
If r.last ϵ $HN(y)$ *do*
$U \leftarrow U\ Z\ \{r, dest\}$
If r.dest ϵ $HN(x)$ *do*
$U \leftarrow U\ Z\ \{r, last\}$
For each $u\ \epsilon\ U$ *do*
If $u\ \epsilon\ U \cap HN(v)$ *do*
$U \leftarrow U - \{u\}$
For each $m\ \epsilon\ MPR'(x)$ *do*
For each $U\epsilon U$ *do*
If $\{m, y\}\epsilon$ *TC such that z is encompassed by m do*
$U \leftarrow U - \{u\}$
if $U \leftarrow \Phi$ *do*
Consider y as a malicious node
Else
Consider y as a trustworthy *MPR*

Using a spurious node, this looks for discrepancies between a WELCOME signal and the system architecture as it is known from previous WELCOME and *TM* messages. However, make sure to double-check each node that the WELCOME message mentions. There are situations in which a node isolation assault is still possible. Think about Figure 3, where y falsely claims that $HN(y) = v$, f, e, and g. *MPR* (y) = "f, h" and *HN2*(x) = "a, b, e, j, l". There are no contradictions that v can find because y does not assert that it is aware of any node in $HN(v)$ except itself (rule No. 1). a, b, e, j, and l are the *MPR*s that were chosen by y to access all of *HN2*(y). Since d is previously approachable by f (rule No. 2) and y does not claim to be aware of all of *HN* (v), in particular b, it is predicted that x would not designate c as one of its *MPR*s (rule No. 3).

Regrettably, if each node in the system declared an extra fake node, all nodes would be recognized as *MPR*s as a result of their false advertisements, and the network would return to Link-State Forwarding. As a result, a technique for restricting false messages must be developed that finds a balance between the requirement to minimize node usage and preserving the network against separation assault.

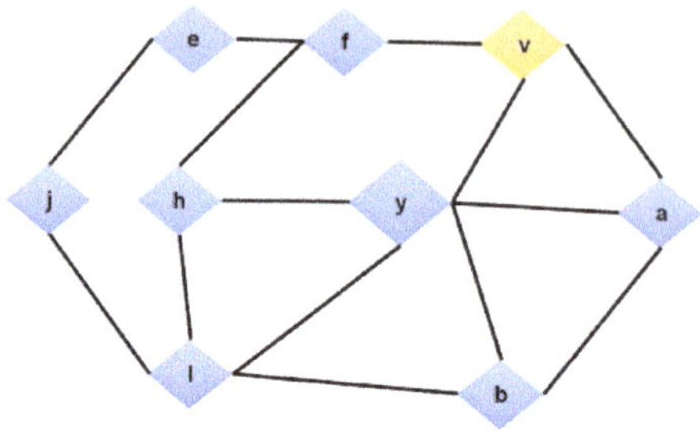

Figure 3. Node attack with no conflicts.

To avoid nodes in the networking from informing the others of misleading data about their connection, we built up a method enabling each node to determine if an attack may be launched via itself. If such a falsehood is feasible, the node creates a spurious node and connects it to the network to stop others from believing they are connected to it. In other words, the nodes themselves are in charge of ensuring that the connection data is accurate since they should prevent others from misusing it. The following provides the limiting method for introducing or eliminating spurious nodes:

- When all nodes in $HN2$ $(v) \in HN$ (v) imply that the separation between y and u is less than three-hops, every node v must incorporate a spurious node;
- $Sv \in HN(v)$;
- New node u promotes Fu by nature before rule 1 is calculated;
- Then, the spurious node is removed when rule (1) is falsifiable;
- Regular inspection must be carried out (every spurious verification period).

There are no nodes in Figure 4 with a separation equal to 3 from any of the nodes $\{y, \mathrm{j}\} \in HN2(\mathrm{b})$. As a result, node c should add a spurious node to the system following rule No. 1 of the fake setting method. Because node y should designate b as an *MPR* to approach Sv, this prevents the assault and safeguards node v. This would be reported as a conflict and in violation of rule No. 2 of the conflict rules. Through this method, the attacks can be prevented, and malicious nodes are identified.

The trust levels of every node in a system, including malevolent nodes, are updated by block transactions. A block will be created by the validating node or a delegation node, which receives all activities. Transactions are distributed by *MPR* nodes under the mechanism used by this method. Every node n will deliver an encoded session (n, transaction) prKeyn, where the secret key of n is used to encode the operation. If the abovementioned process reveals a malevolent node, it will be given a low Trust value (TV) and removed from the system. Even though a node is not an enemy, one node could mistakenly attribute a negative rating to it. Transactions including malevolent node data must first be verified by neighbors before being forwarded to the delegation node to avoid this problem. Because hackers may assert that two neighborhoods of a target are their counterparts in a node attack (NA), the intruder's data and any discrepancies they create must be notified by two neighborhoods. The suspect's secret key is used to encode the target ID (v), assailant ID (y), and Reporting Attempt (discovered discrepancies) in a response signal (v, y, Report Attack) prKeyv that is transmitted. Because the malignant welcome data contain the suspect's two-hop neighborhood, this signal is delivered by piggybacking onto it until it is within two hops of the recipient.

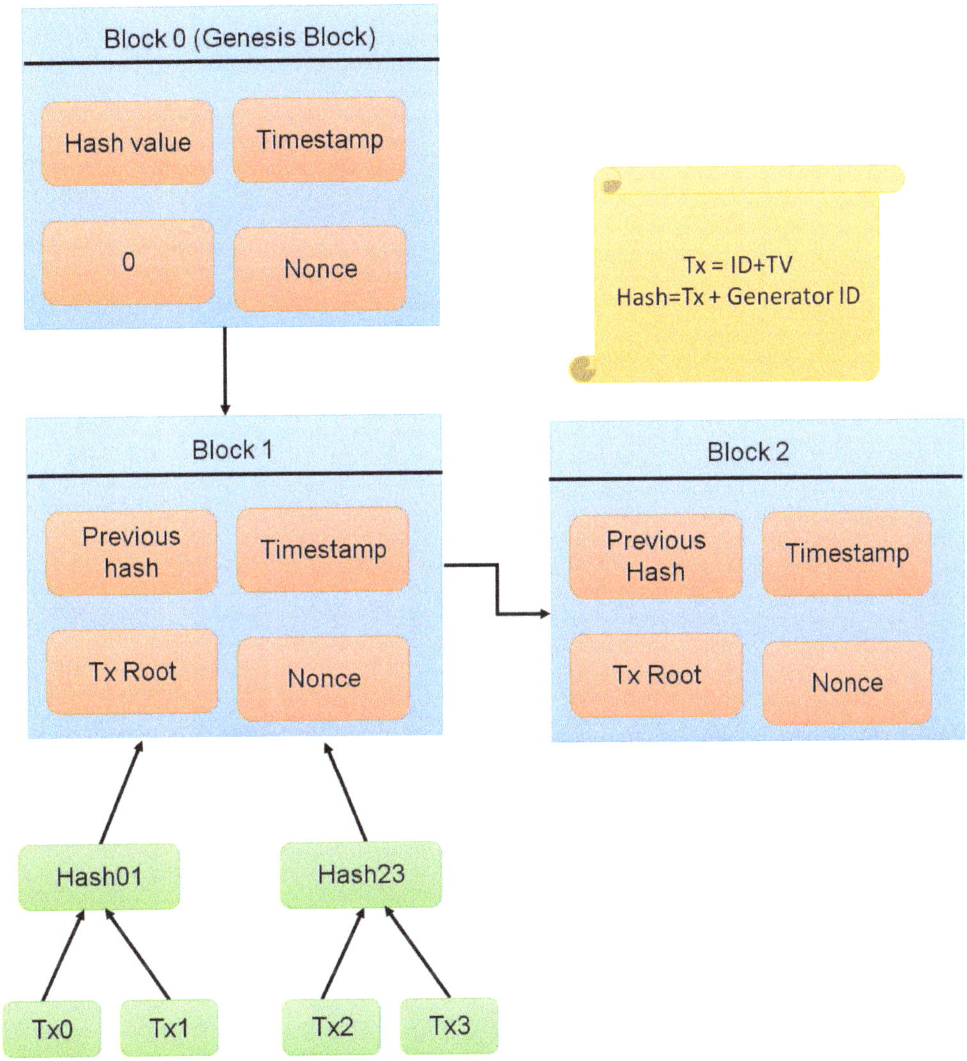

Figure 4. Sample Block Configurations.

If the surrounding nodes accept the transaction, it will respond (i AckReport) prKeyi, validating the transaction. It is hard to receive consensus from all endpoints since the hacker might also incorporate the spurious nodes. Furthermore, the node asking for permission can alternatively be an intruder attempting to identify a reliable node. As a result, the transaction is approved if at least half of the neighbors who received the intruder's Welcome approve. Additionally, even if the intruder states that they are its neighborhood, saying "accept" suggests that they have no link to them. As a result, each node evaluates whether or not they agree using the same criterion. Nodes that have TVs greater than q are the only ones that can transmit non-attacking standard TV transactions. The delegation node will tally the nodes participating in a specific transaction's vote. The delegation will choose the transaction order and create a block depending on the quantity. As a result, using *MPR* nodes across the network, the delegates will disseminate the new block (dl, Block) prKeydl. Every node responds with a verification signal (n, BlockAck) prKeyn after receiving the block from all other nodes. Every node links the new block to a localized blockchain if the most of other nodes approve it. Through this procedure, the suggested method will identify the malicious node and eliminate the node and its attack in the MWCN transmission.

3.2.2. Block Configuration

When building a block, it is important to specify the data that will be contained within it as well as how the delegate node will configure it. In a blockchain system, the pool's transactions are compiled into a block and chained throughout the network because it offers immutability. A hash value (SHA-256 algorithm) is attached to the block in a blockchain and is directly derived from the transaction data. As a result, the hash value will alter even a minor modification in the data. A data update in one block might cause all the other blocks in a blockchain to become disorganized since the hash of the previous block will be incorporated as data in the current block for chaining. There is only one format that the block hash accepts (e.g., a hash signature starting with 10 consecutive zeros). The term "nonce" refers to a piece of data that complies with this criteria. Until a valid hash signature is obtained, the nonce value is continuously modified.

Blocks in a MANET trust blockchain are made up of block transaction data and the aforementioned metadata (timestamp, hash of the transaction, delegate ID, and the nonce). To ensure non-repudiation for the block transactions offered by any nodes, the transaction generator ID, the TVs recommended by the transaction generator, and the delegate ID will all be included when a transaction is hashed.

The first block in the blockchain, known as a "genesis block" (blockchain jargon), is defined as an empty list of transactions when the network is created. Figure 4 displays a sample arrangement for a block.

3.2.3. Block Maintenance

There are two sorts of nodes in a blockchain environment: full nodes, which maintain the blockchain, and lite nodes, which mostly rely on full nodes for information but do not maintain the whole blockchain. We included this idea in our environment as well by the nature of MANETs. A new node will have access to the blockchain data whenever it joins the network. As seen in Figure 2, a node should initially join the network as a light node, which allows it to only download the block's header. A new node can nevertheless produce transactions (attacker detection/TV calculation) in the network even though it will initially function as a light node. To relay block headers until the new node becomes a full node, the network's host node will act as a temporary full node.

4. Results and Discussion

This section displays the findings of the graphical assessments of the efficacy of the suggested and existing strategies. Using the suggested TVDCSN approach, malicious node elimination and intrusion avoidance are carried out. The performance indicators for evaluation include detection accuracy, attack prevention, security, network overhead, and computation time. The suggested TVDCSN is used to compare the performance of the Transfer learning (TL), AdaBoost Regression Classifier (ABRC), malicious intrusion data mining algorithm (MIDTA), and dynamic reputation algorithm (DRA).

4.1. *Detection Accuracy (%)*

Accurately identifying malicious nodes in a wireless communication network is the definition of detection accuracy. The malicious node will reduce the network's communication speed, which would reduce the network's service time. It is necessary to identify these wireless communication nodes. Figure 5 displays the detection accuracy of malicious nodes using both existing and suggested methods. It shows that the proposed approach is effective in detecting precise malicious nodes. Table 1 displays the results for the detection accuracy.

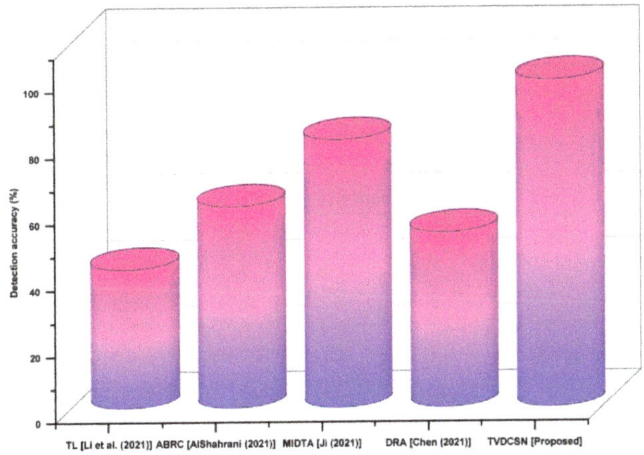

Figure 5. Proposed and existing methods of detection accuracy.

Table 1. Values of proposed and existing methods of detection accuracy.

Methods	Detection Accuracy (%)
TL (Li et al. (2021))	42
ABRC (AlShahrani (2021))	61
MIDTA (Ji (2021))	81
DRA (Chen (2021))	53
TVDCSN (Proposed)	99

4.2. Attack Prevention (%)

During the process of transmitting sensitive information through the MWCN, the network is subject to several attacks. Numerous vulnerable attackers that want to steal sensitive information are the ones who carry out these attacks. In the transmission process, the prevention of attacks is vital. The attack prevention employing both recommended and existing approaches is shown in Figure 6. The attack prevention results are shown in Table 2. It demonstrates how well the suggested strategy works to prevent attacks in MWCN.

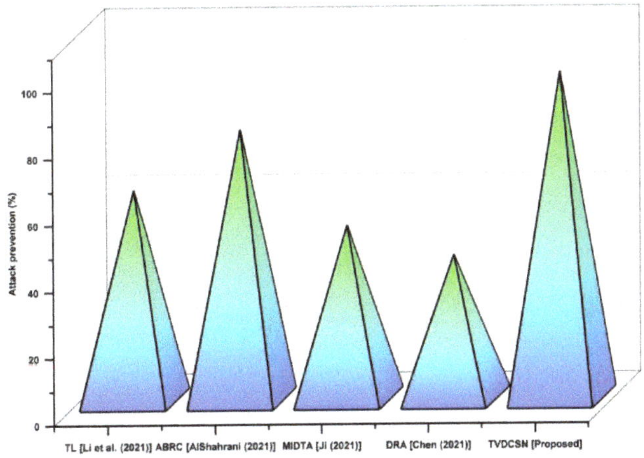

Figure 6. Proposed and existing methods of attack prevention.

Table 2. Values of proposed and existing methods of attack prevention.

Methods	Attack Prevention (%)
TL (Li et al. (2021))	63
ABRC (AlShahrani (2021))	81
MIDTA (Ji (2021))	52
DRA (Chen (2021))	43
TVDCSN (Proposed)	98

4.3. Security (%)

It is essential to have security because it protects sensitive data from being compromised by malicious cyber activity and ensures that the network can be relied upon and is functional at all times. Various security measures are used in effective network security plans to shield people and companies from ransomware and digital threats. Figure 7 shows the security utilizing both the recommended and existing techniques. This demonstrates that the strategy that was proposed is an effective one for providing security. The outcomes for the security are shown in Table 3.

The formula for network security NS = P + Pr + Pe + M + T; NS—Network security, P—policy, Pr—procedure, Pe—people, M—management, and T—technology. The effective collection of data to test and evaluate situational awareness and treat assessment tools for cyber security will be made possible by this adaptable simulation modeling framework.

4.4. Network Overhead (Bits)

Any unlawful use of services such as data, processing, storage, and bandwidth is referred to as network overhead in computing. To hold the additional data required to transport specific information from its source to its recipient, more assets are required. Figure 8 depicts the network overhead of the suggested and current strategies. It shows that the recommended solution has minimal overhead, which enhances the wireless communication network. Table 4 displays the overhead values. The below equation illustrates the comparison of network overhead.

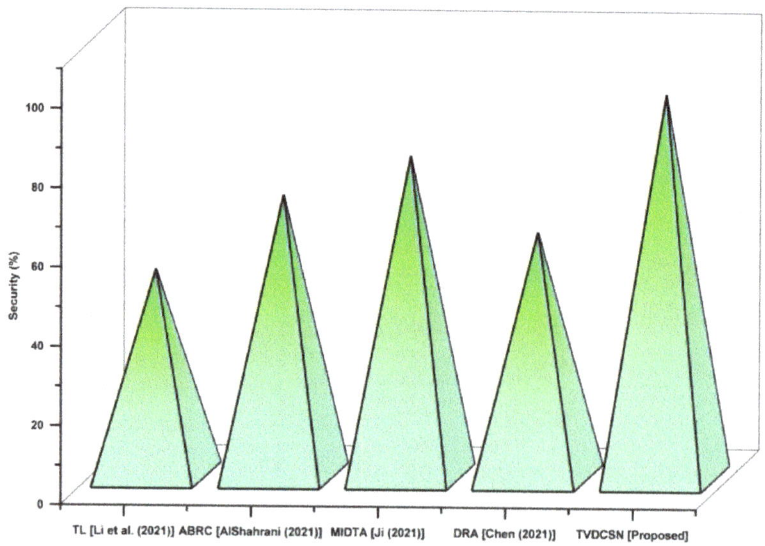

Figure 7. Proposed and existing methods of security.

Table 3. Values of proposed and existing methods of security.

Methods	Security (%)
TL (Li et al. (2021))	52
ABRC (AlShahrani (2021))	71
MIDTA (Ji (2021))	81
DRA (Chen (2021))	62
TVDCSN (Proposed)	97

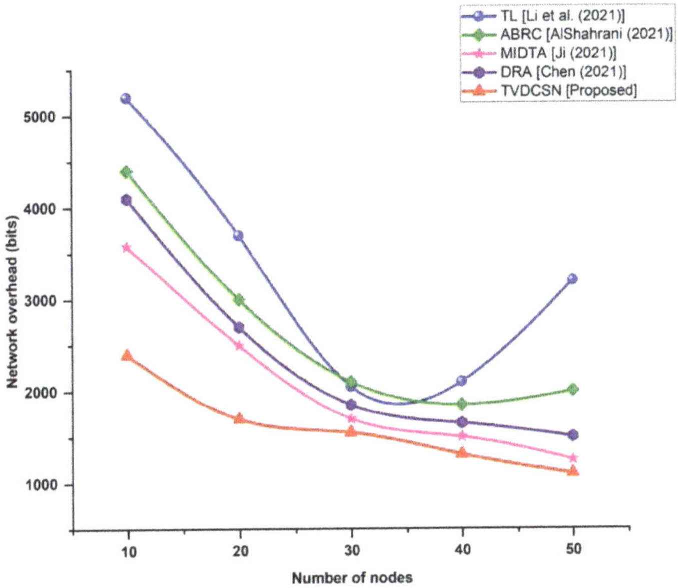

Figure 8. Proposed and existing methods of network overhead.

Table 4. Values of proposed and existing methods network overhead.

Number of Nodes	Network Overhead (Bits)				
	TL (Li et al. (2021))	ABRC (AlShahrani (2021))	MIDTA (Ji (2021))	DRA (Chen (2021))	TVDCSN (Proposed)
10	5200	4400	3580	4100	2400
20	3700	3000	2500	2700	1700
30	2050	2100	1700	1850	1550
40	2100	1850	1500	1650	1310
50	3200	2000	1250	1500	1100

In that situation, O = 2l, where l is the number of connections that calculates the number of overhead networks, O.

4.5. Computation Time (%)

Computation is the amount of time required to accomplish a calculation (also known as "execution periods"). It is a fundamental efficiency criterion that professionals in the fields of software engineering and science have used to evaluate a method's effectiveness. Figure 9 displays the computation times for the suggested and traditional methodologies. Table 5 displays the values of calculation time. It indicates that the suggested strategy operates effectively and rapidly.

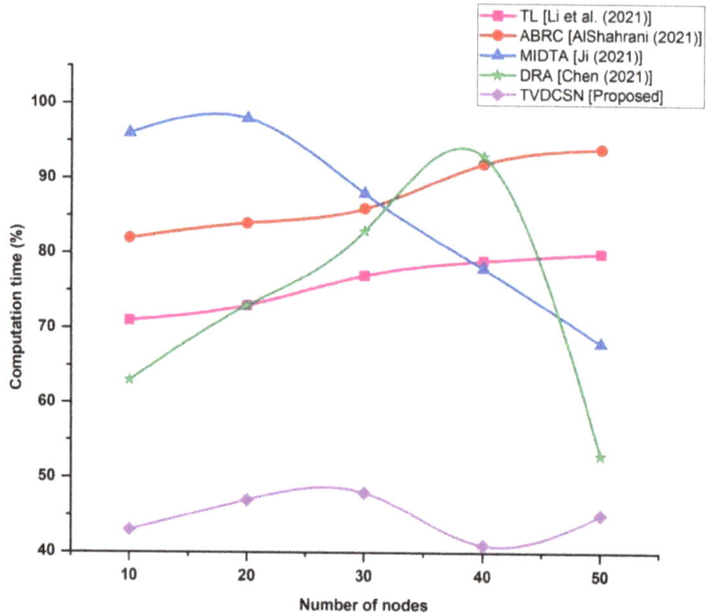

Figure 9. Proposed and existing methods of computation time.

4.6. Block Latency

Block latency is improved even more if an attack detector node serves as the delegate since less communication is needed to send attack information to the delegate. The block duration and transaction ratio are significantly lower when collusive attacks take place in a network. The block generation latency, determined based on attack transactions, can be depicted in Figure 10.

Table 5. Values of proposed and existing methods computation time.

Number of Nodes	Computation Time (%)				
	TL (Li et al. (2021))	ABRC (AlShahrani (2021))	MIDTA (Ji (2021))	DRA (Chen (2021))	TVDCSN (Proposed)
10	71	82	96	63	43
20	73	84	98	73	47
30	77	86	88	83	48
40	79	92	78	93	41
50	80	94	68	53	45

$$Attack\_ratio = \frac{number\_of\_attacker\_in\_the\_network}{number\_of\_nodes\_in\_the\_netowrk} \quad (1)$$

For instance, if two different attackers initiate attacks simultaneously in two different locations, two assault transactions will be included in a block, increasing the effectiveness of the suggested technique. The attack ratio measurement is shown in Equation (1).

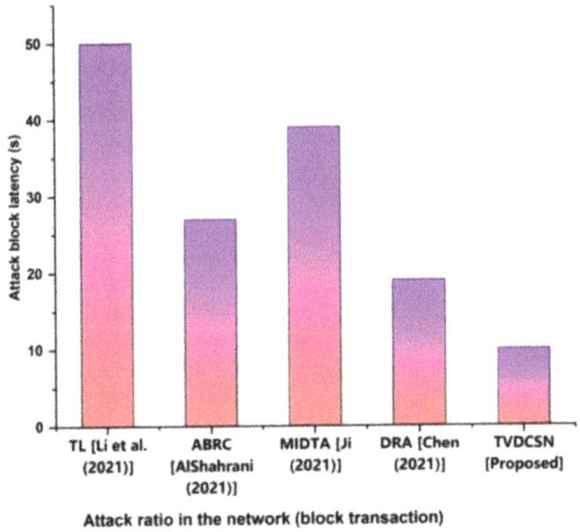

Figure 10. Block Generation Latency Based on Attack Transactions.

5. Discussion

In the area of WCN data analysis, wireless transmission system assault detection is crucial. In unsupervised wireless systems, link forecasting is a challenging issue that can be effectively handled by transfer learning (TL). A link prediction approach relying on the dispersion functional fitting technique of the area diagonal term group is utilized to gather more precise and comprehensive data in the target area [21]. Several security failures have occurred lately as a result of the unfavorable growth of automation. Services are maximized with increased network lifespans to resist those safety dangers and intrusions, particularly for hacking attempts. Artificial intelligence depended on innovation and has advanced to resist intrusions. Deep learning (DL) depended on a categorization strategy for detecting cyber-attacks provided in this article [22,23]. The intrusion detection system using the suggested AdaBoost Regression Classifier (ABRC) uses a deep learning structure. The presented ABRC with DL architecture is implicated in the assessment of network security assault. The privacy of private details in wireless technology cannot be guaranteed because invasive information in the transmission process readily affects wireless private interaction networks. The malicious intrusion data mining algorithm (MIDMA) presented in this study [24,25] is founded on valid large information from wireless personal interaction systems. The main point of malicious infiltration data is repeatedly obtained using the grouping technique, and its predicted participation is determined. The inherent complexity of wireless communication networks makes it difficult to identify rogue nodes using standard approaches, which creates several safety threats in the network setting. In this research, a dynamic reputation algorithm-based technique for detecting rogue wireless transmission nodes is proposed [26,27]. The above methods take a long time to identify and detect malicious activity with less accuracy and fail to effectively prevent attacks.

6. Conclusions

In a "wireless communication network" where the communication of information is fully automated by utilizing electromagnetic waves, like radio waves, which are typically instituted in the physical layer of the system, one of the most significant methods for transferring data between nodes without utilizing wires is used. In the area of data transfer, wireless communication systems have made significant progress to date. This is because they are easy to operate, affordable, and have sufficient bandwidth. The security risks to wirelessly transferred data have risen even if the safety and bandwidth gaps between different kinds of networks have decreased as a result of ongoing advancements in wireless communication innovation. The MWCN has to take measures to reduce the number of security-related issues. As a result, we offered the blockchain-based modal, Transaction Verification Denied conflict with spurious node (TVDCSN) methodology, to be used in MWCN because of the inefficiency of the traditional methods for identifying malicious nodes and preventing attacks. The efficacy of the proposed system is assessed using a variety of performance characteristics, including detection accuracy, attack prevention, security, network overhead, computation time, and average block latency. The proposed method's efficacy is compared with that of conventional techniques such as Transfer learning (TL), AdaBoost Regression Classifier (ABRC), Malicious Intrusion Data Mining Algorithm (MIDTA), and Dynamic Reputation Algorithm (DRA). These assessment results demonstrate the effectiveness of the suggested approach in MWCN for detecting malicious nodes and preventing attacks. Even if an attacker moves around and attacks different nodes from different places, the network will still be safe. No information or time is lost, and the overall level of complexity goes down. Additionally, because of collaborative detection, each node is much less responsible for its actions. The more nodes there are in a network, the less each one is responsible for detecting. In the future, optimization strategies may be introduced into the system to enhance its performance. The proposed scheme will be put to the test with different routing protocols in a wireless communication network.

Author Contributions: Conceptualization, R.R.C.; Methodology, R.R.C.; Software, N.L.S.C.; Validation, N.L.S.C.; Investigation, N.L.S.C.; Resources, A.B.; Data curation, A.B.; Writing—original draft, G.H.L. and F.F.; Writing—review & editing, G.H.L., F.F. and R.N.; Supervision, R.N. All authors have read and agreed to the published version of the manuscript.

Funding: This research received no external funding.

Conflicts of Interest: The author declares no conflict of interest.

References

1. Chowdhury, M.Z.; Shahjalal, M.; Ahmed, S.; Jang, Y.M. 6G Wireless Communication Systems: Applications, Requirements, Technologies, Challenges, and Research Directions. *IEEE Open J. Commun. Soc.* **2020**, *1*, 957–975. [CrossRef]
2. Chowdhury, M.Z.; Shahjalal, M.; Hasan, M.K.; Jang, Y.M. The Role of Optical Wireless Communication Technologies in 5G/6G and IoT Solutions: Prospects, Directions, and Challenges. *Appl. Sci.* **2019**, *9*, 4367. [CrossRef]
3. Jin, L.; Hu, X.; Lou, Y.; Zhong, Z.; Sun, X.; Wang, H.; Wu, J. Introduction to wireless endogenous security and safety: Problems, attributes, structures and functions. *China Commun.* **2021**, *18*, 88–99. [CrossRef]
4. Parvathy, K.; Rajalakshmi, S. A Review on Network Layer Attacks in Wireless Sensor Networks. *Int. J. Comput. Sci. Eng.* **2021**, *9*, 45–48.
5. Srinivasu, P.N.; Bhoi, A.K.; Nayak, S.R.; Bhutta, M.R.; Woźniak, M. Blockchain Technology for Secured Healthcare Data Communication among the Non-Terminal Nodes in IoT Architecture in 5G Network. *Electronics* **2021**, *10*, 1437. [CrossRef]
6. Ramasamy, L.K.; KP, F.K.; Imoize, A.L.; Ogbebor, J.O.; Kadry, S.; Rho, S. Blockchain-Based Wireless Sensor Networks for Malicious Node Detection: A Survey. *IEEE Access* **2021**, *9*, 128765–128785. [CrossRef]
7. Ramesh, S.; Yaashuwanth, C.; Prathibanandhi, K.; Basha, A.R.; Jayasankar, T. An optimized deep neural network based DoS attack detection in wireless video sensor network. *J. Ambient. Intell. Humaniz. Comput.* **2021**, 1–14. [CrossRef]
8. Joon, D.; Chopra, K. Hybrid Deep Learning Prediction Model for Blackhole Attack Protection in Wireless Communication. *Nveo-Nat. Volatiles Essent. Oils J. | NVEO* **2021**, *8*, 10228–10243.
9. Mughaid, A.; AlZu'bi, S.; Alnajjar, A.; AbuElsoud, E.; Salhi, S.E.; Igried, B.; Abualigah, L. Improved dropping attacks detecting system in 5g networks using machine learning and deep learning approaches. *Multimedia Tools Appl.* **2022**, 1–23. [CrossRef]
10. Marriwala, N.; Punj, H.; Panda, S.; Kaur, I.; Rathore, D. An Authentication Based Approach for Prevention of Spectrum Sensing Data Falsification Attacks in Cognitive Radio Network. *Wirel. Pers. Commun.* **2022**, *124*, 119–145. [CrossRef]
11. Hanif, M.; Ashraf, H.; Jalil, Z.; Jhanjhi, N.Z.; Humayun, M.; Saeed, S.; Almuhaideb, A.M. AI-Based Wormhole Attack Detection Techniques in Wireless Sensor Networks. *Electronics* **2022**, *11*, 2324. [CrossRef]
12. Yang, H.; Zhang, X.; Cheng, F. A Novel Algorithm for Improving Malicious Node Detection Effect in Wireless Sensor Networks. *Mob. Networks Appl.* **2021**, *26*, 1564–1573. [CrossRef]
13. Pang, B.; Teng, Z.; Sun, H.; Du, C.; Li, M.; Zhu, W. A Malicious Node Detection Strategy Based on Fuzzy Trust Model and the ABC Algorithm in Wireless Sensor Network. *IEEE Wirel. Commun. Lett.* **2021**, *10*, 1613–1617. [CrossRef]
14. Kandel, R. Provenance-Based Malicious Node Detection in Wireless Sensor Network Using Bloom Filter. Doctoral Dissertation, Pulchowk Campus, Lalitpur, Nepal, 2021.

15. Kumar, M.; Mukherjee, P.; Verma, K.; Verma, S.; Rawat, D.B. Improved Deep Convolutional Neural Network Based Malicious Node Detection and Energy-Efficient Data Transmission in Wireless Sensor Networks. *IEEE Trans. Netw. Sci. Eng.* **2021**, *9*, 3272–3281. [CrossRef]
16. Jagadeesan, S.; Manjula, R.; Johnpeter, T. December. In A Cross-Layer based Hidden Marko (C-HMM) Model for Detection and Prevention of Malicious Attacks in Wireless Ad-hoc Networks. In Proceedings of the 2021 IEEE International Conference on Mobile Networks and Wireless Communications (ICMNWC), Tumkur, India, 3–4 December 2021; IEEE: Piscataway, NJ, USA; pp. 1–5.
17. Gowri, S.; Srinivasulu, S.; Jabez, J.; Vimali, J.S.; Sivasangari, A. Discovery of localized malicious attack in wireless networks. In *Smart Computing Techniques and Applications*; Springer: Berlin/Heidelberg, Germany, 2021; pp. 207–216. [CrossRef]
18. Sharma, T.; Mohapatra, A.K.; Tomar, G. SDBMND: Secure Density-Based Unsupervised Learning Method with Malicious Node Detection to Improve the Network Lifespan in Densely Deployed WSN. *Wirel. Commun. Mob. Comput.* **2022**, *2022*, 1–15. [CrossRef]
19. Ali, F.; El-Sappagh, S.; Islam, S.R.; Ali, A.; Attique, M.; Imran, M.; Kwak, K.-S. An intelligent healthcare monitoring framework using wearable sensors and social networking data. *Futur. Gener. Comput. Syst.* **2021**, *114*, 23–43. [CrossRef]
20. Li, S.; Pang, J.; Wu, Q.; Yao, N.; Yuan, W. Transfer learning based attack detection for wireless communication networks. *Concurr. Comput. Pr. Exp.* **2021**, *33*, e6461. [CrossRef]
21. AlShahrani, B.M.M. Classification of cyber-attack using Adaboost regression classifier and securing the network. *Turk. J. Comput. Math. Educ. (TURCOMAT)* **2021**, *12*, 1215–1223.
22. Ji, K. Malicious Intrusion Data Mining Algorithm of Wireless Personal Communication Network Supported by Legal Big Data. *Wirel. Commun. Mob. Comput.* **2021**, *2021*, 1–7. [CrossRef]
23. Chen, J. An identification method of malicious nodes in wireless communication based on dynamic reputation algorithm. *Int. J. Inf. Commun. Technol.* **2021**, *19*, 343–355. [CrossRef]
24. Bargarai, F.A.M.; AbdulAzeez, A.M.; Tiryaki, V.M.; Zeebaree, D.Q. Management of Wireless Communication Systems Using Artificial Intelligence-Based Software Defined Radio. *Int. J. Interact. Mob. Technol. (iJIM)* **2020**, *14*, 107–133. [CrossRef]
25. Beritelli, F.; Capizzi, G.; Sciuto, G.L.; Napoli, C.; Tramontana, E.; Wozniak, M. Reducing interferences in wireless communication systems by mobile agents with recurrent neural networks-based adaptive channel equalization. In *Photonics Applications in Astronomy, Communications, Industry, and High-Energy Physics Experiments*; SPIE: Washington, DC, USA, 2015; Volume 9662, pp. 497–505. [CrossRef]
26. Kouhalvandi, L.; Shayea, I.; Ozoguz, S.; Mohamad, H. Overview of evolutionary algorithms and neural networks for modern mobile communication. *Trans. Emerg. Telecommun. Technol.* **2022**, *33*, e4579. [CrossRef]
27. Janjua, K.; Shah, M.A.; Almogren, A.; Khattak, H.A.; Maple, C.; Din, I.U. Proactive forensics in IoT: Privacy-aware log-preservation architecture in fog-enabled-cloud using holo-chain and containerization technologies. *Electronics* **2020**, *9*, 1172. [CrossRef]

electronics

Article

The Concept regarding Vehicular Communications Based on Visible Light Communication and the IoT

Eduard Zadobrischi [1,2]

[1] Integrated Center for Research, Development and Innovation in Advanced Materials, Nanotechnologies, and Distributed Systems for Fabrication and Control, Stefan cel Mare University of Suceava, 720229 Suceava, Romania; eduard.zadobrischi@usm.ro

[2] Department of Computers, Electronics and Automation, Stefan cel Mare University of Suceava, 720229 Suceava, Romania

Abstract: The most controversial technology—visible light communication—is becoming increasingly promising in the field of wireless networks, being ideal for many indoor and outdoor applications. This article proposes VLC methods and architectures capable of providing high security in vehicles and in their communications with the environment or other cars in traffic. The architectures proposed involve the inclusion of ambient lighting equipment and systems and indoor and outdoor lighting systems, such as headlights, traffic lights, and stoplights. Securing data within vehicular networks and validating them through multiple layers of filtering at the level of the physical PHY layer would drastically strengthen the position of VLC. They are the only source of information through which direct contact is maintained with the other entities in the network. The evaluations and proposals presented here are highly viable and deserve future consideration in light of the results obtained in the practical steps carried out in the research process.

Keywords: in-vehicle communication; inter-vehicle communication; optical communication; security wireless; visible light communication; wireless optical communication

Citation: Zadobrischi, E. The Concept regarding Vehicular Communications Based on Visible Light Communication and the IoT. *Electronics* **2023**, *12*, 1359. https://doi.org/10.3390/electronics12061359

Academic Editors: Tao Huang, Shihao Yan, Guanglin Zhang, Li Sun, Tsz Hon Yuen, YoHan Park and Changhoon Lee

Received: 28 January 2023
Revised: 7 March 2023
Accepted: 9 March 2023
Published: 12 March 2023

1. Introduction

Visible light communication (VLC) represents an important component of optical wireless communication (OWC) and has brought many challenges to the research community, as well as those attracted to this field [1]. VLC could become an extremely remarkable technology because, in addition to being used for lighting, it can also be used for data communication between devices, users, and the outside environment. This approach is important in terms of the benefits it can bring, as well as in terms of its huge potential for future development across extremely vast areas [2]. VLC is different from the technologies we know. It can be developed at the level of pre-existing lighting infrastructure or at the level of equipment containing LEDs, offering the opportunity for the mass development of a fast and cost-effective network [3]. According to the specialized literature, the basic principle of VLC is that the data are transported using an optical carrier without leading to higher energy consumption, which is another advantage of this technology [4]. Most studies conclude that LED light, in conjunction with the data transmission process, is becoming more and more common in our society and employ the lens of avoiding health risks due to exposure to emissions and radiation, as VLC is one of the greenest technologies. The specialized literature demonstrates the high potential of VLC technology and elucidates several of its aspects through the lens of its standardization by competent organizations, including the IEEE [5]. In terms of energy efficiency, VLC uses LED light to transmit data, and this is known as a low-power-consumption factor. Wi-Fi low power is designed as a restrained form of energy consumption but, compared to VLC, it is much more expensive. Channel bandwidth represents a constraint for the number of data packets that can be transmitted over a certain channel. Wi-Fi has a much higher channel bandwidth than

VLC, but low-power Wi-Fi operates in the 2.4 GHz or 5 GHz frequency bands and has bandwidths of 20 MHz or more. Although it is a technology brought back to the public after a period of evanescence, VLC reappeared with prototypes developed and the first standardization, recognized with the acronym IEEE 802.15.7, was achieved in 2011 and later benefited from new updates [6]. VLC is showing an upward trend and represents an opportune moment for today's society, branching out into more and more fields. In the early days of the technology, it was used to make high-speed wireless connections, as it is extremely suitable for broadband internet. In this field, VLC technology has proven its capabilities: it can ensure data transfer at speeds of several gigabits per second and, in ideal cases and laboratory tests, the technology can also establish connections that reach transfer values of over 100 Gb/s [7,8]. These aspects make VLC an extremely promising candidate for systems based on technologies such as 5G or 6G. VLC is suitable for most fields due to its ability to reuse space and small communication cells. Therefore, 5G and 6G technologies could achieve much higher transfer rates than known before; low latencies, even below <1 ms; and extremely wide coverage [9,10]. In accordance with the extremely wide distribution of LED lighting sources, in addition to applications related to its information- and energy-transfer capacities, VLC could also be used in Internet of Things-type applications, see Figure 1. Great progress could be made in the transition towards Industry 4.0 or 5.0 through the application of wireless communication in production lines and automation. VLC's simplified implementation, cost efficiency, flexibility, and versatility would help in significantly scaling these processes, and it can be declared an ideal technology. Automation and robotization processes could use VLC for communication, control, management, and location tasks, and identification of equipment could also be controlled with it [11].

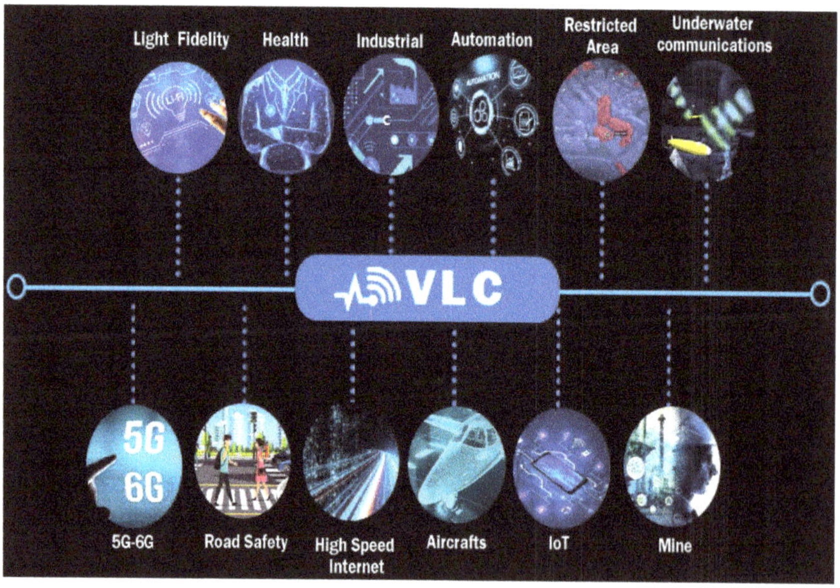

Figure 1. Illustration of the most representative scenarios and applications that use visible light communication.

Perhaps the most representative and popular field of use for VLC technologies is in road and safety and the design of applications dedicated to vehicular communications. This field also receives increased attention in light of the loss of human life involved, as road accidents are the second leading cause of death worldwide. Implementations in this direction are more and more numerous, proving that VLC is extremely reliable and can provide adequate resistance against noise. Communication distances can exceed 200 m and latencies meet the requirements of some vehicle-to-vehicle communication applications [12]. Analyzing all the research and the specialized literature, perhaps the most important aspect for today's society is that VLC is safe for the human body and for other equipment that can be influenced by or is sensitive to interference [13], and VLC is recommended even in RF-restricted areas. VLC can be used in medical procedures, transportation, logistics, security, oil rigs, nuclear power plants, highly confined areas, and even in aquatic research. VLC technology offers many unique benefits, including very high bandwidth and high transfer rates, in addition to the green zone aspects already presented [14]. This technology is increasingly being exploited by research groups, as well as in the private environment, to open up new fields and explore applications that could fix certain pressing problems of our society.

The effectiveness of current networks for communication between vehicles has been demonstrated, as well as their use in communication and control in autonomous cars, but aspects related to safety and information protection have been neglected. Therefore, this study focused on the analysis, presentation, and development of an architecture capable of providing a high degree of security in the process of communication between vehicles and between vehicles and infrastructure through the distribution of light in the indoor and outdoor environment [15]. The application of the solution is oriented toward both the user and the infrastructure or vehicles. The large number of systems produced by research groups so far proves the usefulness of this technology. The experimental evaluation process and the implementation were carried out in different stages, and the concepts were determined at the architectural level but without the implications related to the hardware and software components through which these processes were carried out. The data security aspect is extremely important for both the user and other traffic participants. The data communicated can be intercepted and, subsequently, the control of autonomous vehicles or on-board systems can undergo changes that may jeopardize the condition of the vehicle and endanger the driver, pedestrians, and other traffic participants [16]. Many of the major challenges currently impeding the implementation of new technologies, such as 5G, can be mitigated by using VLC [17,18]. The most important point is that VLC provides an alternative by not having a limitation in the radio frequency spectrum, which is already loaded and limited, and VLC even has a capacity more than 10,000 times higher than that of RF [19,20]. As the VLC spectrum remains unregulated and unlicensed, it can be considered an extremely important solution from a bandwidth perspective, capable of mitigating the limitations of the RF spectrum. New approaches and an increase in the degree of security for VLC, as well as development of an implementation method, are imperative. The most important contributions of the article are the proposal for a network architecture for future implementation in relation to vehicular communication and the enhancement of data security through multiple connections based on primary authentication keys and the parameterization of information using unique IDs. Section 2 reviews the methodology, outlines a proposal, while Section 3 is related to the implementation, and describes some of the results. Section 4 includes further discussion of the experimental results, and Section 5 is dedicated to the conclusion and future approaches.

2. Methodology and Design Parameters

Based on experience in the field of optical communication, research groups have consistently focused on adding new functions and generating related applications for VLC, including in relation to the IoT and vehicular communication. Several papers have discussed the use of VLC as part of various wireless technologies, but very few have focused on the IoT and road safety applications. In [21], architectures were proposed for VLC systems employing the IoT and its integration in the dark, using orthogonal frequency division modulation (OFDM) to overcome the identified limits [22]. In [23], VLC-over-UART-type systems were proposed that used the bit error rate and system evaluation. Another research paper [24] elucidated the potential of Li-Fi and its capacity for use in outdoor lighting, stating that it may represent a new backbone in the field of wireless communications. The activation of 5G wireless access using Li-Fi technology was addressed in [25] based on OFDM, demonstrating speeds of 200–300 kbps. Many of the challenges related to VLC, as well as the potential it has for industrial applications, were presented in [26], along with other concepts. Another review can be found in [27] that discusses various aspects and contributions, as well as providing an ensemble presentation, including aspects related to the optical IoT (OIoT). One new approach in V2V technology is the use of light fidelity (Li-Fi), which is an alternative medium in data transmission. The capacity of this technology to send data over an optical medium wirelessly using light-emitting diodes that propagate the signal makes it very promising. In the case of Li-Fi technology, data are extracted from the vehicle and spread via headlights or stoplights to other traffic participants or infrastructure, but there are many challenges related to bandwidth and data latencies. As shown in Figure 2, Li-Fi systems are composed of luminous media (LEDs) that transmit data and information, and the receiving system is based on a photo-detector that processes the data and analyzes the obtained signal. It is imperative to implement systems of this type because the actions that traffic participants take are based on information obtained from other vehicles and involve short durations of time and extremely low-validity data. Thus, in the case of systems of this type, GPS and Wi-Fi units are not necessary because Li-Fi technology can use interface or PIC controllers to emit tiny pulses of sound, which can penetrate barriers and be employed with straight roads or those of the T-junction type [24].

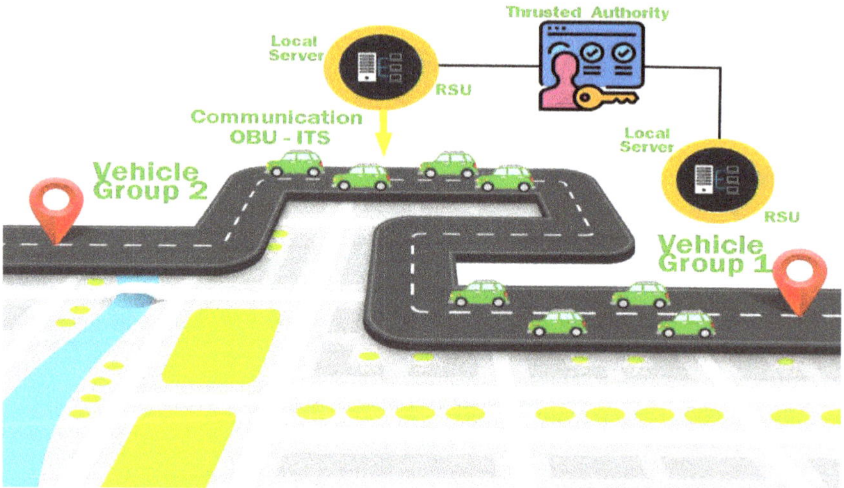

Figure 2. Illustration regarding the fields of communication within VANETs.

V2V technology can accurately calculate the moment T of a collision and highlight its severity when used synchronously with a laser detector or laser rangefinder (LRF). The guarantee that this is a highly viable system comes from the accumulation of adjustable vehicle functions, which can allow the generation of protocols and procedures to expedite the activation of pre-crash systems or airbags even before collisions occur [28]. Communication between vehicles is dependent on the distance of the convergence or divergence between them because the density of cars on roads is involved in the first process of the information transmission mechanism. In the case of congested roads where the traffic density shortens the distance between vehicles, the communication process takes place in a platoon-type network with vehicles separated by small distances, and hybrid implementation of VLC-RF is necessary [29]. In the case of visible light, data can be transmitted with a single data-stream instance extremely quickly [30]. In addition to what has been presented above, there are also aspects of the topic related to the exploitation of THz bands dedicated to vehicular networks, which have intrinsic properties. As millimeter-wave technology moves toward commercial implementation, it is clear that the terahertz (THz) band is the next frontier of communications. Summaries of all the RF techniques are provided in Table 1, where we show how they can be used under different traffic density conditions, and in Table 2, where other related work is addressed.

Table 1. Network systems and approaches in relation to coverage.

Traffic Networks	RF Systems	Network Approaches
Low-density and sparse network	Radio cognitive network Short-range radio	Low
Vehicular network-type for densities	Short-range radio millimeter waves	High
Use of high-density network millimeter waves	Network based on VLC radio	High
Use of higher frequency bands (THZ)	Network based on 5G	High

Table 2. Existing approaches and prospects for development.

VLC Applicability	VLC/Li-Fi Design and Architecture	VLC Approaches and Proposals	VLC IoT Challenges	Solution Presented	Future Perspectives
VLC systems [31]	✓	✓	✓	✓	✓
Vehicular systems [32]	✓	✓	✓	✓	✓
IoT [33]	✓	✓	✓	✓	✓
Safety systems [34]	✓	✓	✓	✓	✓
V2V and Li-Fi [35]	✓	✓	✓	✓	✓
IoT [36]	✓	✓	X	✓	✓
OIoT [37]	✓	X	✓	X	X
Industrial applications [38]	✓	✓	✓	✓	✓
IoT [39]	✓	X	✓	✓	✓
5G and IoT [40]	✓	✓	✓	X	✓
6G [41]	✓	✓	✓	X	✓

The proposed solution could make major contributions to the emergency transmission of priority messages, the avoidance of road accidents, and the safety and security of data. In all these processes, it is extremely important to also consider the adoption of vehicular ad hoc network (VANET) technology, which can guarantee the safety of vehicles and transmits information through central roadside units (RSUs) or electronic control units (ECUs) that can pre-secure data with up to six encryption cycles [42]. With vehicular ad hoc networks (VANETs), it is possible to manage multiple vehicles that have on-board units or roadside units, as illustrated in Figure 2. Further measures at the security level could involve Euclidean distance calculation components, which can provide data on the distance between vehicles and RSUs or on the occurrence of adverse events at the edge of the road surface. Therefore, the protocols used to secure data could be based on event detection crawling and information filtering procedures, sending the data only through repetitive loops to the RSU-type units or concatenating the input data with the output data to encrypt them. If an accident is detected, the system sends the information to nearby vehicles and, through a filtering process that also uses an advanced driver-assistance system (ADAS), implements assistance processes, even providing traffic updates. Data are pre-swapped and routed through band-switching to ensure security and privacy, then initialized and keyed into the cloud. Any sudden change in the amount of data or any data modification result in a software trigger that processes each routine and compares it with the additional sets [43,44].

2.1. Li-Fi Communication System Proposal

The most important advantage that this new Li-Fi technology brings to the field of communications is greater security through the lower radius of the coverage area, as well as data encoding. As a consequence of limiting the coverage area, VLC cannot penetrate opaque surfaces or obstacles, even when geographically limited. In addition, VLC systems can employ connectivity based on unique IDs to encrypt the information in a format that can only be decoded with an adapted receiver. Therefore, this approach is extremely important in terms of the security and integrity of data communication, both for users and within vehicular networks. In the design process for a VLC system, the transmitter is not necessarily the central component, although it is important for the communication process, but an extremely volatile and important part of the system rests on the shoulders of the receiver. This idea was deduced from the specialized literature [45,46] (see Figure 3).

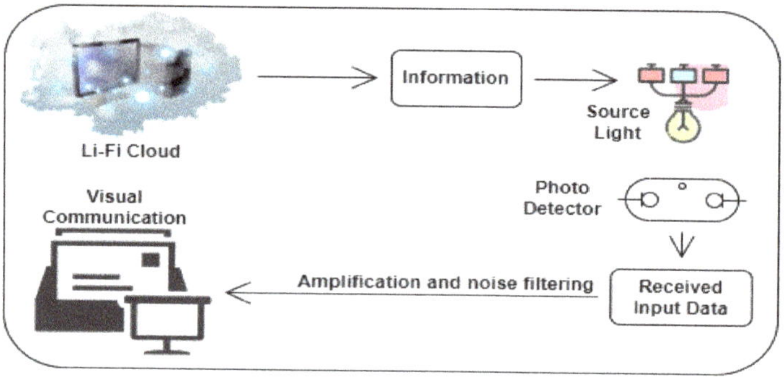

Figure 3. Illustration presenting Li-Fi technology and its general utility.

The IEEE 802.15.7 standard provides additional functions, some of which are complementary to lighting devices, and, in this case, they do not negatively influence the process [47]. Even if these systems have to be able to provide both lighting and data communication, they must not induce a flickering effect perceptible to the human eye, and it is necessary to implement functions capable of diminishing the light intensity if this is required. Hardware and software solutions have been found for this problem. Figure 4 shows a complete VLC diagram, as well as a way to secure the communication. The diagram includes an ARM Cortex M7 microcontroller component with a frequency of 1008 MHz, which is the central element around which the entire system gravitates. The basic function of the microcontroller is to transmit the data and transform/demodulate them to obtain a continuous stream of information/bitstream [48,49].

Thus, the microcontroller facilitates the processing, encoding, modulation, and contouring of bit matrices to expose them and transmit them further. Through the prism of its versatility, its performance and data security can be substantially improved, including by using on–off keying (OOK) at the emission and modulation side. The improved security process is also based on the central anti-flicker aspect; the VLC transmitter runs a code based on unique IDs and a run-length-limited (RLL) code, which can overlap the logic levels "1" and "0" at the same light intensity. Encrypting data by using security keys with unique IDs assigned to each data matrix also increases the data transmission speed and the instantiation capacity; in some cases, the speeds are around 250–300 kb/s [50]. To validate the security process carried out, the GVLC comparator structured based on the message intent iterates a log with the purpose of validating the receiver as a part of the system; everything is undertaken based on unique IDs. Subsequently, the message header provides the VLC receiver with information regarding the modulation technique, coding, transfer rate, and length of the message, aspects that ultimately validate the communication and security process. The instance frame contains all the transmitted data, and it is followed by a stop and validate header, the purpose of which is to inform the VLC receiver that the data have arrived.

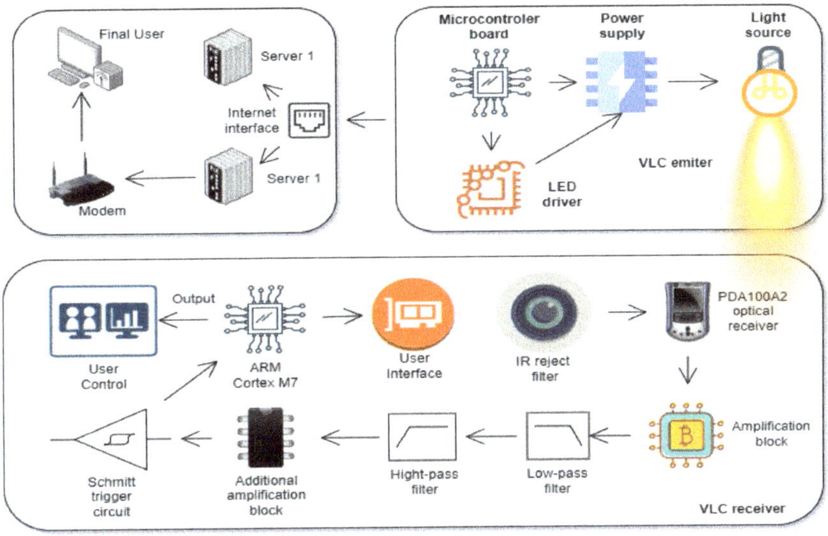

Figure 4. Architecture of the proposed visible light communication system at the road infrastructure level.

As shown by communications tests, the intermittency of the signals between data frames also facilitates the implementation of an additional security protocol. The top microcontroller component can generate data and transmit them to other devices, but, at the same time, it can be interfaced with other devices through CAN, I2C, or SPI ports, which increases the safety and veracity of the information [51]. Regarding the level of frame data and the data quantity, solutions related to the cadence of the data transmission can be established with the help of an LED driver. The generation of the light beam that contains that data is related to the dynamics of the environment and the data quantity; the light contains the data that must be provided to the receiver, and the data take a path through the optical channel in free space (see Figure 4).

To obtain a more robust structure, this research focused on the development of a system capable of providing information and connectivity with any other device. For the reception side, as can be seen, the collector optical system had a processing component and a processing unit. These components integrate optical filters that adjust the signal to noise ratio (SNR) and remove unwanted components from the optical spectrum. For the optical detection part, a PDA component and PIN photodiodes connected to direct transimpedance circuits were used. For the processing blocks, bandpass filters were introduced with certain cutoff frequencies determined by the spectral densities. The data encoding and decoding process are carried out using quadratic triggers and other types of triggers, and the final signal is ultimately analyzed and received by the ARM Cortex M4-type controller. The final data arrive at the last unit and can be accessed by the end user [52,53].

2.2. Proposal for Software Infrastructure and Architectural Components

In accordance with the prospects for future use of Li-Fi networks for V2V interactions, the proposed algorithm validates and authenticates users in vehicles based on the queries it makes within the internal nomenclature, which includes all IDs and encryption codes related to users. The interaction between the device and the vehicle involves a database located at the level of the vehicle ECU infrastructure. The risk of information leakage is minimal due to the degree of encryption and the transmission of information through light, the identifiers being encrypted at the level of the internal stack without generating an exact reference to a specific device and constantly reinitialized every time they are reintegrated into the system. Regarding the security protocol for VLC-based networks such as Li-Fi, the process by which this occurs takes places in several stages. Devices are verified before connecting to the network and, if a device is in range, it receives a query asking for an access code and its verification leads to the first process. When the code is valid, the data are checked for identifiers and for whether there have been any previous logins. When the device authenticates and authorizes itself, the data are encrypted and sent within the network, with the traffic on that network being constantly monitored. In addition to these aspects, there are functions dedicated to the additional protection of the network from possible external attacks or penetrations.

Therefore, encrypting and transmitting data through a Li-Fi-type network can be achieved using a symmetric encryption algorithm. A symmetric encryption algorithm is based on the use of unique keys that encrypt and decrypt data. In a Li-Fi network, these keys are generated and distributed across the network to all devices. One example of an algorithm based on symmetric encryption is the advanced encryption standard (AES). According to specialist studies, it is considered one of the most viable and powerful algorithmic structures in terms of symmetric encryption, and it is used in more and more security standards and, now, in VLC [54].

The structure in this case could be implemented according to the following steps:

(a) The header is initialized and a cryptographic key is generated for each ID (i.e., for each device on the network or each device that should connect to it);
(b) The AES algorithm is used to encrypt the data and transmit them via the Li-Fi network;
(c) The unique cryptographic keys are then sent to the connected devices to enable the decryption of the received data. Their form being that of the MD5 hash function, the unique identifier generated for each vehicle in the network can be inserted into their headers;
(d) The keys are used to decrypt the data when they are received by the connected device.

The existence of new technologies dedicated to cryptography has a direct connection with quantum cryptography, which is advancing quickly around the world. Specifically, quantum cryptography involves quantum key distribution (QKD) and subsequent redistribution of a cryptographic key. Later, the degree of security can be proven by using new instantiations on the bases of computational complexity and processing with emerging quantum computers [55]. In QKD, the quantum key can be exchanged between network users in the form of light to increase security. When quantum sequences are iterated, they are measured post-processing to generate identical keys on both sides of the network. A first step in this direction is the QuINSiDa project, which is the first to incorporate a "QKD over Li-Fi"-type system [56]. This aspect makes QKD data transfer possible, which is more widely used in communications between buildings and offices. The project aims to demonstrate that a quantization-based data communication network can be flexible in its secure backbone infrastructure and can make the step to the vehicular area. In summary, the project aims to realize wireless data communication in a point-to-point scenario but, at the same time, simultaneously secure all individual communication channels through quantum keys [57].

An extremely important aspect in any network is maintaining the confidentiality of the generated key at all times and, in cases where new external connections are introduced, generating authentication keys after a certain time interval. The security of wireless networks and their data security routines are based on encrypted connections centered on protocols for data transmission, such as WPA2 or WP3. Another aspect of security is related to the use of firewalls, which limit access to networks and, at the same time, do not let external entities connect. In addition, two- or three-step authentication systems can be used to increase safety.

For our proposal regarding securing Li-Fi data communication networks for vehicles, we considered aspects related to the creation of an encryption and decryption algorithm to limit access to the infrastructure created. Li-Fi networks offer a high degree of security because this technology is bidirectional and can only be accessed through certain decoding procedures with dedicated systems. There are no generally valid programming languages that can be used to create security protocols in Li-Fi networks or to define a security standard to date. Therefore, this approach is extremely important for the scientific environment and, as a result of the experience accumulated in this field, presents a viable alternative through which this technology will soon be able to branch out into more fields. Even if technological advances favor communication based on the 802.11 p/a standard, 4G, 5G, and even 6G, the complementarity and usefulness of VLC are undeniable [58,59].

The encryption and formation of a security protocol at the physical level in a data network based on VLC could have the following structure (Algorithm 1).

Algorithm 1: Pseudocode for the encryption process for data at the physical level of the network.

```
from vlc_supp
import SEC1_SUPP
vlc = SEC1_SUPP(lifi0)
vlc.set_network('usr', 'pass', Li-Fi1#!2)
vlc.enable_network()
```

In Algorithm 1, an attempt was made to create a much stronger encryption process compared to those known from the much more widespread networks that encrypt data within wireless networks. In the previously presented case, security keys are outlined in the function header, after which the imports are undertaken and the user and password are validated sequentially as the first iterator.

In Algorithm 2, a firewall-type procedure is outlined that can manage the network more efficiently, restrict access to it, and prevent external attacks. Towards the end, the created traffic network and its port are also highlighted. A final step in accomplishing the process of securing a network is authentication and the creation of a way to validate previous data. Therefore, in Algorithm 3, all the data from the encrypted validation files are imported, along with the user input data to be filled in by the handler, and the password is maintained at the same time as the credentials. All the data are saved in a nomenclator in the VLC database and requested for access and validation through iterative instance comparators. In the last stage, the algorithm decides whether access is allowed depending on the degree of portability of the user and the password.

Algorithm 2: Pseudocode for the import and filtering of access data for the generated address.

```
from ipvlctables import Iptablesvlc
ipt = Ipvlctables()
ipt.block_all_trafficvlc()
ipt.allow_traffic(100.100.1.0/88)
```

Algorithm 3: Pseudocode for final validation and authentication in the created communication process.

```
from passlibvlc.hash import sha256_crypt
usr = input("Enter user: ")
pass = input("Enter pass: ")
stored_passvlc = "hashed_pass_from_databasevlc"
if shabvlc_crypt.verify(pass, stored_pass):
print("Access granted LiFI.")
else:
print("Access denied to Li-Fi.")
```

The proposed architecture provides, using several LEDs, an authentication ID regarding the location and identification data for the vehicle, these being managed with cryptography [60,61]. The network undertakes the distribution of the authentication ID and the lighting sources, a process that increases the degree of security through power lines. Extremely high scalability can be achieved through the efficient management of IDs, highlighting how the use of VLC in a direct approach with new technologies can be extremely interesting. Therefore, ID management guarantees the validity of the IDs and offers a control mechanism through which the necessary data can be obtained. The combination of the existing infrastructure with the IDs generated through validation within the existing nomenclature with preset IDs interchangeable between vehicles offers a new security policy for optical communications.

3. Implementation and Results

The proposal was tested using various methods and tools based on the Linux operating system capable of intercepting data or connections, such as BackTrack and WireShark. These methods' connections and their traffic management were analyzed. These aspects are important and each type of amendment was staged, which was the purpose of the study, starting from the unstructured ones and then the structured, external, and internal ones. In vehicular networks in particular, we can experience unstructured threats from other users without a high level of training; these practices are undertaken only out of pure curiosity and their method of operation is extremely easy to identify. These types of penetration are carried out by users who know the methods of operation and the vulnerabilities of networks and later develop scripts capable of disrupting access to them.

Vehicular network security and communications between infrastructure, pedestrians, cars, and intelligent traffic systems are extremely pressing topics for today's society. Attacks from the outside can be initiated at the level of intelligent transportation systems (ITSs) by capable individuals who gain access to the entire infrastructure, generating chaos and panic in addition to pursuing extremely well-defined goals of controlling certain areas of activity. Many of the attacks from outside target issues that are closely related to bank fraud, personal information, and the mining of confidential data. Analyzing the subject in detail, persons with hidden intentions could take control of autonomous vehicles, as well as intelligent traffic systems. Attacks of an internal type, however—and at this moment it is much too early to take these aspects into account—have more to do with the accuracy and degree of security established by existing users and the way they set up their accounts.

Analysis of the Security Process and the Threats to Which the Network Is Exposed

When there are security and privacy concerns, in order to ensure reliable communication between the sender and the recipient, we need to perform certain tests that can give us feedback on the VLC's compliance with the requirements imposed by other wireless systems. Therefore, the system proposed in this study aims to provide protection against external connections and rejects data assignment to other users outside the network or who may compromise the network. The proposed system does not fully behave like a commercial Internet network but, as shown by the simulation process, it provides the most important features: authenticity, confidentiality, integrity, and availability.

The authenticity feature aims to limit the introduction of messages into the communication channel that may disturb the receivers and prevent them from transmitting messages. The privacy feature imposes limits on data access to prevent disclosure of communication routes or routes created between senders and receivers. The integrity feature maintains the accuracy of content throughout its transmission from source to destination. If available, it prevents authentication from being given to unauthorized users, while for others, it requires access keys, in addition to the username and password. For such a system based on VLC to meet the mandatory requirements, several critical issues must be addressed in the final implementation. In the case of much more established communications, the network layer assumes all responsibilities of protecting the data and keeping them private from all points of view, including the legal and the commercial perspectives. In terms of the VLC channel, it can be vulnerable to attacks within vehicular networks, and the confidentiality of the data transmitted between the vehicles may be endangered. The measurements and tests presented here were carried out to track how a VLC system can be protected from various types of attacks in the network, such as flooding attacks, poisoning attacks, and cache attacks. These types are the ones that could endanger the integrity of the system, and this study proves that the security breaches in the case of VLC are much more critical than in the case of standard VANET communications (Figure 5).

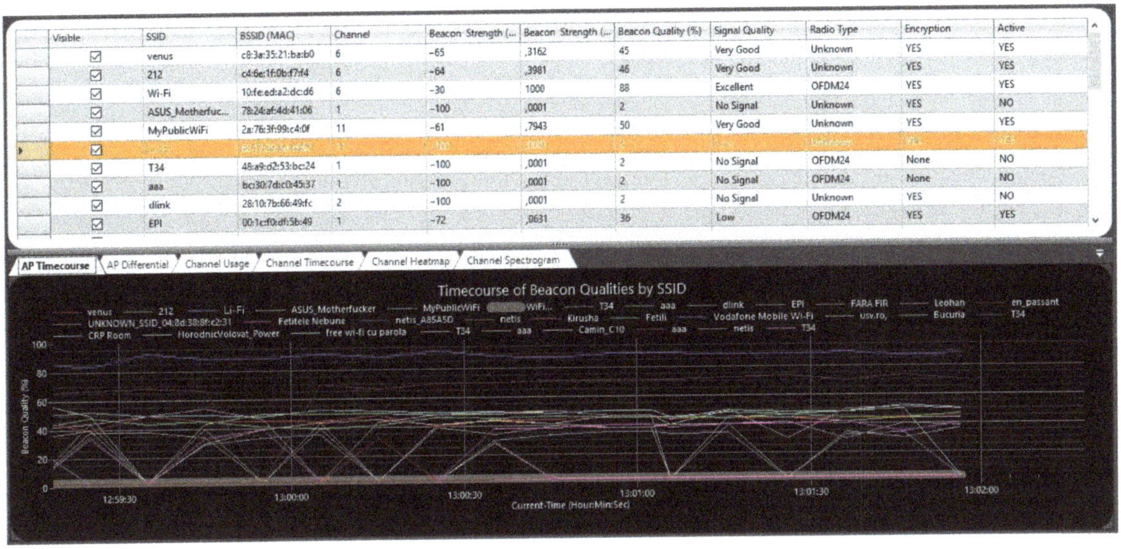

Figure 5. Analyses regarding the risk to which the infrastructure is exposed.

According to the research carried out, the sources of and exposure to external risks can only come about when there are failures in the physical infrastructure or the system itself. Matrix data approaches and evolutionary determinants of attacks within the physical layers have been considered and these threats to VLC do not compromise data security and integrity. To penetrate a VLC network or a communication system based on this technology, dedicated modules and receivers are needed, and if the communication system behaves like a classic wireless network, the penetration procedure is difficult and the contamination time is relatively long (see Figures 6 and 7).

Regarding the connectivity of such a network, the insertion of data packets capable of providing a perspective on the network through their iteration was also considered. Therefore, cascading data templates were used, and these were split into multiple data matrices that randomly populated the network. A premature conclusion regarding their capacity and accuracy should not be drawn, but, regarding the main aspect of security, there were many indications that confirmed that the data were in a network capable of providing them with a high degree of protection. As shown by Figure 8, no technical problems were encountered, and the data packet penetration process, which was constantly monitored, could not be derailed.

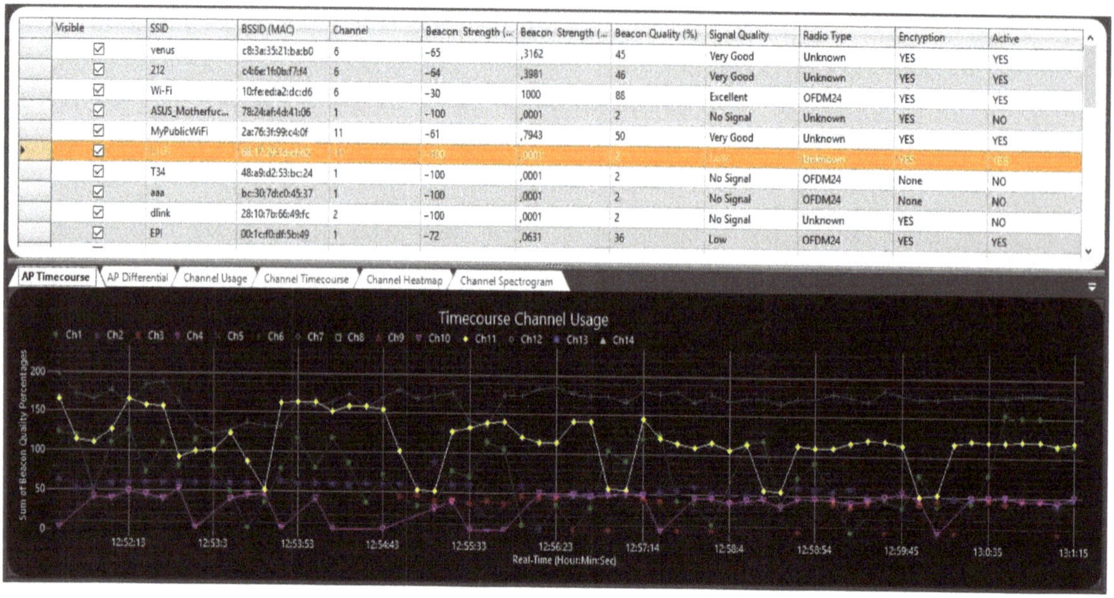

Figure 6. Report on the use of the communication channel in relation to the other active networks.

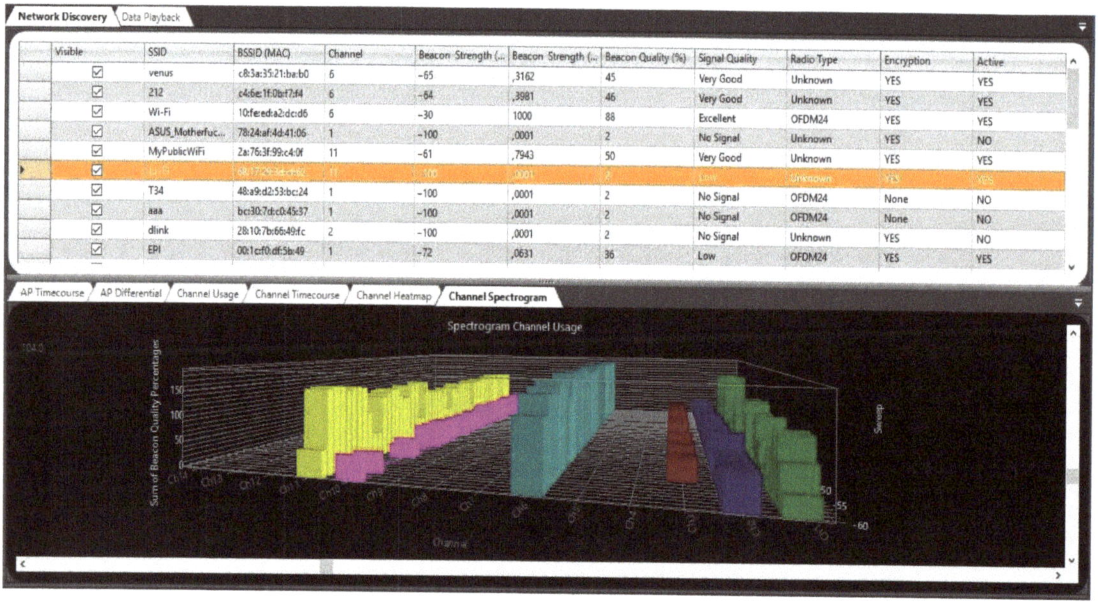

Figure 7. Spectrogram of the created communication channel.

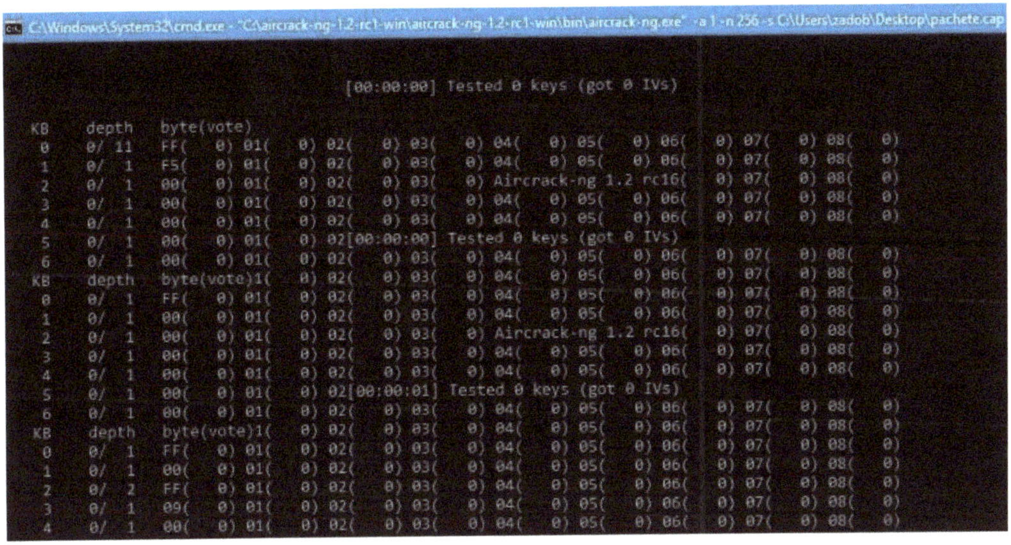

Figure 8. Communication testing with a dataset created in order to validate its reliability.

4. Discussion

Following the analysis of the proposal offered as a communication and security alternative, new directions were generated, especially concerning the use of VLC-RF and the two systems' integration in lighting systems, both in public and in vehicle lighting systems, to achieve communication of permanent data and in complete safety. These new approaches can address extremely pressing problems in today's society. It should not be overlooked that pollution and congestion are causes of traffic and mismanagement. The purpose of the proposal was to highlight the usefulness of VLC in systems other than the standard ones while, at the same time, indicating the high degree of security offered by the new standard compared to the existing ones. The communication through visible light employed in the proposed approach is performed in the PHY layers, resulting in unidirectional UDP connections in the first instance. The tests are in the early stages and have not passed the first stages where addresses are generated and packets with digital samples and minimal processing blocks are sent. Various parameters, sample rates, data rates, and modulation schemes have been studied, but no conclusions have yet been drawn, as these digital samples are routed internally without processing. An outline of a GNURadio-type processing block can be proposed that targets a future direction of being able to modulate the transmission bandwidth in both directions of the optical channel. When the intensity of the transmitted light is detected by a receiver and converted into an electrical current, unforeseen effects can occur. In this case, a file-type protection board on the transmitter–receiver path that can demodulate the received carrier signal is imperative. Thus, the data in Table 3 were extracted from the first stage of analysis of the presented proposal. The sources were retrieved and identified as the main data providers within the network, while the nodes represented the control units with destinations and sources. Figure 8 presents the degree of security of the data exposed in the information transmission process, showing that they do not deviate in the process they follow, nor do they present certain violations.

Table 3. Testbed routing.

Source	Next Node	Destination Node	Mask Address	Interface Connection
100.100.1.0	100.100.1.1	100.100.1.0	255.255.255.0	ethvlc0
100.100.2.0	100.100.2.2	100.100.2.0	255.255.255.0	ethvlc0
100.100.3.0	100.100.3.3	100.100.3.0	255.255.255.0	ethvlc0
100.100.4.0	100.100.4.4	100.100.4.0	255.255.255.0	ethvlc0
100.100.5.0	100.100.5.5	100.100.5.0	255.255.255.0	ethvlc0
100.100.6.0	100.100.6.6	100.100.6.0	255.255.255.0	ethvlc0
100.100.7.0	100.100.7.7	100.100.7.0	255.255.255.0	ethvlc0
100.100.8.0	100.100.8.8	100.100.8.0	255.255.255.0	ethvlc0
100.100.9.0	100.100.9.9	100.100.9.0	255.255.255.0	ethvlc0
100.100.10.0	100.100.10.10	100.100.10.0	255.255.255.0	ethvlc0
100.100.11.0	100.100.11.11	100.100.11.0	255.255.255.0	ethvlc0

Visible light communication is considered extremely important because it is a new physical environment that promises to alleviate the pressure hovering over the use of the RF spectrum. This tool is becoming more and more common, highlighting the performance of VLC in the case of end-to-end network integrations. The purpose of this article was to highlight the diverse applications of VLC, its complementarity with RF, and, in particular, to process of emergence through which it can have an important role within the same system. The VLC model was validated by the tests and the measurements undertaken, but there is a need for the independence of VLC to be finalized, and questions arise as to how it can be used with Wi-Fi and RF, encouraging proposals for hybrid networks at scale.

Instrumenting binaries with additional code sequences can be used to achieve a higher degree of routing by passing each instruction into the buffer dedicated to validation and generation of execution. Implementation at the architecture or prototype levels demonstrates leaps in tracking binary macs. Arguably, this implies the detection of unknown exploits from the previous parsing that trigger new routines in the buffer and create tags for system-wide validation.

Therefore, any type of attack against the networks outlined on board vehicles can be successfully mitigated because such attacks cannot be backed by classic exploitation techniques, penetration tests, buffer overflows, packet injection, or fake routines. Through such an approach, it is possible to ensure that the information is and will remain private and the data are kept within the ECUs until the moment of validation by the issuer and confirmation that the data can travel the unidirectional route. The major problem is created when binary tools need a larger number of binary tags; these must be extracted from multiple sources until MAC addresses can have one-byte characteristics with the ability to expose distinct tags for IDs, they must obtain contains eight characteristics generated for each. It is possible to limit the volume of data and the sources, but the goal is to secure a small dataset, which is a priority for optical communication dedicated to road safety and vehicular communication. A brake pressed suddenly in a major emergency triggers a request from a certain distance, and the existing cryptographic security process inserts labels for each VLC code and transmits them to the other sensors to verify the veracity of the data. The role of tags is to validate data from several sources: at least three tagging sources for each court validating code, cause code, or sub-cause code. These codes are already developed with a nomenclature dedicated to the traffic codes that the V2X–DSRC networks have, and they are called CAM and DENM messages. The presented solution is at an early stage and requires intensive study of the codes that the ECUs generate and analysis of data flows and the protocols they use, as well as identification of efficient ways of processing data in a relatively short time. The process cannot be used by all existing vehicles but may be an extremely important feature for future approaches.

5. Conclusions

When resource-intensive computing services employing big data are used and they contain location data or large-scale derivations, query requests also arise. They are encrypted and transmitted to vehicles to efficiently manage the processes carried out by the RSUs. The roadside control and management units can calculate the shortest routes to the desired destinations, a feature that transforms the network into a continuous flow of variants. In addition to these features, distance and location data are only captured at the user's physical level of inquiry and are shown in standard CPSs. The transfer of information and its sharing between vehicles is undertaken through a different environment that can control each physical level and a layer of the sharable service query. When the data are collected, they are adjusted with on-board units capable of branching the information into distinct stages and iterative processes through which the filtering of the usual information from that of control and management is distinctly achieved. Regardless of the area in which communications are used or their type, keeping personal information confidential and encrypting it are of paramount importance. The future is extremely promising for the application and development of such systems but, at the same time, the accumulation of factors and the dangers to which individuals and users are exposed can create more identity and authentication intrusions. Location and destination monitoring may make new systems veritable maps of possible targets for malicious individuals. Another approach that the scientific community can consider is that of generating fake data for the external environment and constantly updating IDs, producing a model capable of maintaining some degree of discretion in terms of safety and security.

Funding: This research received no external funding.

Acknowledgments: This work was supported by a grant of the Ministry of Research, Innovation and Digitization, CNCS-UEFISCDI, project number PN-III-P1-1.1-TE-2021-1371, within PNCDI III.

Conflicts of Interest: The author declares no conflict of interest.

References

1. Chowdhury, M.Z.; Hossan, M.T.; Islam, A.; Jang, Y.M. A Comparative Survey of Optical Wireless Technologies: Architectures and Applications. *IEEE Access* **2018**, *6*, 9819–9840. [CrossRef]
2. Matheus, L.E.M.; Vieira, A.B.; Vieira, L.F.M.; Vieira, M.A.M.; Gnawali, O. Visible Light Communication: Concepts, Applications and Challenges. *IEEE Commun. Surv. Tutor.* **2019**, *21*, 3204–3237. [CrossRef]
3. WHO. Global Status Report on Road Safety 2021 (World Health Organization, 2021), 138. 2021. Available online: https://www.who.int/violence_injury_prevention/road_safety_status/2021/en/ (accessed on 12 January 2023).
4. Amani, A.S.; Hesham, A.E.Z.; Sadek, A.A.-S. Secure and intelligent road trafc management system based on rfd technology. In Proceedings of the World Symposium on Computer Applications and Reserch (WSCAR), Cairo, Egypt, 12–14 March 2016; pp. 41–46.
5. Jungnickel, V.; Uysal, M.; Serafimovski, N.; Baykas, T.; O'Brien, D.; Ciaramella, E.; Ghassemlooy, Z.; Green, R.; Haas, H.; Haigh, P.A.; et al. A European view on the next generation optical wireless communication standard. In Proceedings of the 2015 IEEE Conference on Standards for Communications and Networking (CSCN), Tokyo, Japan, 28–30 October 2015; pp. 106–111.
6. Ifada, E.; Surajudeen-Bakinde, N.T.; Faruk, N.; Abubakar, A.; Mohammed, O.O.; Otuoze, A.O. Implementation of a data transmission system using Li-Fi technology. In Proceedings of the 2nd International Conference of the IEEE Nigeria Computer Chapter (NigeriaComputConf), Zaria, Nigeria, 14–17 October 2019; pp. 1–7.
7. *IEEE Std 802.15.7-2018 (Revision of IEEE Std 802.15.7-2011)*; IEEE Standard for Local and Metropolitan Area Networks-Part 15.7: Short-Range Optical Wireless Communications. IEEE: Piscataway, NJ, USA, 2019; pp. 1–407.
8. Shi, D.; Zhang, X.; Shi, L.; Vladimirescu, A.; Mazurczyk, W.; Cabaj, K.; Meunier, B.; Ali, K.; Cosmas, J.; Zhang, Y. On Improving 5G Internet of Radio Light Security Based on LED Fingerprint Identification Method. *Sensors* **2021**, *21*, 1515. [CrossRef]
9. Maia, G.; Aquino, A.L.L.; Viana, A.C.; Boukerche, A.; Loureiro, A.A.F. HyDi: A hybrid data dissemination protocol for highway scenariosin vehicular ad hoc networks. In Proceedings of the second ACMinternational Symposium on Design and Analysis of Intelligentvehicular Networks and Applications, Paphos, Cyprus, 21–25 October 2012; pp. 15–122.
10. Alraih, S.; Shayea, I.; Behjati, M.; Nordin, R.; Abdullah, N.F.; Abu-Samah, A.; Nandi, D. Revolution or Evolution? Technical Requirements and Considerations towards 6G Mobile Communications. *Sensors* **2022**, *22*, 762. [CrossRef]
11. Desai, V. Design and implementation of GSM and GPS based vehicle accident detection system. *Int. J. Technol. Sci.* **2014**, *1*, 306–3014.

12. Abdelhady, A.M.; Amin, O.; Chaaban, A.; Shihada, B.; Alouini, M. Spectral-Efficiency—Illumination Pareto Front for Energy Harvesting Enabled VLC Systems. *IEEE Trans. Commun.* **2019**, *67*, 8557–8572. [CrossRef]
13. Abdillah, M.I.; Darlis, D. Distance measurement implementation for VLC-based V2V communication on motorbike platooning. In Proceedings of the International Conference of Engineering Technology Entrepreneurship, Bandung, India, 13–15 August 2019.
14. Căilean, A.-M.; Dimian, M.; Popa, V. Noise-Adaptive Visible Light Communications Receiver for Automotive Applications: A Step Toward Self-Awareness. *Sensors* **2020**, *20*, 3764. [CrossRef]
15. Delgado-Rajo, F.; Melian-Segura, A.; Guerra, V.; Perez-Jimenez, R.; Sanchez-Rodriguez, D. Hybrid RF/VLC Network Architecture for the Internet of Things. *Sensors* **2020**, *20*, 478. [CrossRef]
16. Alin-Mihai, C.; Barthélemy, C.; Luc, C.; Suat, T.; Yasser, A.; JeanMarc, B. Visible light communications: Application to cooperation between vehicles and road infrastructures. In Proceedings of the IEEE Intelligent Vehicles International Symposium, Alcalá de Henares, Spain, 3–7 June 2012.
17. Marabissi, D.; Mucchi, L.; Caputo, S.; Nizzi, F.; Pecorella, T.; Fantacci, R.; Nawaz, T.; Seminara, M.; Catani, J. Experimental Measurements of a Joint 5G-VLC Communication for Future Vehicular Networks. *J. Sens. Actuator Netw.* **2020**, *9*, 32. [CrossRef]
18. de Oliveira, M.A.; Lima, E.S.; Cunha, M.S.P.; Abreu, M.; Arismar Cerqueira, S. RGB-based VLC system using 5G NR standard. *Opt. Commun.* **2021**, *481*. [CrossRef]
19. Hammouda, M.; Akın, S.; Vegni, A.M.; Haas, H.; Peissig, J. Link Selection in Hybrid RF/VLC Systems Under Statistical Queueing Constraints. *IEEE Trans. Wirel. Commun.* **2018**, *17*, 2738–2754. [CrossRef]
20. Khalifeh, A.F.; AlFasfous, N.; Theodory, R.; Giha, S.; Darabkh, K.A. An experimental evaluation and prototyping for visible light communication. *Comput. Electr. Eng.* **2018**, *72*, 248–265. [CrossRef]
21. Cevik, T.; Yilmaz, S. An overview of visible light communication systems. *arXiv* **2015**, arXiv:1512.03568. [CrossRef]
22. Chen, Q.; Wen, H.; He, J.; Deng, R.; Chen, M.; Zhou, Z.; Ma, J.; Zong, T. Performance enhancement of OFDM-VLC system using joint preprocessing scheme. *Opt. Commun.* **2019**, *451*, 111–115. [CrossRef]
23. Chen, C.-W.; Wang, W.-C.; Wu, J.-T.; Chen, H.-Y.; Liang, K.; Wei, L.-Y.; Hsu, Y.; Hsu, C.-W.; Chow, C.-W.; Yeh, C.-H.; et al. Visible light communications for the implementation of internet-of-things. *Opt. Eng.* **2016**, *55*, 060501. [CrossRef]
24. Mohsin, M.J.; Murdas, I.A. Performance analysis of an outdoor Li-Fi system-based AO-OFDM architecture under different FSO turbulence and weather conditions. *Optik* **2023**, *273*, 170427. [CrossRef]
25. Abdallah, W.; Boudriga, N. Enabling 5G wireless access using Li-Fi technology: An OFDM based approach. In Proceedings of the 2016 18th International Conference on Transparent Optical Networks (ICTON), Trento, Italy, 10–14 July 2016; pp. 1–6. [CrossRef]
26. Durgun, M.; Gokrem, L. VLC4WoT: Visible Light Communication for Web of Things. *KSII Trans. Internet Inf. Syst. (TIIS)* **2020**, *14*, 1502–1519. [CrossRef]
27. Yu, T.-C.; Huang, W.-T.; Lee, W.-B.; Chow, C.-W.; Chang, S.-W.; Kuo, H.-C. Visible Light Communication System Technology Review: Devices, Architectures, and Applications. *Crystals* **2021**, *11*, 1098. [CrossRef]
28. Khan, L.U. Visible light communication: Applications, architecture, standardization and research challenges. *Digit. Commun. Netw.* **2017**, *3*, 78–88. [CrossRef]
29. Hoffmann, M.; Kryszkiewicz, P.; Kliks, A. Frequency Selection for Platoon Communications in Secondary Spectrum Using Radio Environment Maps. *IEEE Trans. Intell. Transp. Syst.* **2022**, *23*, 2637–2650. [CrossRef]
30. Renzler, T.; Stolz, M.; Watzenig, D. Feudalistic Platooning: Subdivide Platoons, Unite Networks, and Conquer Efficiency and Reliability. *Sensors* **2022**, *22*, 4484. [CrossRef]
31. Căilean, A.-M.; Beguni, C.; Avătămăniței, S.-A.; Dimian, M.; Popa, V. Design, Implementation and Experimental Investigation of a Pedestrian Street Crossing Assistance System Based on Visible Light Communications. *Sensors* **2022**, *22*, 5481. [CrossRef]
32. Căilean, A.-M.; Avătămăniţei, S.-A.; Beguni, C.; Popa, V.; Dimian, M. Experimental Demonstration of a 188 meters Infrastructure-to-Vehicle Visible Light Communications Link in Outdoor Conditions. In Proceedings of the 2021 IEEE Sensors Applications Symposium (SAS), Sundsvall, Sweden, 23–25 August 2021; pp. 1–6. [CrossRef]
33. Beguni, C.; Căilean, A.-M.; Avătămăniței, S.-A.; Zadobrischi, E.; Stoler, R.; Dimian, M.; Popa, V.; Béchadergue, B.; Chassagne, L. In-Vehicle Visible Light Communications Data Transmission System Using Optical Fiber Distributed Light: Implementation and Experimental Evaluation. *Sensors* **2022**, *22*, 6738. [CrossRef] [PubMed]
34. Kadam, K.; Dhage, M.R. Visible Light Communication for IoT. In Proceedings of the 2016 2nd International Conference on Applied and Theoretical Computing and Communication Technology (iCATccT), Bangalore, India, 21–23 July 2016; pp. 275–278. [CrossRef]
35. Ismail, S.N.; Salih, M.H. A review of visible light communication (VLC) technology. In *AIP Conference Proceedings*; AIP Publishing LLC: Melville, NY, USA, 2020; p. 020289.
36. Katz, M.; O'Brien, D. Exploiting novel concepts for visible light communications: From light-based IoT to living surfaces. *Optik* **2019**, *195*, 163176. [CrossRef]
37. Teli, S.R.; Zvanovec, S.; Ghassemlooy, Z. Optical internet of things within 5G: Applications and challenges. In Proceedings of the 2018 IEEE International Conference on Internet of Things and Intelligence System (IOTAIS), Bali, Indonesia, 1–3 November 2018; IEEE: Piscataway, NJ, USA, 2018; pp. 40–45.
38. Almadani, Y.; Plets, D.; Bastiaens, S.; Joseph, W.; Ijaz, M.; Ghassemlooy, Z.; Rajbhandari, S. Visible Light Communications for Industrial Applications—Challenges and Potentials. *Electronics* **2020**, *9*, 2157. [CrossRef]

39. Beguni, C.; Zadobrischi, E.; Avătămăniței, S.-A.; Căilean, A.-M. Experimental Demonstration of a Visible Light Communications Crosswalk Assistance System. In Proceedings of the 2022 International Symposium on Electronics and Telecommunications (ISETC), Timisoara, Romania, 10–11 November 2022; pp. 1–4. [CrossRef]
40. Yang, H.; Zhong, W.-D.; Chen, C.; Alphones, A. Integration of Visible Light Communication and Positioning within 5G Networks for Internet of Things. *IEEE Netw.* **2020**, *34*, 134–140. [CrossRef]
41. Katz, M.; Ahmed, I. Opportunities and challenges for visible light communications in 6G. In Proceedings of the 2020 2nd 6G wireless summit (6G SUMMIT), Levi, Finland, 17–20 March 2020; IEEE: Piscataway, NJ, USA, 2020; pp. 1–5.
42. Lee, Y.U. Secure Visible Light Communication Technique Based on Asymmetric Data Encryption for 6G Communication Service. *Electronics* **2020**, *9*, 1847. [CrossRef]
43. Webber, J.; Mehbodniya, A.; Teng, R.; Arafa, A.; Alwakeel, A. Finger-Gesture Recognition for Visible Light Communication Systems Using Machine Learning. *Appl. Sci.* **2021**, *11*, 11582. [CrossRef]
44. Rehman, S.U.; Ullah, S.; Chong, P.H.J.; Yongchareon, S.; Komosny, D. Visible Light Communication: A System Perspective-Overview and Challenges. *Sensors* **2019**, *19*, 1153. [CrossRef]
45. Sharma, P.K.; Ryu, J.H.; Park, K.Y.; Park, J.H.; Park, J.H. Li-Fi based on security cloud framework for future IT environment. *Hum. Cent. Comput. Inf. Sci.* **2018**, *8*, 23. [CrossRef]
46. Perwej, Y. The Next Generation of Wireless Communication Using Li-Fi (Light Fidelity) Technology. *J. Comput. Netw. (JCN)* **2017**, *4*, 20–29. [CrossRef]
47. Cailean, A.M.; Dimian, M. Impact of IEEE 802.15.7 Standard on Visible Light Communications Usage in Automotive Applications. *IEEE Commun. Mag.* **2017**, *55*, 169–175. [CrossRef]
48. Sejan, M.A.S.; Rahman, M.H.; Aziz, M.A.; Kim, D.-S.; You, Y.-H.; Song, H.-K. A Comprehensive Survey on MIMO Visible Light Communication: Current Research, Machine Learning and Future Trends. *Sensors* **2023**, *23*, 739. [CrossRef]
49. Shi, J.; Niu, W.; Ha, Y.; Xu, Z.; Li, Z.; Yu, S.; Chi, N. AI-Enabled Intelligent Visible Light Communications: Challenges, Progress, and Future. *Photonics* **2022**, *9*, 529. [CrossRef]
50. Aydin, B.; Duman, Ç. Comparison of OOK-RZ and 4-PPM performances in Li-Fi systems using LED arrays. *Opt. Laser Technol.* **2022**, *153*, 108247. [CrossRef]
51. Stoicuta, O.; Riurean, S.; Burian, S.; Leba, M.; Ionica, A. Application of Optical Communication for an Enhanced Health and Safety System in Underground Mine. *Sensors* **2023**, *23*, 692. [CrossRef]
52. Gao, D.; Zhang, J.; Wang, F.; Liang, J.; Wang, W. Design and simulation of ultra-thin and high-efficiency silicon-based trichromatic PIN photodiode arrays for visible light communication. *Opt. Commun.* **2020**, *475*, 126296. [CrossRef]
53. Li, J.; Zou, P.; Ji, X.; Guo, X.; Chi, N. High-speed visible light communication utilizing monolithic integrated PIN array receiver. *Opt. Commun.* **2021**, *494*, 127027. [CrossRef]
54. Qiu, J.; Zhang, L.; Li, D.; Liu, X. High security chaotic multiple access scheme for visible light communication systems with advanced encryption standard interleaving. *Opt. Eng.* **2016**, *55*, 066121. [CrossRef]
55. Zia-Ul-Mustafa, R.; Boroujeni, S.S.; Guerra-Yanez, C.; Ghassemlooy, Z.; Minh, H.L.; Zvanovec, S. Quantum Key Distribution for Visible Light Communications: A Review. In Proceedings of the 2022 13th International Symposium on Communication Systems, Networks and Digital Signal Processing (CSNDSP), Porto, Portugal, 20–22 July 2022; pp. 589–594. [CrossRef]
56. Osama, E.; Mohsen, R. Wireless Quantum Key Distribution in Indoor Environments. *J. Opt. Soc. Am. B* **2018**, *35*, 197–207. [CrossRef]
57. Lu, Q.; Wang, F.; Huang, K.; Wu, X.; Wang, S.; He, D.; Yin, Z.; Guo, G.; Chen, W.; Han, Z. Quantum key distribution over scattering channel. *arXiv* **2021**, arXiv:2109.12282. [CrossRef]
58. Muhammad, T.; Marcos, K. Performance evaluation of IEEE 802.11p, LTE and 5G in connected vehicles for cooperative awareness. *Eng. Rep.* **2022**, *4*, e12467. [CrossRef]
59. Plascencia, E.; Guan, H.; Chassagne, L.; Barrois, O.; Shagdar, O.; Căilean, A.-M. A Comprehensive Investigation on Multi-User Interference Effects in Vehicular Visible Light Communications. *Sensors* **2023**, *23*, 2553. [CrossRef] [PubMed]
60. Wang, Y.; Chen, H.; Jiang, W.; Li, X.; Chen, X.; Meng, X.; Tian, P.; Sun, B. Optical encryption for visible light communication based on temporal ghost imaging with a micro-LED. *Opt. Lasers Eng.* **2020**, *134*, 106290. [CrossRef]
61. Liao, T.-L.; Chen, C.-Y.; Chen, H.-C.; Chen, Y.-Y.; Hou, Y.-Y. Realization of a Secure Visible Light Communication System via Chaos Synchronization. *Math. Probl. Eng.* **2021**, *2021*, 6661550. [CrossRef]

electronics

Article

Discriminating WirelessHART Communication Devices Using Sub-Nyquist Stimulated Responses

Jeffrey D. Long, Michael A. Temple * and Christopher M. Rondeau

Department of Electrical and Computer Engineering, US Air Force Institute of Technology, Wright-Patterson AFB, Dayton, OH 45433, USA
* Correspondence: michael.temple@afit.edu

Abstract: Reliable detection of counterfeit electronic, electrical, and electromechanical devices within critical information and communications technology systems ensures that operational integrity and resiliency are maintained. Counterfeit detection extends the device's service life that spans manufacture and pre-installation to removal and disposition activity. This is addressed here using Distinct Native Attribute (DNA) fingerprinting while considering the effects of sub-Nyquist sampling on DNA-based discrimination. The sub-Nyquist sampled signals were obtained using factor-of-205 decimation on Nyquist-compliant WirelessHART response signals. The DNA is extracted from actively stimulated responses of eight commercial WirelessHART adapters and metrics introduced to characterize classifier performance. Adverse effects of sub-Nyquist decimation on active DNA fingerprinting are first demonstrated using a Multiple Discriminant Analysis (MDA) classifier. Relative to Nyquist feature performance, MDA sub-Nyquist performance included decreases in classification of $\%C_\Delta \approx 35.2\%$ and counterfeit detection of $\%CDR_\Delta \approx 36.9\%$ at SNR = −9 dB. Benefits of Convolutional Neural Network (CNN) processing are demonstrated and include a majority of this degradation being recovered. This includes an increase of $\%C_\Delta \approx 26.2\%$ at SNR = −9 dB and average CNN counterfeit detection, precision, and recall rates all exceeding 90%.

Keywords: convolutional neural network; CNN; counterfeit detection; device fingerprinting; distinct native attribute (DNA); information and communications technology; multiple discriminant analysis; MDA; WirelessHART; wireless communications security

Citation: Long, J.D.; Temple, M.A.; Rondeau, C.M. Discriminating WirelessHART Communication Devices Using Sub-Nyquist Stimulated Responses. *Electronics* **2023**, *12*, 1973. https://doi.org/10.3390/electronics12091973

Academic Editors: Tao Huang, Shihao Yan, Guanglin Zhang, Li Sun, Tsz Hon Yuen, YoHan Park, Changhoon Lee and Cheng-Chi Lee

Received: 8 February 2023
Revised: 11 April 2023
Accepted: 17 April 2023
Published: 24 April 2023

1. Introduction

The development of new electronic, electrical, and electromechanical device technologies supporting critical information and communications technology systems will continue for decades to come. The deployment and availability of new devices provides certain benefits for expanding interconnectivity capabilities within the critical information and information technology arena. This expansion has heightened awareness and increased concerns associated with maintaining operational integrity and resiliency within the information and communications technology supply chain [1,2]. The adverse effects caused by a loss of operational integrity or resiliency range from increased inconvenience (degraded, inefficient, or intermittent service) at one extreme to premature lifecycle termination (removal from service) at the other extreme.

Supply chain integrity concerns are not unique within the information and communications technology community and are shared among other service communities that rely on electronic communications. Activities within these other service communities vary widely but are generally embodied within critical infrastructure, internet of things, industrial internet of things, and/or fourth industrial revolution frameworks [3–6]. Regardless of the framework, the use of digital communications requires that integrity assurance measures be taken during all phases of the device's technical lifespan (service life) [5].

The demonstration emphasis here is on pre-deployment protection (i.e., counterfeit detection) applied within the near-cradle phase of the device's technical lifespan. This includes pre-deployment manufacturing and distribution protection.

Lifespan assurance is addressed here using Radio Frequency (RF)-based Distinct Native Attribute (DNA) fingerprinting to provide reliable pre-deployment detection of counterfeit devices. Such protection can be considered during manufacturing, following manufacturing, and/or at any point in the supply chain as the device makes its way into service. A form of active DNA fingerprinting is considered here that uses fingerprint features extracted from externally stimulated responses of non-operating, non-operably connected WirelessHART communication devices. The operational and technical motivations for making this choice are presented in Sections 1.1 and 1.2, respectively.

1.1. Operational Motivation

The operational motivation for considering Wireless Highway Addressable Remote Transducer (WirelessHART) field device discriminability is generally unchanged from that put forth in prior related works [5,7–10]. It is even reasonable to argue that the motivation today remains stronger than ever given that (1) the number of fielded WirelessHART devices has reached into the tens-of-millions [11], and (2) hundreds of thousands of WirelessHART devices are manufactured annually and enter the supply chain [12]. Since its initial introduction WirelessHART has been well-received in European and North American industries given that [3,9,11,12]:

- It operates using the legacy wired HART protocol and users can take maximum advantage of prior experience, training, tool purchases, etc.;
- The deployment, installation, and maintenance cost are considerably reduced since no additional infrastructure cabling is generally required;
- There is considerable network architecture flexibility and expansion is easily accommodated using additional field devices or by connecting other nearby networks;
- The time required to commission (bring into service and put online) new devices takes hours versus days thanks to efficient pre-deployment benchtop programing.

It has been noted that a five-device WirelessHART network provides "sufficiently redundant operation" [12] and flexibility to support general industrial network architectures [11] and the communications lifeline between critical infrastructure elements [13]. Thus, the consideration and demonstration of counterfeit detection using N_{Dev} = 8 hardware devices here is not overly simplified and has appropriate applicability to small-scale networks supporting information and communications technology applications.

The concerns with maintaining operational integrity and resiliency in critical information and communications technology systems [1,2] are not new and related protection criteria was established early in November 2009 by the Society of Automobile Engineers under SAE-AS6462 guidelines [14]. This guidance targeted aerospace applications and was adopted by the US Defense Logistics Agency in May 2014 [15]—they subsequently reaffirmed SAE-AS6462 relevance in April 2020. To cope with an expanding supply chain attack space, the SAE-AS6462 guidelines were subsequently updated to the most recent AS-5553D revision in March 2022 [16]. The evolution to AS-5553D includes the recognition of expanded applicability to all organizations (beyond aerospace) that procure "parts and/or systems, subsystems, or assemblies, regardless of type, size, and product provided". AS-5553D aptly notes that mitigation of adverse counterfeit effects is "risk-based" and steps taken to mitigate these effects "will vary depending on the criticality of the application, desired performance and reliability of the equipment/hardware".

1.2. Technical Motivation

Identifying counterfeit devices early in their lifecycle is crucial to ensuring that operational integrity and resiliency are maintained. Near-cradle counterfeit detection activity within the technical cradle-to-grave protection strategy [5] has been considered using fundamentally different RF-based approaches for various electronic, electrical, and electromechanical devices. Representative active stimulation methods include:

- RF-based fingerprinting that uses interrogated responses of intentionally embedded onboard structures emplaced during manufacture [6,17,18]. These methods embedded structures at the integrated circuit level to impart unique RF fingerprint features when stimulated. The stimulated features are extracted and used to track and verify device identity as it traverses the supply chain (manufacturer, distributor, installer);
- DNA-based fingerprinting that exploits inherently present uniqueness resulting from device component, sub-assembly, and/or manufacturing process differences [9,19–21]. These methods exploit stimulated features that are distinct (unique from device-to-device), native (instilled during manufacture), and collectively embody device hardware/operating attributes (power consumption, mode, status, etc.).

The work in [8] was the first to consider the discrimination of four -Siemens AW210 [22] and four Pepperl + Fuchs Bullet [23] WirelessHART adapters using active DNA fingerprinting. These demonstrations were motivated by earlier passive DNA fingerprinting works in [5] that used the same adapters. Passive DNA fingerprints are generated from devices that are operably connected and perform their normal by-design communication function. Subsequent demonstrations in [9] used the same WirelessHART adapters with passive DNA fingerprinting processes from [5,7] and the active DNA fingerprinting process from [8].

The work in [9] provides the main motivation for demonstrations performed here with a goal of improving overall computational efficiency and enhancing the operational transition potential. Several options were considered in [9], including the application of conventional signal processing (down-conversion and filtering) and factor-of-5 decimation of the active DNA stimulated responses from [8]. This provided (1) an effective sample rate reduction from 1 Giga Samples per second (GSps) to 200 Mega Samples per second (MSps) and (2) a corresponding decrease in the number of pulse time domain samples (1,150,000 to 230,000) used for DNA fingerprint generation.

1.3. Relationship to Prior Research

Numerous RF fingerprinting methods have been considered as a means to discriminate electrical, electronic, and electromechanical components, and to improve operational reliability and security. For brevity, a detailed summary of RF fingerprinting methods is not included in this paper and the reader is referred to [24]. The authors in [24] have done a commendable job of categorizing various RF fingerprinting methods that use physical layer features to discriminate transmission sources. While the overall end-to-end identification process for the various methods can vary considerably, the main task of signal collection and digitization is largely the same and aimed at capturing signals that contain "useful features" to enable reliable identification. What is not immediately evident in [24] and the various fingerprinting works noted therein, is the Nyquist sampling conditions and how satisfying them does or does not impact the ability to extract useful features. This is not saying that these prior works did not consider Nyquist conditions, but rather that details for this consideration are not explicitly detailed in the works.

A majority of the works in [24] are believed to be based on discriminating features extracted from Nyquist sampled signal responses. This is a consequence of the researchers (1) considering conventional digital signal processing techniques that include consideration for Nyquist sampling conditions, or (2) using collected signals and/or methods from related work(s) where Nyquist sampling criteria were maintained. Satisfying Nyquist criteria enables receiver systems to reliably reconstruct the transmitted signal of interest and perform their intended by-design function (communicate, navigate, track, etc.). Nyquist criteria include sampling the signal of interest at a rate equal to, or greater than, the maximum system operating frequency—as operating frequency increases so does the amount of sampled data and required computational resources. The desire to minimize the amount of sampled data has motivated extensive research over the past decade. These works demonstrate acceptable signal reconstruction using a reduced number of samples without satisfying Nyquist criteria [25–28]—these are but a few representative works from a search using sub-Nyquist, undersampling, and compressive sensing terms.

Given the lack of detailed discussion on Nyquist sampling conditions in the RF fingerprinting works noted in [24], the authors believe that the work presented here is perhaps the first to consider a direct comparison of fingerprint discrimination performance with (1) fingerprint features generated under both Nyquist and sub-Nyquist conditions, (2) using the same collected device responses, and (3) a given classifier architecture. While work remains to consider sub-Nyquist conditions for other signal types and fingerprinting methods, results here suggest that deviating from Nyquist sampling constraints is a viable option and fingerprinting can be performed without regard for preserving by-design signal information.

1.4. Paper Organization

The remainder of this paper is organized as follows. The Demonstration Methodology is presented in Section 2 which provides selected details for relevant processes used to generate the demonstration results. This includes details for Experimental Collection and Post-Collection Processing in Section 2.1, Nyquist Decimation in Section 2.2, Sub-Nyquist Decimation in Section 2.3, Time Domain DNA Fingerprint Generation in Section 2.4, and Multiple Discriminant Analysis (MDA) in Section 2.5. Section 2.5 includes two sub-sections that provide details for Device Classification in Section 2.5.1 and Device ID Verification in Section 2.5.2. Details for Convolutional Neural Network (CNN) Discrimination is provided in Section 2.6. This includes implementation details for the one-dimensional CNN (1D-CNN) architecture in Section 2.6.1 and the two-dimensional CNN (2D-CNN) architecture in Section 2.6.2. Section 3 provides the Device Discrimination Results and includes MDA Classification Performance in Section 3.1 and CNN Classification Performance in Section 3.2. The final Counterfeit Discrimination Assessment results are presented in Section 3.3 and the paper concludes with the Summary and Conclusions presented in Section 4.

2. Demonstration Methodology

This section summarizes the experimental demonstration steps used to generate the classification results presented in Section 3. These steps include:

- Experimental Collection and Post-Collection Processing in Section 2.1: this includes a summary of processing details from [8] for obtaining the post-collected WirelessHART $s_{PC}(t)$ responses used here for Nyquist and sub-Nyquist decimation prior to DNA fingerprint generation. Selected details are provided for SFM waveform stimulus generation, WirelessHART device under test hardware, device under test response collection, and pre-fingerprint generation processing.
- Nyquist Decimation in Section 2.2: this includes the use of a theoretically selected NDecFac = 5 decimation factor based on conventional signal processing aimed at preserving by-design signal information. This process was first considered in [9] and was revisited here for completeness in making a Nyquist versus sub-Nyquist comparative performance assessment. In this case, decimation of $s_{PC}(t)$ by the NDecFac = 5 factor was preceded by down-conversion (D/C) to near-baseband and BandPass (BP) filtering. The uniform frequency spacing of the SFM sub-pulses and overall SFM waveform bandwidth are maintained.
- Sub-Nyquist Decimation in Section 2.3: this includes the use of an empirically selected NDecFac = 205 decimation factor that is applied without regard for preserving by-design signal information. Empirical selection details are provided and include (1) consideration of community feedback received as part of [20] proceedings, and (2) the desire to maintain both the number of SFM tones present and their spectral domain relationship. The NDecFac = 205 factor provided the desired sample rate reduction and computational efficiency increase.

- Time Domain DNA Fingerprint Generation in Section 2.4: this includes selected details for the time domain DNA fingerprint generation process adopted from prior related work in [8,9]. The adopted process has steadily evolved through numerous demonstrations in wireless communications applying time domain DNA fingerprinting to multiple modulation types and having similar classification objectives.
- Multiple Discriminant Analysis (MDA) Discrimination in Section 2.5: this includes a description of MDA model development and MDA-based device classification (discrimination). The confusion matrix construction is presented and calculation of the average cross-class percent correct classification (%C) metric is defined.
- Convolutional Neural Network (CNN) Discrimination in Section 2.6: this includes the CNN architectures selected for demonstration and the layer constructions used for performing (1) one-dimensional CNN (1D-CNN) processing with Time-Domain-Only (TDO) and Frequency-Domain-Only (FDO) samples, and (2) two-dimensional CNN (2D-CNN) processing with Joint-Time-Frequency (JTF) samples.

2.1. Experimental Collection and Post-Collection Processing

Stimulated WirelessHART response signals were originally collected and post-collection processed for demonstration activity in [8]. An overview of the collection setup is provided in Figure 1 and is based on integrated circuit anti-counterfeiting work in [18]. There are three main hardware components, including (1) a Keysight N5222B network analyzer [29] for generating the SFM input stimulus $s_{IN}(t)$, (2) a LeCroy WaveMaster 825Zi-A oscilloscope [30] for collecting the device under test output response $s_{OUT}(t)$, and (3) the WirelessHART device under test.

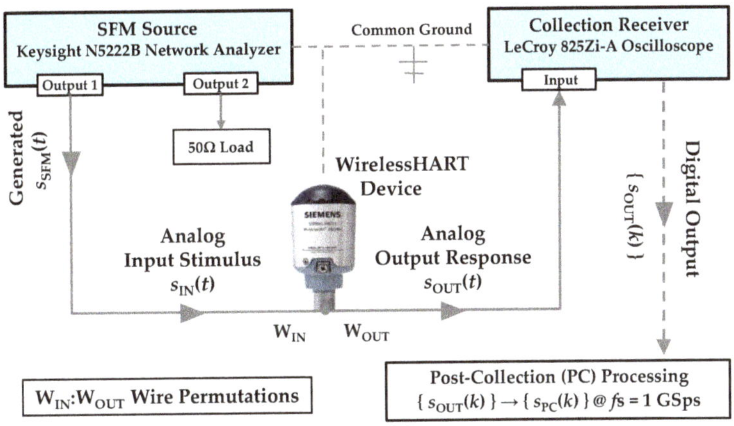

Figure 1. Active DNA fingerprinting setup used for collecting WirelessHART device responses. Post-collection processing applied prior to DNA fingerprinting.

Table 1 shows details for the four Siemens [22] and four Pepperl + Fuchs [23] WirelessHART adapters considered. The N_{Dev} = 8 adapters are identified as the D1, D2, ..., and D8 devices for demonstration. Although the device labeling of Siemens AW210 and Pepperl + Fuchs Bullet devices makes it appear that they are from two different manufacturers, it was previously determined in [5] that these devices are actually from the same manufacturer. The devices were distributed under two different labels with dissimilar serial number sequencing (a result of company ownership transition). Thus, the device discrimination being considered is the most challenging, that is the like-model and intra-manufacturer case using identical hardware devices that vary only by serial number.

Table 1. Selected details for N_{Dev} = 8 WirelessHART adapters used for demonstration.

Device ID	Device Label	Serial Number
D1	Siemens AW210	003095
D2	Siemens AW210	003159
D3	Siemens AW210	003097
D4	Siemens AW210	003150
D5	Pepperl + Fuchs Bullet	1A32DA
D6	Pepperl + Fuchs Bullet	1A32B3
D7	Pepperl + Fuchs Bullet	1A3226
D8	Pepperl + Fuchs Bullet	1A32A4

The N5222B source parameters were set to produce the SFM stimulus signal $s_{SFM}(t)$ that was input as $s_{IN}(t)$ to 1-of-5 available adapter wires that are denoted as W^j_{IN} for $j \in \{1, 2, \ldots, 5\}$. The SFM parameters were empirically set to maximize the source and device under test electromagnetic interaction with a goal of increasing discriminable information. The post-collected SFM response characteristics from [8] included (1) a total of $N_{SFM} = 9$ sub-pulses, (2) sub-pulse duration of $T_\Delta = 0.125$ ms for a total SFM pulse duration of $T_{SFM} = 1.125$ ms, and (3) sub-pulse spectral spacing of $f_\Delta = 5$ MHz yielding an SFM pulse bandwidth of $W_{SFM} \approx 50$ MHz that approximately spans 400 MHz $< f <$ 450 MHz. Each $s_{OUT}(t)$ response received by the 825Zi-A oscilloscope was digitized, stored, and its corresponding output sample sequence $\{s_{OUT}(t)\}$ used for fingerprint generation.

As indicated in Figure 1, the SFM stimulus is applied to a given W^j_{IN} wire ($j \in \{1, 2, \ldots, 5\}$) and the output response $s_{OUT}(t)$ is collected from 1 of 4 remaining wires. The output collection wire is denoted as W^k_{OUT} for $k \in \{1, 2, \ldots, 5\}$, $k \neq j$. Thus, there are a total of 20 order-matters $W^j_{IN} : W^k_{OUT}$ permutations available for active DNA fingerprinting. The collections from [8] were used here for demonstration and included W^j_{IN} being the device input power wire and W^k_{OUT} being the HART communication signaling wire.

2.2. Nyquist Decimation

Initial computational complexity reduction activity using the post-collected pulses from [8] was performed as part of work detailed in [9]. However, details of the theory-based Nyquist decimation process were omitted from [9] due to page constraints. Selected details are now included here to highlight differences between Nyquist decimated and the sub-Nyquist decimated processing detailed in Section 2.3. Processing of post-collected s_{PC}(t) pulses with Nyquist decimation is shown in Figure 2. The processing includes conventional signal processing of down-conversion (D/C), near-baseband BandPass (BP) filtering, decimation, and estimation of various powers and SNRs included.

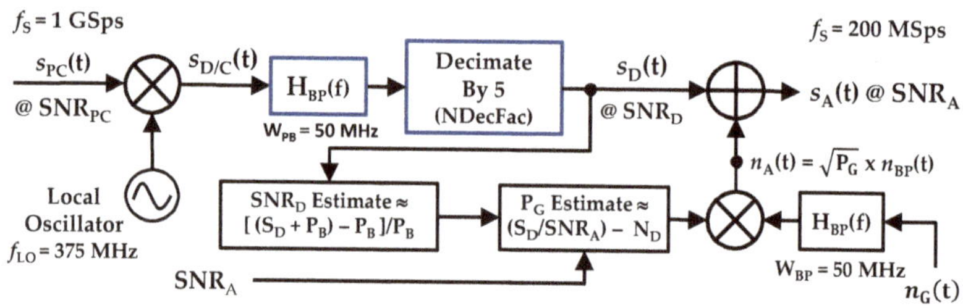

Figure 2. Overall down-conversion, filtering, Nyquist decimation, and SNR scaling processes used to generate the desired analysis s_A(t) for DNA fingerprinting.

The so-called "proper" decimation that is used here is consistent with Matlab's down-sample function and includes every NDecFac sample being retained and all others discarded. The desired effects of this decimation include (1) an effective sample rate reduction by a factor of 1/NDecFac (computational complexity reduction), and (2) retention of as-collected sample values and inherent source-to-device electromagnetic interaction effects (discriminable fingerprint information retention).

Figure 2 shows how the post-collected device response $s_{PC}(t)$ is (1) Down-Converted (D/C) to near-baseband using a local oscillator frequency of f_{LO} = 375 MHz, (2) BandPass (BP) filtered at the D/C center frequency of $f_{D/C}$ = 425 − 375 = 50 MHz using a 16th-order Butterworth filter having a passband of W_{BP} = 50 MHz, and (3) decimated by NDecFac = 5 to produce the decimated $s_D(t)$—this decimation factor choice was based on being the highest decimation factor that can be used while ensuring that Nyquist criteria is maintained. Thus, each of the WirelessHART $s_{PC}(t)$ responses at a sample rate of f_S = 1 GSps (N_{PC} = 1,150,000 post-collected time domain samples per pulse) are converted to have an f_S = 200 MSps rate (N_{Dec} = 230,000 decimated time domain samples per pulse) prior to fingerprint generation. This processing was performed for all of the N_{Pls} = 1132 pulses that were collected and post-collection processed for each of the N_{Dev} = 8 WirelessHART adapters (D1, D2, ..., D8) listed in Table 1.

The time domain effects of Figure 2 Nyquist decimation processing is illustrated for a representative WirelessHART $s_{PC}(t)$ signal is in Figure 3. These plots are for the case where there is no like-filtered AWGN SNR scaling ($SNR_A = SNR_{PC}$). The Region of Interest (ROI) samples for DNA fingerprint generation are highlighted in Figure 3 as well. Apart from ROI sample index number changes required to ensure the pulse ROI duration remains unchanged following decimation, the time domain amplitude effects of sample decimation are minimally discernable.

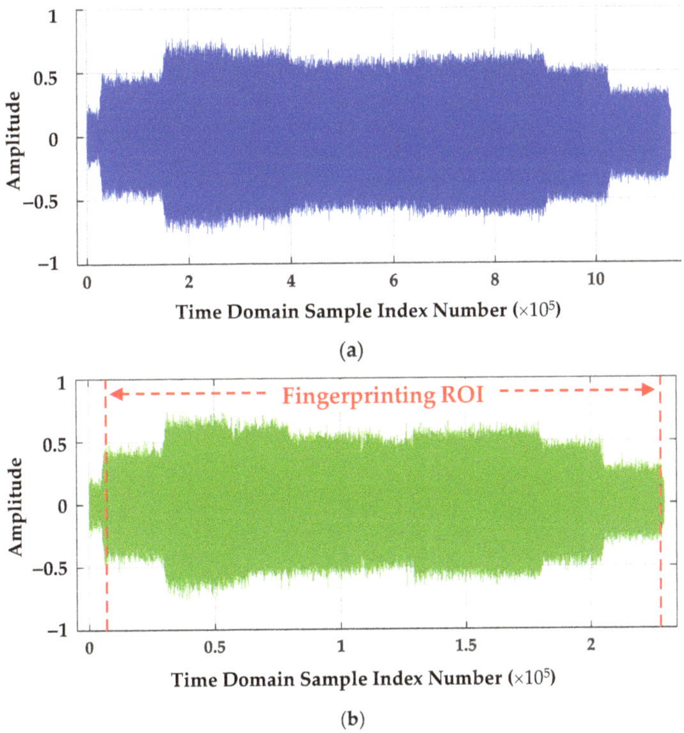

Figure 3. Time domain amplitude responses for (**a**) a representative post-collection processed pulse at f_S = 1 GSps and (**b**) the corresponding NDecFac = 5 Nyquist decimated pulse at f_S = 200 MSps showing the DNA fingerprinting ROI sample range.

The impact of Figure 2 Nyquist decimation processing is most evident in the frequency domain power spectral densities shown in Figure 4. This figure shows power spectral density (PSD) overlays for the (1) input $s_{PC}(t)$ response (far right red), (2) down-converted $s_{D/C}(t)$ response (far left green plus middle blue), and (3) final down-converted, bandpass filtered, and decimated $s_D(t)$ response (far left green) used for the analysis $s_A(t)$ generation. The impulse response (green dashed line) of the post-D/C bandpass filter W_{BP} is shown for reference. As desired for Figure 2 processing, the spectral content of $s_D(t)$ is displaced but structurally unchanged from the input $s_{PC}(t)$.

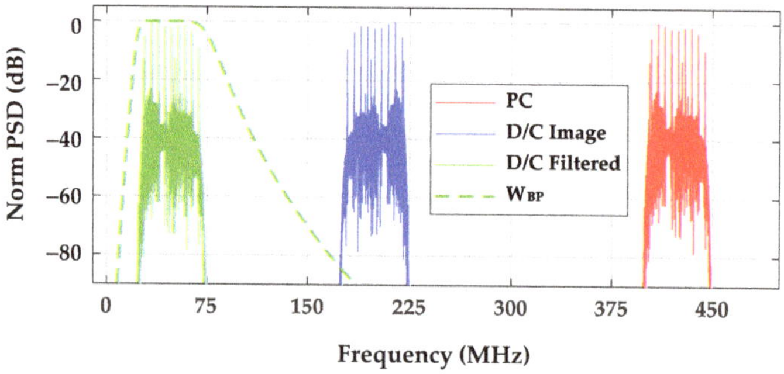

Figure 4. Overlay of Post-Collected (PC) input, Down-Converted (D/C) Image, D/C Filtered, and BandPass filtered (W_{BP}) impulse response (dashed line).

2.3. Sub-Nyquist Decimation

The main computational complexity reduction activity using post-collected WirelessHART pulses is referred to herein as sub-Nyquist decimation. Relative to the Nyquist decimation detailed in Section 2.2, the goal involves further reduction in the number of time domain samples in $s_{PC}(t)$ used for classifier training and testing. The overall processing for sub-Nyquist decimation is illustrated in Figure 5. The indicated NDecFac = 205 decimation factor was empirically chosen and implemented through "proper" decimation. The choice of NDecFac = 205 was motivated by community feedback relative to the presentation made in support of [20]. This feedback included suggestions that "a minimum sample rate reduction of 200" should be considered to make the DNA fingerprinting method more attractive for adoption and operational implementation. The final choice of NDecFac = 205 was based on observing the decimated spectral responses and ensuring that both the number of SFM tones and the order of the tones were maintained. The process included "proper" decimation of $s_{PC}(t)$ signals from [8] such that every 205th sample in the collections were retained and all others are discarded.

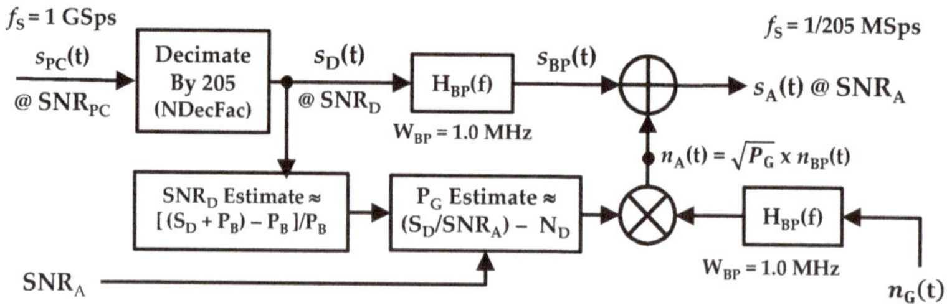

Figure 5. Overall sub-Nyquist decimation, estimation, filtering, and SNR scaling processes used to generate the desired analysis $s_A(t)$ that is input to the DNA fingerprinting process.

Nyquist sampling conditions of f_S = 1 GSps > 2 × f_{Max} = 2 × 425 MHz = 950 MHz were satisfied for the original post-collected $s_{PC}(t)$ signals in [8]. Thus, application of the empirically chosen NDecFac = 205 proper decimation factor effectively yields sub-Nyquist sampled signals for DNA fingerprinting. As illustrated throughout the remainder of this subsection using the same representative SFM response pulse used for Section 2.2, the NDecFac = 205 sub-Nyquist decimation of $s_{PC}(t)$ results in (1) the desired reduction in the number of samples used for fingerprint generation and classification, (2) an effective sample rate reduction by a factor of 1/NDecFac, and (3) inherent down-conversion, bandwidth compression, and increased background noise power in the spectral domain.

The sub-Nyquist decimation of SFM response signals was performed using an empirically chosen NDecFac = 205 decimation factor. The post-collection processing in [8] resulted in N_{PC} = 1,150,000 samples per SFM pulse at a sample frequency of f_S = 1 GSps. Thus, for the empirically chosen NDecFac = 205 factor, the sub-Nyquist decimated SFM pulses used here included a total of N_{Dec} = 1,150,000/205 = 5610 samples at a decimated sample rate of f_{SDec} = 1GSps/205 ≈ 4.88 MSps. The overall sub-Nyquist decimation process in Figure 5 was applied to a total of N_{Pls} = 1132 pulses that were collected and post-collection processed for each of the N_{Dev} = 8 WirelessHART devices being considered.

The effect of sub-Nyquist time domain sample decimation is illustrated in the amplitude responses shown in Figure 6. The ROI samples used for DNA fingerprint generation are highlighted as well. As implemented in [8], the post-collected SFM pulses are comprised of N_{SFM} = 9 sub-pulses, with (1) the duration of each sub-pulse being T_Δ = 0.125 ms and contributing to an overall SFM pulse duration of T_{SFM} = 9 × 0.125 ms ≈ 1.125 ms, and (2) the sub-pulses sequentially occurring at a uniform frequency spacing of $f_\Delta \approx$ 5 MHz and contributing to an overall SFM pulse bandwidth of $W_{SFM} \approx$ 50 MHz. The $T_{SFM} \approx$ 1.125 ms pulse duration includes to a total of N_{SPC} = (1.125 ms × 1 GSps) ≈ 1,125,000 time domain samples in the post-collected pulse responses (Figure 6a) and N_{SDec} = (1.125 ms × 4.88 MSps) ≈ 5490 time domain samples in the decimated responses (Figure 6b) used for DNA fingerprinting.

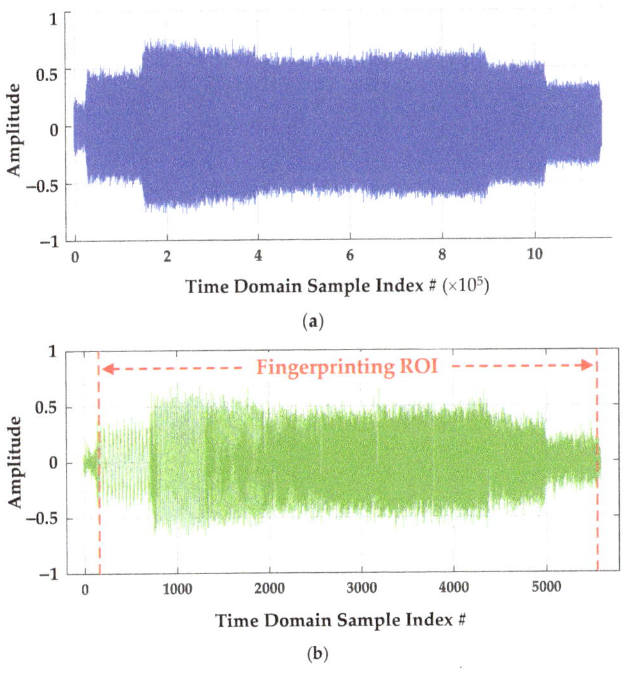

Figure 6. Time domain amplitude responses for (**a**) a representative post-collection processed pulse at f_S = 1 GSps and (**b**) the corresponding NDecFac = 205 sub-Nyquist decimated pulse response f_{SDec} = 1/205 GSps showing the DNA fingerprinting ROI sample range.

The corresponding power spectral density (PSD) responses for sub-pulses contained in Figure 6 pulses are shown overlaid in Figure 7. The non-decimated SFM pulse bandwidth indicated in Figure 7a (dashed lines) is $W_{PC} \approx 50.0$ MHz and spans a frequency range of $400 < f < 450$ MHz. For the post-collected $f_{Max} = 450$ MHz and $f_{SPC} = 1$ GSps, the Nyquist criteria of $f_{SPC} = 1\ \text{GSps} \geq 2 \times 450\ \text{MSps} = 900$ MHz is satisfied for post-collection processed SFM pulses from [8]. The corresponding NDecFact = 205 decimated SFM pulse bandwidth in Figure 7b (dashed lines) is $W_{Dec} \approx 1.0$ MHz and spans a frequency range of approximately $87 < f < 1150$ KHz. For the decimated $f_{Max} = 1150$ KHz and the decimated $f_{SDec} \approx 4.88$ MSps, the Nyquist criteria of $f_{SDec} \approx 4.88\ \text{MSps} \geq 2 \times 1150\ \text{KHz} \approx 2.3$ MHz is not satisfied for decimated SFM pulses and sub-Nyquist DNA fingerprinting is performed. Comparison of Figure 7a,b highlights the earlier noted down-conversion and bandwidth compression of $s_{PC}(t)$ resulting from sub-Nyquist NDecFac = 205 decimation.

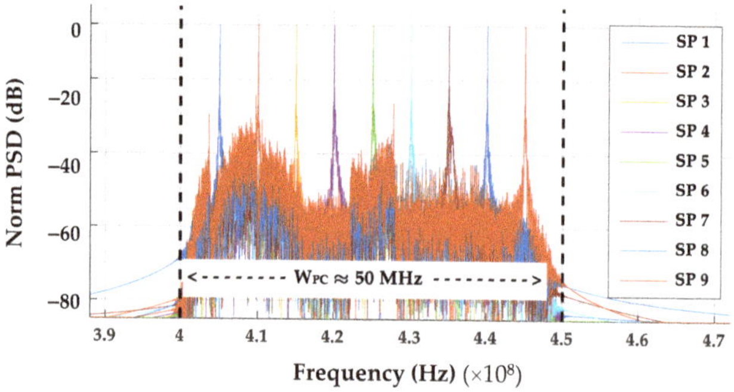

(a)

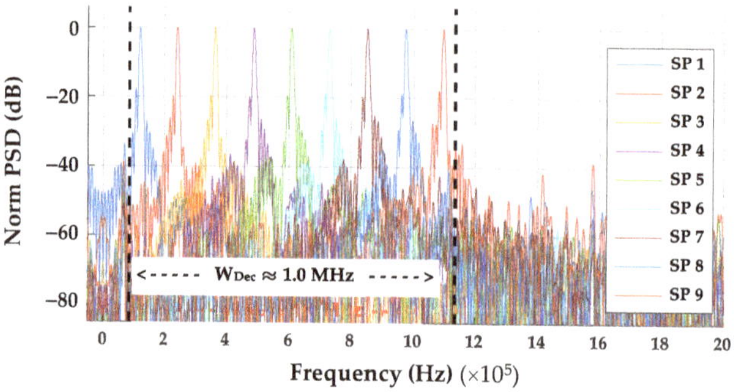

(b)

Figure 7. Overlay of individual SFM sub-pulse (SP) power spectral density (PSD) responses for pulses in Figure 6 with approximate SFM bandwidths bounded by the vertical dashed lines. (**a**) Post-collected SFM sub-pulse (SP) responses spanning $W_{SFM} \approx 50$ MHz. (**b**) Corresponding NDecFac = 205 sub-pulse (SP) responses spanning $W_{Dec} \approx 1$ MHz.

The power spectral densities for the composite SFM pulses are provided in Figure 8. Of note in comparing the non-decimated post-collected (PC) and decimated (Dec) responses in this figure are the average estimated background noise powers (N_{PC} and N_{Dec}) shown in the captions. The $N_{PC} \approx 5.54 \times 10^{-5}$ (W/MHz) background noise power for Figure 8a was calculated as the average of seven noise powers estimated in seven adjacent ideal $W_{PC} = 50.0$ MHz filters (red dashed line regions) spanning $0 < f < 350$ MHz. The decimated $N_{Dec} \approx 3.56$ (W/MHz) background noise power for Figure 8b was calculated as the average noise power in a single ideal $W_{Dec} \approx 1.0$ MHz filter (red dashed line region) spanning $13.5 < f < 23.5$ MHz. Considering the ratio of N_D/N_{PC} noise powers, the difference in post-collected and sub-Nyquist decimated background noise powers ($N_{B\Delta}$) is given by $N_{B\Delta} \approx 10 \times \log10[3.56/(5.54 \times 10^{-5})] \approx 48.1$ dB. This is the previously noted increased background noise power level resulting from sub-Nyquist decimation.

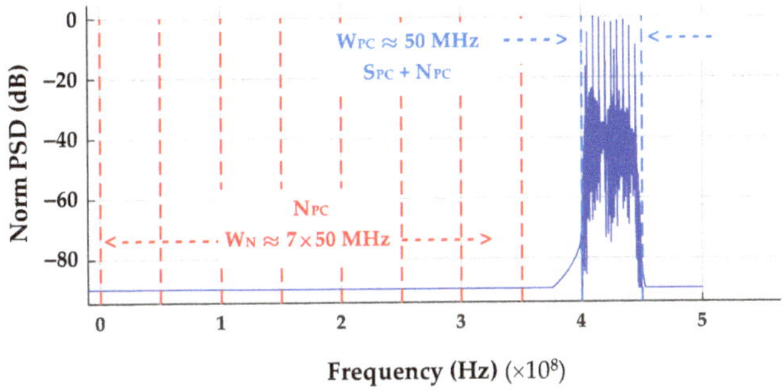

(a)

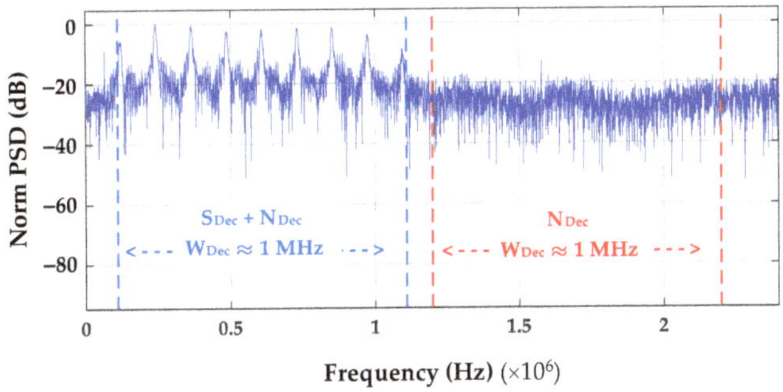

(b)

Figure 8. Composite SFM power spectral density responses showing spectral regions used to estimate average background noise powers. (**a**) Post-collected (PC) composite SFM pulse power spectral density with $W_{PC} \approx 50$ MHz. Estimates made within W_{PC} include $S_{PC} + N_{PC} \approx 78.74$, $N_{PC} \approx 5.54 \times 10^{-5}$ and $SNR_{PC} \approx 61.53$ dB. (**b**) Corresponding NDecFac = 205 decimated pulse power spectral density with $W_{SFM} \approx 1$ MHz. Estimates made within W_{SFM} include $S_{Dec} + N_{Dec} \approx 69.92.74$, $N_{Dec} \approx 3.56$ and $SNR_{Dec} \approx 12.71$ dB.

The overall pre-fingerprint generation processing with decimation, filtering, Signal-to-Noise Ratio (SNR) estimation and analysis SNR (SNR_A) scaling is illustrated in Figure 5. Using the post-collected SFM signal $s_{PC}(t)$ and desired analysis SNR_A as inputs, the steps for generating analysis signal $s_A(t)$ at the desired SNR_A include:

- Properly decimating $s_{PC}(t)$ by the selected NDecFac factor to obtain $s_D(t)$. The result is bandpass filtered with a $W_{Dec} \approx 1.0$ MHz filter to produce $s_{BP}(t)$.
- Using $s_D(t)$ to estimate the combined decimated signal and decimated noise ($S_D + N_D$) power using the signal-plus-background samples (see Figure 8b) that approximately span $W_{Dec} \approx 1.0$ MHz.
- Estimating background noise power P_{Dec} using noise-only region samples (see Figure 8b) that approximately span $W_{Dec} \approx 1.0$ MHz. The assumption here is that the estimated P_{Dec} noise power in this region is the same as the P_{Dec} noise power present in the estimated ($S_{Dec} + P_{Dec}$) power.
- Estimating $s_D(t)$ signal-only power S_{Dec} using $S_{Dec} \approx (S_{Dec} + P_{Dec})$ and the P_{Dec} noise power estimated in the previous step.

- Estimating the required like-filtered Additive White Gaussian Noise (AWGN) power as $P_G \approx (S_{Dec}/SNR_A)$. P_{Dec} using the desired SNR_A and the S_{Dec} and P_{Dec} power estimates from the two previous steps.
- Generating AWGN $n_G(t)$ and BandPass (BP) filtering it with the $W_{Dec} \approx 1.0$ MHz used to produce $n_{BP}(t)$. The result is power-scaled by P_G to produce the noise analysis signal given by $n_A(t) = \sqrt{P_G} \times n_{BP}(t)$.
- The final analysis signal $s_A(t)$ at the desired SNR_A is formed as $s_A(t) = s_{BP}(t) + n_A(t)$ and input to the DNA fingerprinting process.

The time and frequency domain effects of Figure 5 processing are shown in Figures 9 and 10 for a NDecFac = 205 decimated $s_D(t)$ SFM pulse at $SNR_{Dec} \approx 12.71$ dB and a desired like-filtered power-scaled analysis $s_A(t)$ at $SNR_A = 0$ dB. These plots were generated for the decimated $s_D(t)$ of the representative SFM pulse used for Figure 8 responses. The estimated $SNR_A \approx 0$ dB shown in Figures 9 and 10 captions was estimated using the $W_{Dec} \approx 1.0$ MHz decimation filter bandwidth. The final ROI samples used for time domain DNA fingerprint generation are highlighted in the bottom plot of Figure 9.

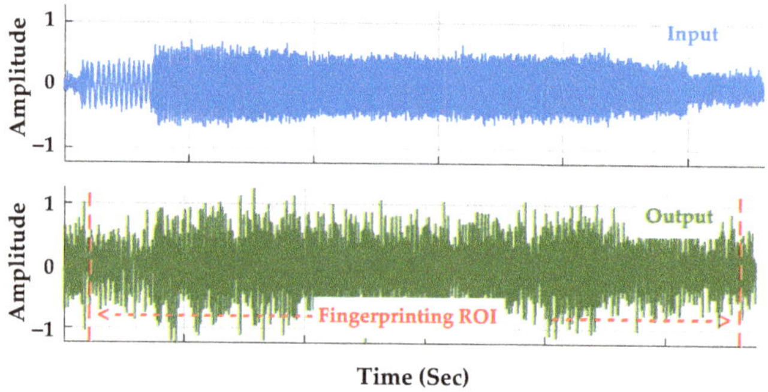

Figure 9. Time domain effects of Figure 5 processing showing (**Top**) an input NDecFact = 205 decimated signal $s_D(t)$ at $SNR_{Dec} \approx 12.71$ dB and (**Bottom**) corresponding output analysis $s_A(t)$ signal at $SNR_A \approx 0$ dB. Non-normalized plots are provided with the same vertical amplitude scale to highlight the effects of SNR degradation due to adding like-filtered AWGN.

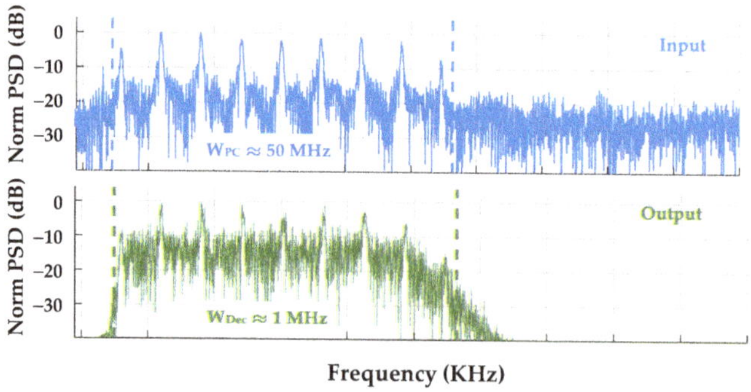

Figure 10. Frequency domain effects of Figure 5 processing showing normalized power spectral density (PSD) for (**Top**) input NDecFact = 205 decimated signal $s_D(t)$ at $SNR_{Dec} \approx 12.71$ dB and (**Bottom**) corresponding bandpass filtered analysis signal $s_A(t)$ at $SNR_A \approx 0$ dB. Normalized plots are provided with the same vertical scale to highlight the effects of within band SNR degradation.

2.4. Time Domain DNA Fingerprint Generation

The time domain DNA generation process used here is a variant of previous passive [5,10,31,32] and more recent active [8,9] DNA-based fingerprinting works. Active

DNA fingerprinting in [8,9] emerged from the earlier passive DNA fingerprinting methods [5,10,31,32] that steadily evolved within the wireless communications arena. Selected elements of time domain fingerprint generation are extracted from [9] and summarized here. The reader is referred to [9] and the other noted works for additional details.

The time domain region of interest samples from the analysis signals (those highlighted in Figure 9 and carried forward into Figure 11) are used to calculate statistical DNA features. For a real-valued sample sequence $\{s_{FP}(n)\}$ the statistical DNA features are calculated for instantaneous (1) *amplitude* response samples given by $M_{FP}(n) = |s_{FP}(n)|$; (2) *phase* response samples given by $\Theta_{FP}(n) = \tan^{-1}[H_{Re}(n)/H_{Im}(n)]$ where $H_{Re}(n)$ and $H_{Im}(n)$ are real and imaginary components of the Hilbert Transform denoted by Hilbert$[s_{FP}(n)]$; and (3) *frequency* response samples given by $\Phi_{FP}(n) = \text{gradient}[\Theta_{FP}(n)]$.

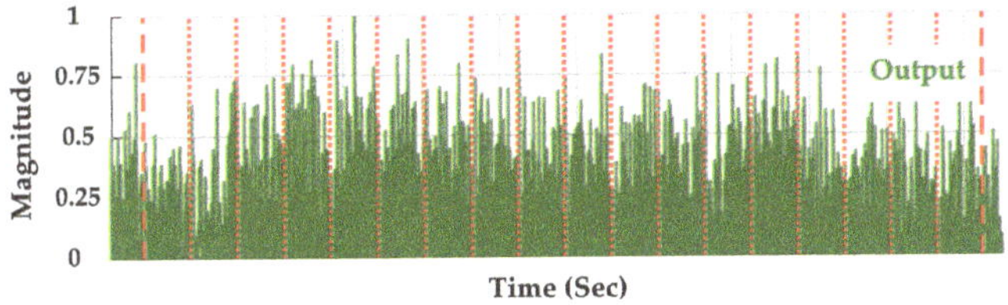

Figure 11. Time domain magnitude response for Figure 9 pulse showing the selected ROI (samples between the red dashed lines) and division into $N_{Srgn} = 18$ subregions (samples between adjacent red dotted lines) used for generating statistical time domain DNA fingerprint features.

Statistical DNA features are calculated using the $N_{Resp} = 3$ instantaneous response sequences of $\{M_{FP}(n)\}$, $\{\Theta_{FP}(n)\}$, and $\{\Phi_{FP}(n)\}$ and N_{Srgn} contiguous subregions of $\{s_{FP}(n)\}$ that span the selected ROI. This is illustrated in Figure 11 which shows the $\{M(n)\}$ magnitude responses for the representative pulse in Figure 9. Considering the calculation of $N_{Stat} = 3$ three statistical features of variance, skewness, and kurtosis [33] using samples within each of the $N_{Srgn} = 18$ subregions, and across the entire ROI as well, the time domain DNA fingerprints included a total of $N_{TD} = (N_{Srgn} + 1) \times N_{Resp} \times N_{Stat} = (19 + 1) \times 3 \times 3 = 171$ features.

2.5. Multiple Discriminant Analysis (MDA) Discrimination

The MDA-based discrimination methodology used here was adopted from prior related work in [5,8,9]. These works exploited DNA features for device discrimination using the same $N_{Dev} = 8$ WirelessHART adapters listed in Table 1 and used here. While providing a motivational basis for active DNA fingerprinting demonstration here, care is taken in making direct comparison of results in [5,8,9] with results provided here—this is reiterated with greater detail in Section 3 results. Regardless, the MDA processing here is fundamentally the same and summary details are presented for completeness. The reader is referred to [5] for additional details and a more complete development of MDA-based device classification.

MDA-based classification (discrimination) assessments were performed using a trained $(\mathbf{W}, \boldsymbol{\mu}_F, \boldsymbol{\sigma}_F, \boldsymbol{\mu}_n, \boldsymbol{\Sigma}_n)$ MDA model—bold variables are used here and henceforth throughout the paper to denote non-scalar vector or matrix quantities. The model components include (1) the MDA projection matrix $\mathbf{W}$ (dimension $N_{Feat} \times (N_{Cls} - 1)$), (2) the input fingerprint mean normalization factor $\boldsymbol{\mu}_F$ (dimension $1 \times N_{Feat}$), (3) the input fingerprint standard deviation normalization factor $\boldsymbol{\sigma}_F$ (dimension $1 \times N_{Feat}$), (4) the projected training class means $\boldsymbol{\mu}_n$ (dimension $1 \times (N_{CLS} - 1)$), and (5) the projected training class covariance $\boldsymbol{\Sigma}_n$ (dimension $(N_{Cls} - 1) \times (N_{Cls} - 1)$).

The classification process includes taking an unknown device fingerprint $\mathbf{F}_{Unk}$ (dimension $1 \times N_{Feat}$) and projecting it with $\mathbf{p}_{Unk} = \left[(\mathbf{F}_{Unk} - \boldsymbol{\mu}_F) \odot \boldsymbol{\sigma}_F^{-1}\right]\mathbf{W}$ into the MDA decision space [5]. The resultant $\mathbf{p}_{Unk}$ (dimension $1 \times N_{Dev} - 1$) is used with a given measure of similarity and a given test statistic (Z_{Unk}) generated. The resultant Z_{Unk} is used for making device classification decisions using threshold comparison. This represents an estimate indicating which 1 of N_{Cls} modeled devices the unknown $\mathbf{F}_{Unk}$ most closely represents. The Z_{Unk} test statistics used here were generated from probability-based Multi-Variate Normal (MVN) measures of similarity given their demonstrated superiority for device fingerprint discrimination [5,9].

2.5.1. Device Classification

Device classification decision results are summarized in a confusion matrix format [34], such as shown in Table 2, for a representative N_{Cls} = 8 model. This matrix shows MDA classifier testing using N_{Tst} = 2830 unknown testing fingerprints per class. Average cross-class percent correct classification (%C) is calculated as the sum of diagonal elements divided by the total number of estimates in the matrix ($N_{Tot} = N_{Tst} \times N_{Cls}$). The bold diagonal entries in Table 2 yield an overall %C = [21,184/(2830 × 8)] × 100 ≈ 93.6 ± 0.3%. This calculation includes a $\pm CI_{95\%} = \pm 0.3\%$ factor representing the 95% Confidence Interval ($CI_{95\%}$) calculated per [35]. The individual per-class testing is like-wise calculated on a row-by-row basis and ranges from a low of $\%C_{Cls}$ = (2459/2830) × 100 ≈ 86.9% (Class 2 and Class 5) to a high of $\%C_{Cls}$ = (2822/2830) × 100 ≈ 99.7% (Class 3).

Table 2. Representative classification confusion matrix for N_{Cls} = 8 class assessment.

		Called Class								
		Class 1	**Class 2**	**Class 3**	**Class 4**	**Class 5**	**Class 6**	**Class 7**	**Class 8**	**$\%C_{Cls} \pm CI_{95\%}$**
Input Class	**Class 1**	2639	0	1	2	0	0	0	188	93.3 ± 0.9%
	Class 2	0	2459	20	0	323	19	9	0	86.9 ± 1.2%
	Class 3	2	0	2822	3	3	0	0	0	99.7 ± 0.2%
	Class 4	18	0	4	2804	0	0	0	4	99.1 ± 0.4%
	Class 5	0	368	2	0	2459	0	1	0	86.9 ± 1.2%
	Class 6	5	18	0	3	35	2612	157	0	92.3 ± 0.9%
	Class 7	3	9	1	5	14	136	2662	0	94.1 ± 0.8%
	Class 8	103	0	0	0	0	0	0	2727	96.4 ± 0.7%

Results in Table 2 show that a majority of the classification error (bold red entries) is attributable to mutual confusion between (1) Class 2 and Class 5, (2) Class 6 and Class 7, and (3) Class 1 and Class 8. The individual per class testing is like-wise calculated on a row-by-row basis and ranges from a low of $\%C_{Cls}$ = (2459/2830) × 100 ≈ 86.9% (Class 2 and Class 5) to a high of $\%C_{Cls}$ = (2822/2830) × 100 ≈ 99.7% (Class 3).

2.5.2. Device ID Verification

As detailed in [5], device ID verification is performed using the trained MDA model ($\mathbf{W}_{Best}$, $\boldsymbol{\mu}_{F}$, $\boldsymbol{\sigma}_{F}$, $\boldsymbol{\mu}_{k}$, $\boldsymbol{\Sigma}_{k}$) with (1) testing fingerprints from an "unknown" device (denoted as D_j for $j = 1, 2, \ldots, N_{Dev}$) and (2) a claimed ID associated with one of the authorized model devices (denoted as D_k for $k = 1, 2, \ldots, N_{Dev}$ and $j \neq k$). For the D_j:D_k ID verification assessment, a given measure of similarity (Z_k) is generated for each unknown fingerprint, compared with the established training threshold (T_k) for device D_k, and a binary accept (e.g., $Z_k \geq T_k$) or reject (e.g., $Z_k < T_k$) decision made. Assuming the unknown device D_k is counterfeit, the desired outcome is a reject decision. The resultant Counterfeit Detection Rate percentage (%CDR) can be simply estimated as the total number of reject decisions divided by the total number of testing fingerprints considered. The reader is referred to [5] for a more formal development of the ID verification process.

The counterfeit detection potential for a given classifier can be estimated using confusion matrix results, such as that provided in Table 3. The classification results in Table 3 are taken from Table 2 confusion matrix and divided into four sub-matrices (quadrants) that effectively reflect performance for an N_{Cls} = 2 classifier. The quadrants are segregated by the dashed lines to highlight elements used for calculating %CDR and the alternate Counterfeit Precision Rate percentage (%CPR) and Counterfeit Recall Rate percentage (%CRR) metrics that are introduced later. The two classes correspond to Class 1 being all Table 1 Siemens devices (D1, D2, D3, and D4) and Class 2 being all Table 1 Pepperl + Fuch devices

(D5, D6, D7, and D8). The mechanics for assessing counterfeit detection potential from a classification confusion matrix are demonstrated with the Siemens devices designated as authentic and the Pepperl + Fuch devices designated as counterfeits.

Table 3. Division of Table 2 classification confusion matrix into N_{Cls} = 2 sub-matrices to highlight elements used for estimating counterfeit detection metrics.

	Called Class							
	Class 1				Class 2			
Input Class 1 (Authentic)	2639	0	1	2	0	0	0	188
	0	2459	20	0	323	19	9	0
	2	0	2822	3	3	0	0	0
	18	0	4	2804	0	0	0	4
Input Class 2 (Counterfeit)	0	368	2	0	2459	0	1	0
	5	18	0	3	35	2612	157	0
	3	9	1	5	14	136	2662	0
	103	0	0	0	0	0	0	2727

The classification results in Table 3 are consistent with the four Pepperl + Fuch devices being previously screened and declared as counterfeit devices. The counterfeit detection rate is estimated using diagonal elements in lower right hand quadrant of Table 3 and is given by %CDR = [(2459 + 2612 + 2662 + 2727)/(4 × 2830)] × 100 ≈ 92.40%. This exceeds the arbitrary performance benchmark of %CDR ≥ 90%. Calculation of this generally less rigorous %CDR metric is consistent with previous DNA works [5,31,32] and motivated by the desire to bolster cross-discipline understanding and appreciation for the work.

It has been suggested that a more rigorous counterfeit detection assessment can be made using hypothesis testing [5,34]. The test here involves counterfeit hypothesis testing with an unknown device (authentic or counterfeit) presenting an identity for a given counterfeit device. In this case, the hypothesis testing outcomes include: (1) a true positive (TP), the unknown counterfeit device is correctly declared counterfeit; (2) a false positive (FP) error, the unknown authentic device is errantly declared counterfeit; and (3) a false negative (FN) error, the unknown counterfeit device is errantly declared authentic. The resultant TP, FP, and FN outcomes are estimated from confusion matrix entries and used to calculate the alternate %CPR and %CRR metrics using [5,34],

$$\%\mathrm{CPR} = \left(\frac{\mathrm{TP}}{\mathrm{TP}+\mathrm{FP}}\right) \times 100, \tag{1}$$

$$\%\mathrm{CRR} = \left(\frac{\mathrm{TP}}{\mathrm{TP}+\mathrm{FN}}\right) \times 100. \tag{2}$$

For the Table 3 confusion matrix, the hypothesis testing outcomes required for calculating the %CPR and %CRP metrics include TP = 2459 + 2612 + 2662 + 2727 = 10,460 (sum of diagonal elements in the lower right hand quadrant), FP = 546 (sum of all elements in the upper right hand quadrant), and FN = 517 (sum all elements in the lower left hand quadrant). These values are input to Equations (1) and (2) to yield the alternate %CPR ≈ 95.04% and %CRP ≈ 95.29% metrics to characterize counterfeit detectability.

2.6. Convolutional Neural Network (CNN) Discrimination

Convolution Neural Network (CNN) processing is used to improve detection, identification, tracking, and classification in numerous application spaces. This is most evident when considering the plethora of more recent 2021–2022 research that has been conducted. These works include image processing centric CNN investigations supporting spatial terrain [36–38], smart grid [39], transfer learning [40], encoding/decoding [41], automatic modulation detection [42], and various electronic/electrical/electromechanical applications [7,43,44]. While [38] is not presented as a survey type paper, it does provide a noteworthy survey and summary with a relatively concise perspective on CNN processing.

Details for the CNN architectures used here are consistent with the basic CNN working principles noted in [38]. In the context of DNA fingerprinting, these principles are generally consistent with other classification problems and include: (1) data acquisition; (2) data exploration; (3) data preparation, decimation, digital filtering, standardization/normalization, and data splitting for training, validation, and testing; (4) CNN model development through hyperparameter selection; (5) model compilation with selected parameters, optimizer type, loss function, and metrics selection; (6) model training, weight updating, and biasing to increase classification performance; and (7) model application using testing fingerprints to make classification %C estimates.

In addition to the active DNA response conditioning and sub-Nyquist decimation in Section 2.3, data standardization and data splitting (training, validation, testing) with labels was required for CNN classification. The standardization included mapping to a Gaussian distribution (zero mean and unit variance) through calculation of a standard **Z**-score ($\mathbf{Z}_{\text{Std}}$) given by

$$\mathbf{Z}_{\text{Std}} = \frac{\mathbf{X} - \mu_{\mathbf{X}}}{\sigma_{\mathbf{X}}} \tag{3}$$

where **X** is the data vector to be scored, $\mu_{\mathbf{X}}$ is the mean of **X**, $\sigma_{\mathbf{X}}$ is the standard deviation of **X** and $\mathbf{Z}_{\text{Std}}$ is the normalized value of **X**. The **X** here includes sequence **X**: $\{x\}$ (time domain or frequency domain elements) that requires normalization aid deep learning and enable faster convergence.

CNN development requires selection of key hyperparameters (tuning) that include the number of neurons, activation function, optimizer, learning rate, batch size, and number of epochs. The CNN architectures considered here differ from traditional machine learning implementations where feature extraction is performed by a human. For the CNN processing here, the CNN plays the primary role in feature extraction with a goal of maximizing classification performance during training. This process includes the use of backpropagation to adjust weights and biases [45]. The CNN input data sizes are adaptable and exhibit immunity to small transformations from input data [46]. The convolutional layer filters are randomly initialized and optimized during training to identify discrimination-rich features.

For all CNN results here, the input data samples correspond to N_{Pls} = 5660 independent preprocessed pulses per device. These were randomly divided into pools containing approximately 80% training (N_{Tng} = 4528), 10% validation (N_{Val} = 566), and 10% testing (N_{Tst} = 566) samples. Each input pulse sample was assigned a unique label corresponding to one of N_{Dev} = 8 device IDs and contained N_{Dec} = 230,000 sub-Nyquist decimated time samples per Section 2.3. The training, validation, and testing pulse samples were used to characterize (1) 1-D CNN performance using time-domain-only and frequency-domain-only features, and (2) 2-D CNN performance using joint-time-frequency features.

All samples for the N_{Dev} = 8 devices are assigned labels such that each device number is represented in a string of eight digits. For example, device D1 is encoded as 10000000, device D2 is encoded as 01000000, device D3 is encoded as 00100000, and so on. This encoding enables the application of dense layer output processing within the CNN using softmax activation. Softmax activation converts a vector of numbers into a vector of probabilities, with the probability estimate being proportional to a relative scale of each value in the vector. In multiclass output classification problems, such as DNA fingerprinting, the last layer is usually the softmax layer. The softmax operation used here is given by [47]

$$\hat{p}_k = \alpha(s(x))_k = \frac{e^{s_k(x)}}{\sum_{j=1}^{N_{\text{Cls}}} e^{s_j(x)}} \tag{4}$$

where N_{CLS} is the number of classes, $s(x)$ is a vector containing the scores of each class for instance x, and $\alpha(s(x))_k$ is the estimated probability that instance x belongs to class k, given the scores for each class for that instance.

2.6.1. 1D-CNN Architecture

The 1D-CNN architecture used for device discrimination with sub-Nyquist signal responses is shown in Figure 12. The core CNN processing was implemented using four hidden layers, including three 1D-CNN layers (1DCNN) with Rectified Linear Unit (ReLu) activation and one pooling layer. The pooling layer finds the most significant abstract features (fingerprints) for the dense output layer for device classification. The 1DCNN layers are not fully connected and require fewer parameters when compared to fully connected layers. The non-fully connected 1DCNN convolutional layers in Figure 12 share weights among neurons whereas within the fully connected layer every output neuron is connected to every input neuron through a specific set of weights.

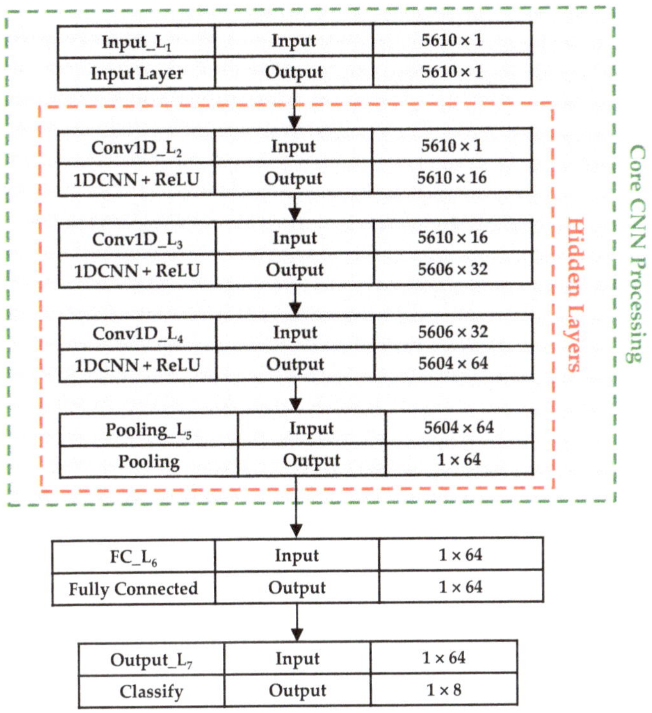

Figure 12. 1D-CNN architecture used for time-domain-only and frequency-domain-only WirlessHART device classification. The gray shaded box includes core hidden layer and input layer processing elements that are common with the 2D-CNN architecture in Figure 13.

Figure 12 shows the 1D-CNN architecture used for device classification using TDO and FDO responses of sub-Nyquist sampled signals. This architecture is based on the generic 1D-CNN architecture detailed in [48]. As shown, the architecture includes seven total layers with the four core CNN processing layers being hidden layers. The 5610 × 1 dimensional input data were processed in the second CNN Conv1D_L$_2$ layer (first convolutional layer) using N_{Cfil} = 16 filters with a kernel size of N_{Krn} = 5 and output a 5606 × 16 feature map. The third CNN Conv1D_L$_3$ layer (second convolutional layer) utilized N_{Cfil} = 32 filters and N_{Krn} = 3 and output a 5604 × 32 feature map. The fourth CNN Conv1D_L$_4$ layer (third and final convolutional layer) utilized N_{Cfil} = 64 filters and N_{Krn} = 3 and a 5602 × 64 feature map. As indicated in Figure 12, all three Conv1D layers use a Rectified Linear Unit (ReLU) activation function.

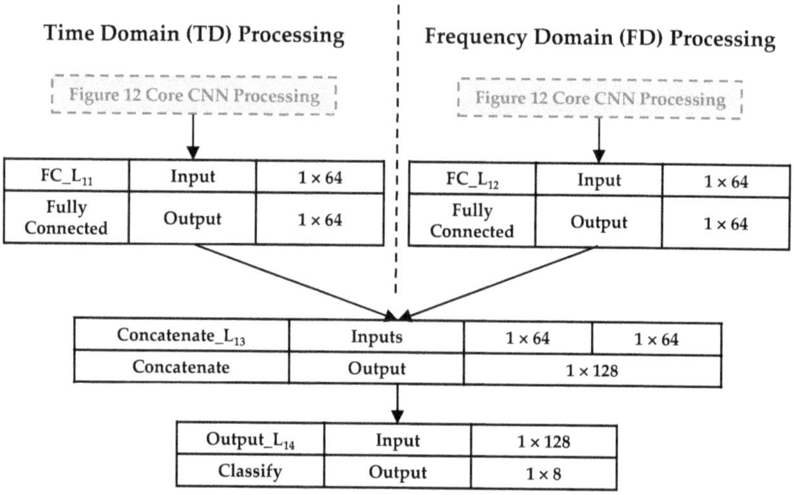

Figure 13. 2D-CNN architecture used for Joint-Time-Frequency (JTF) domain WirelessHART device classification. The two Core CNN Processing blocks are independent and functionally equivalent to those shown in Figure 12.

The feature map output of the final hidden Conv1D_L_4 convolutional layer in Figure 12 is input to the fifth CNN Pooling_L_5 layer. This layer performs global average pooling to accentuate feature rich information used for subsequent device classification. The Pooling_L_5 layer output is input to a fully connected FC_L_6 layer where optimization is performed to enhance class scoring and classification accuracy in the final Output_L_7 layer. The Output_L_7 results are used to form the classification confusion matrix detailed in Section 2.5 and estimate the %C and %CDR percentages.

The algorithm pseudocode for implementing the 1D-CNN processing in Figure 12 is presented in Algorithm 1. As detailed in the code, CNN processing employs dropout and kernel regularization to address issues associated with overfitting and to accelerate data processing. As indicated in Line 1, the learning process was implemented with an N_{Lrn} = 0.001 learning rate, N_{Epc} = 40 epochs, and a mini-batch size of N_{MB} = 32.

Algorithm 1. Algorithm pseudocode for implementing 1D-CNN processing.

1: **CNN** (trainX, trainY, validationX, validationY, testX, testY,
learningrate = 0.001, epoch = 40, batchsize = 32):
2: inputs = shape (datapoints, dimension = 1)
3: model ← Conv1D (filters = 16, kernels = 5, activation = ReLU) (input)
4: model ← Conv1D (filters = 32, kernels = 3,
activation =ReLU) (model)
5: model ← Conv1D (filters = 64, kernels = 3,
activation = ReLU) (model)
6: model ← GlobalAvergagePooling (model)
7: model ← Flatten() (model)
8: model ← Dropout(0.20) (model)
9: model ← Dense (neurons = 8, activation = softmax,
kernel_regularizer =
regularizers.L1L2 (l1 = 1×10^{-5}, l2 = 1×10^{-4}) (model)
10: model.compile (loss = categorical_crossentropy,
optimizer = Adam, learningrate)
11: model. Fit (trainX, trainY, validationX, validationY, epoch, batchsize)
12: Accuracy = model.evaluate (testX, testY)
13: **return** Accuracy

2.6.2. 2D-CNN Architecture

Figure 13 shows the 2D-CNN architecture used for JTF-based classification. This architecture includes replication of the core 1D-CNN processing layers in Figure 12 with time domain and frequency domain data input separately. The time domain Pooling_L_5^{TD} and frequency domain Pooling_L_5^{FD} layer outputs are independently processed within the fully connected FC_L_{11} and FC_L_{12} layers, respectively. These fully connected layer outputs are merged within the Concatenate_L_{13} layer before final Output_L_{14} classification occurs. The impact of this 2D-CNN JTF processing on the final classification performance is determined by analyzing confusion matrix %C and %CDR percentage estimates.

3. Device Discrimination Results

Classification performance of MDA models representing all N_{Cls} = 8 devices is first considered in Section 3.1. These results are provided to (1) highlight the effects of Nyquist decimation detailed in Section 2.2 and sub-Nyquist decimation detailed in Section 2.3, and (2) to establish a baseline for subsequent CNN performance results in Section 3.2 that highlight the benefits of CNN processing. As with prior related DNA-based discrimination works [5,7–9], classification performance analysis is focused on the %C vs. SNR region neighboring an arbitrary performance benchmark of %C = 90%. Section 3.1 MDA and Section 3.2 CNN classification confusion matrices are used with the ID verification process in Section 2.5.2 to generate the counterfeit detection assessment results in Section 3.3.

3.1. MDA Classification Performance

MDA classification results are presented in Figure 14 for the N_{CLS} = 8 class discrimination of the N_{Dev} = 8 WirelessHART adapters in Table 1. Results are presented using fingerprints for WirelessHART signals with no decimation (•), Nyquist decimate-by-5 decimation (•), and sub-Nyquist decimate-by-205 decimation (•). Note that the no decimation (•) results are visually obscured by the overlaid Nyquist decimation (•) results—based on $CI_{95\%}$ confidence intervals the no decimation and Nyquist decimate-by-5 results are statistically equivalent for all SNR considered.

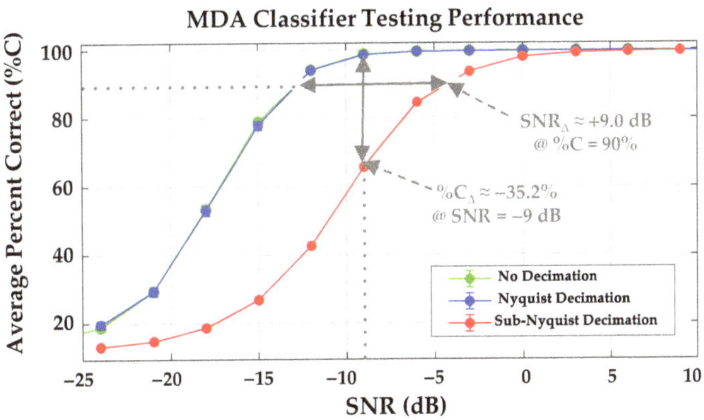

Figure 14. MDA classification using fingerprints for WirelessHART signals with no decimation (•), NDecFac = 5 Nyquist decimation (•) and NDecFac = 205 sub-Nyquist decimation (•). The sub-Nyquist SNR (SNR_Δ) and %C ($\%C_\Delta$) degradations are highlighted at the dotted line values.

By comparison with the statistically equivalent no decimation (•) and Nyquist decimate-by-5 decimation (•) results, the sub-Nyquist decimate-by-205 decimation (•) results are considerably poorer. Considering the %C = 90 arbitrary benchmark region, poorer performance is reflected in degradation metrics that include (1) a decrease in %C ($\%C_\Delta$) that is calculated as $\%C_\Delta \equiv \%C_{Dec} - \%C_{NonDec} \approx 63.2\% - 98.4\% \approx -35.2\%$ at SNR = −9 dB, and (2) an increase SNR (SNR_Δ) calculated as $SNR_\Delta = SNR_{Dec} - SNR_{NonDec} \approx -3.96 + 12.98 \approx$

+9.02 dB at %C = 90%. These degradations are highlighted in Figure 14 at the dotted line values.

3.2. CNN Classification Performance

CNN classification results are presented in Figure 15 for the N_{Dev} = 8 WirelessHART adapters in Table 1. This figure shows classification performance of the 1D-CNN Time-Domain-Only (TDO), 1D-CNN Frequency-Domain-Only (FDO) and 2D-CNN Joint Time-Frequency (JTF) architectures overlaid on an expanded region of the MDA %C vs. SNR results in Figure 14. Considering the sub-Nyquist performance results in Figure 15, the 2D-CNN JTF (▲) architecture performance is best overall and includes:

- The %C = 90% benchmark being achieved for SNR ≥ -9 dB;
- A major share of MDA degradation being recovered. This includes the indicated (a) $\%C_{\Delta} \equiv \%C_{CNN} - \%C_{MDA} \approx 91.8\% - 65.6\% \approx +26.2\%$ improvement at SNR = −9 dB, and (b) $SNR_{\Delta} \equiv SNR_{CNN} - SNR_{MDA} \approx -9 - (-4.5) \approx -5.0$ dB improvement at %C ≈ 92%;
- A marginal sub-Nyquist (▲) versus Nyquist (▼) average performance trade-off loss of $\%C_{\Delta} \approx -5.6\%$ across the -15 dB $\leq$ SNR ≤ 0 dB range—considerably more tolerant when considering the MDA $\%C_{\Delta} \approx -35.2\%$ loss noted in Figure 14.

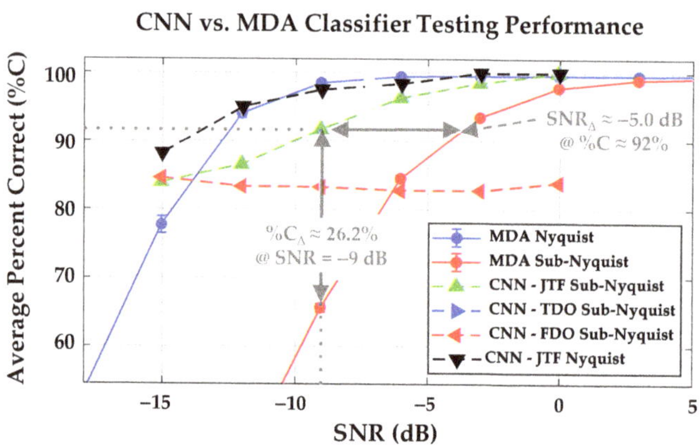

Figure 15. MDA vs. CNN classification highlighting the benefits of CNN processing. The 2D-CNN JTF (▲) $\%C_{\Delta}$ and SNR_{Δ} benefits are highlighted at the dotted line values and represent recovery of MDA classification degradation resulting from sub-Nyquist response decimation.

3.3. Counterfeit Discrimination Assessment

The estimated %CDRs with $\pm CI_{95\%}$ intervals for Figure 14 MDA classification results are summarized in Table 4 for three selected SNR. These %CDRs were calculated using confusion matrices and the estimation process detailed Section 2.5.2. Comparing the No Decimation and Nyquist Decimated estimates in Table 4, there is (1) no statistical difference in %CDR for the SNR = −15 dB and SNR = −9 dB conditions, and (2) less than 1% difference in %CDR for Nyquist decimation at SNR = −3 dB conditions. As reflected in the $\%CDR_{\Delta}$ differences in Table 4, there is considerable sub-Nyquist decimation degradation.

The estimated %CDRs with $\pm CI_{95\%}$ intervals for Figure 15 CNN sub-Nyquist classification results are summarized in Table 5 for three selected SNR. These were calculated using classification confusion matrices for results in Figure 15 and the estimation process detailed in Section 2.5.2. Based on the $\pm CI_{95\%}$ intervals, CNN %CDR performance of (1) FDO is the poorest for all SNR, (2) TDO and JTF are statistically equivalent for SNR = −15 dB and SNR = −9 dB, and (3) JTF is marginally better than TDO by $\%CDR_{\Delta} \approx 2\%$ at SNR = −3 dB. In light of minimizing computational complexity, it could be argued that the 1D-CNN TDO architecture may be preferred over the 2D-CNN JTF architecture for operational implementation if the $\%CDR_{\Delta} \approx 2\%$ performance trade-off is not tolerable.

Table 4. Estimated %CDRs with $\pm CI_{95\%}$ intervals for MDA classification results in Figure 14. The Nyquist decimated versus Sub-Nyquist decimated $\%CDR_{\Delta}$ differences are provided in the bottom row for comparison and highlight the degrading effects of sub-Nyquist decimation.

	SNR (dB)			
	−15.0	**−9.0**	**−3.0**	**Average**
No Decimation	86.4 ± 1.41%	99.1 ± 0.39%	99.7 ± 0.23%	95.1%
Nyquist Decimated	87.0 ± 1.39%	99.2 ± 0.37%	98.8 ± 0.45%	95.0%
Sub-Nyquist Decimated	28.4 ± 0.83%	62.3 ± 0.89%	92.4 ± 0.49%	61.0%
$\%CDR_{\Delta}$	−56.6%	−36.9%	−6.4%	−33.3%

Table 5. Estimated %CDR with $\pm CI_{95\%}$ for Figure 15 CNN classification results. All results for sub-Nyquist decimation with MDA %CDRs taken from Table 4 and reintroduced for comparison.

	SNR (dB)			
	−15.0	**−9.0**	**−3.0**	**Average**
CNN TDO	83.1 ± 1.54%	92.3 ± 1.10%	97.3 ± 0.67%	90.9%
CNN FDO	79.5 ± 1.66%	82.3 ± 1.57%	82.8 ± 1.55%	81.5%
CNN JTF	82.2 ± 1.58%	91.5 ± 1.15%	99.2 ± 0.37%	91.0%
MDA	28.4 ± 0.83%	62.3 ± 0.89%	92.4 ± 0.49%	61.0%
JTF vs. MDA $\%CDR_{\Delta}$	+53.8%	+29.2%	+6.8%	+29.9%

The corresponding sub-Nyquist MDA results from Table 4 are also provided in Table 5 for comparison. As indicated, the CNN JTF classifier outperforms the MDA classifier by a considerable margin and achieves the arbitrary %CDR > 90% benchmark for all SNR ≥ -9 dB. The CNN JTF classifier improvement relative to MDA is reflected in the $\%CDR_{\Delta} = \%CDR_{JTF} - \%CDR_{MDA}$ percentages in the bottom row. Collectively considering $\%CDR_{\Delta}$ for the three represented SNR, the CNN JTF classifier provides an average improvement of $\%CDR_{\Delta} \approx 29.9\%$ in counterfeit detection performance relative to the MDA classifier, while achieving the %CDR > 90% benchmark for all SNR ≥ -9 dB.

The final counterfeit assessment results are presented in Table 6 to enable performance comparison between the generally less rigorous %CDR detection metric and the alternate more rigorous hypothesis testing %CPR precision and %CRR recall metrics calculated using Equations (1) and (2), respectively. These results show that the cross-SNR average CNN counterfeit detection, precision, and recall rates all exceed 90%.

Table 6. Comparison of estimated counterfeit detection (%CDR), precision (%CPR), and recall (%CRR) metrics for best-case Figure 15 results using the 2D-JTF CNN with sub-Nyquist features.

	SNR (dB)			
	−15.0	**−9.0**	**−3.0**	**Average**
%CDR	82.2 ± 1.58%	91.5 ± 1.15%	99.2 ± 0.37%	91.0%
%CPR	87.4 ± 1.37%	93.9 ± 0.99%	99.2 ± 0.37%	93.5%
%CRR	85.4 ± 1.45%	92.9 ± 1.06%	99.5 ± 0.29%	92.6%

4. Summary and Conclusions

This work was motivated by the need to achieve reliable detection of counterfeit electronic, electrical, and electromechanical devices being used in critical information and communications technology applications. The counterfeit mitigation goal is to ensure that operational integrity and resiliency objectives are maintained [1,2]. WirelessHART is among the key communications technologies requiring protection and the current motivation for protecting WirelessHART systems is generally unchanged from prior related work [5,7–10]. One could argue that the motivation today is even stronger than ever given the number of fielded WirelessHART devices is approaching tens of millions [11] and hundreds of thousands of WirelessHART devices enter the supply chain annually [12]. Counterfeit device detection is addressed with a goal of enhancing the operational transition potential of previously demonstrated active DNA fingerprinting methods [8,9]. The goal is addressed in light of increased computational efficiency (decreased computational complexity) and increased counterfeit detection rate objectives.

Computational efficiency can generally be improved by reducing the total number of processed signal samples. This reduction is easily accomplished through sample decimation which is generally applied with a goal of retaining information—this is generally assured when the Nyquist sampling constraint is enforced. Retaining signal information is not a DNA fingerprinting requirement and thus an aggressive NDec = 205 sample decimation was applied to the WirelessHART adapter responses from [8]—this pushed the spectral information content well-below the Nyquist constraint. This resulted in an effective sample rate reduction (1 GSps to 200 MSps) and the desired reduction in the total number of samples (1,150,000 to 230,000) being processed.

The sub-Nyquist decimate-by-205 sampled responses were used for DNA-based Multiple Discriminant Analysis (MDA) and Convolutional Neural Network (CNN) classification. Counterfeit device classification and detectability was performed using eight commercial WirelessHART communication adapters [7–9]. The MDA classifier performance provided a baseline for highlighting (1) the overall degrading effects of sub-Nyquist sampling, and (2) detectability improvements that are realized using the CNN classifier. Relative to using Nyquist-compliant DNA fingerprint features, MDA performance using DNA features from sub-Nyquist sampled WirelessHART responses included *decreases* of $\%C_\Delta \approx 35.2\%$ and $\%CDR_\Delta \approx 36.9\%$ in classification and counterfeit detection at SNR = -9 dB. Corresponding CNN classifier performance using the same sub-Nyquist sampled responses was considerably better with a majority of the MDA degradation being recovered. This included best case CNN performance with a 2D Joint Time-Frequency (JTF) CNN architecture providing increases of $\%C_\Delta \approx 26.2\%$ and $\%CDR_\Delta \approx 29.2\%$ at SNR = -9 dB. For the full range of -15 dB $\leq$ SNR ≤ -3 dB average CNN performance included $\%CDR_\Delta \approx 29.9\%$, with corresponding detection, precision and recall rates all exceeding 90% for SNR ≥ -9 dB.

Author Contributions: Conceptualization, M.A.T. and C.M.R.; Data curation, M.A.T.; Formal analysis, J.D.L., M.A.T. and C.M.R.; Investigation, J.D.L.; Methodology, J.D.L., M.A.T. and C.M.R.; Project administration, M.A.T. and C.M.R.; Resources, M.A.T. and C.M.R.; Supervision, M.A.T.; Validation, J.D.L. and M.A.T.; Graphic Visualization, J.D.L. and M.A.T.; Writing—original draft, M.A.T.; Writing—review and editing, J.D.L., M.A.T. and C.M.R. All authors have read and agreed to the published version of the manuscript.

Funding: This research was funded in part by support funding received from the Spectrum Warfare Division, Sensors Directorate, U.S. Air Force Research Laboratory, Wright-Patterson AFB, Dayton OH, during U.S. Government Fiscal Years 2019–2022.

Institutional Review Board Statement: Not applicable.

Informed Consent Statement: Not applicable.

Data Availability Statement: The experimentally collected WirelessHART data used to obtain results were not approved for public release at the time of paper submission. Requests for release of these data to a third party should be directed to the corresponding author. Data distribution to a third party will be made on a request-by-request basis and are subject to public affairs approval.

Acknowledgments: The views and conclusions contained in this document are those of the authors and should not be interpreted as representing the official policies, either expressed or implied, of the United States Air Force or the U.S. Government. This paper is approved for public release, Case Number 88ABW-2023-0065.

Conflicts of Interest: The authors declare no conflict of interest. The funders had no role in the design of the study; in the collection, analyses, or interpretation of data; in the writing of the manuscript, or in the decision to publish the results.

Abbreviations

The following abbreviations are used throughout the manuscript:

%C	Average Cross-Class Percent Correct Classification
AWGN	Additive White Gaussian Noise
%CDR	Counterfeit Detection Rate Percentage
%CPR	Counterfeit Precision Rate Percentage
%CRR	Counterfeit Recall Rate Percentage
$CI_{95\%}$	95% Confidence Interval
CNN	Convolutional Neural Network
1D-CNN	One Dimensional CNN
2D-CNN	Two Dimensional CNN
DNA	Distinct Native Attribute
FDO	Frequency Domain Only
GSps	Giga-Samples Per Second
ID	Identity/Identification
JTF	Joint Time-Frequency
MDA	Multiple Discriminant Analysis
MHz	Megahertz
MSps	Mega-Samples Per Second
PC	Post-Collected
PSD	Power Spectral Density
RF	Radio Frequency
SFM	Stepped Frequency Modulated
SNR	Signal-to-Noise Ratio
SNR_{Dec}	Decimated Signal-to-Noise Ratio
SNR_A	Analysis Signal-to-Noise Ratio
TD	Time Domain
TDO	Time-Domain-Only
HART	Highway Addressable Remote Transducer

References

1. Cyber Security and Infrastructure Agency (CISA). Assessment of the Critical Supply Chains Supporting the U.S. Information and Communications Technology Industry: Overview of Executive Order 14017—America's Supply Chains. 2021. Available online: https://www.dhs.gov/publication/assessment-critical-supply-chains-supporting-us-ict-industry (accessed on 7 February 2023).
2. U.S. Department of Commerce; U.S. Department of Homeland Security. Assessment of the Critical Supply Chains Supporting the U.S. Information and Communications Technology Industry. Available online: https://www.dhs.gov/sites/default/files/2022-02/ICT%20Supply%20Chain%20Report_2.pdf (accessed on 7 February 2023).
3. FieldComm Group. *WirelessHART: Proven and Growing Technology with a Promising Future*; Global Control; FieldComm Group: Austin, TX, USA, 2018. Available online: https://tinyurl.com/fcgwirelesshartglobalcontrol (accessed on 7 February 2023).
4. Majid, M.; Habib, S.; Javed, A.R.; Rizwan, M.; Srivastava, G.; Gadekallu, T.R.; Lin, J.C.W. Applications of Wireless Sensor Networks and Internet of Things Frameworks in Industry Revolution 4.0: A Systematic Literature Review. *Sensors* **2022**, *22*, 2087. [CrossRef] [PubMed]
5. Rondeau, C.M.; Temple, M.A.; Betances, J.A.; Schubert Kabban, C.M. Extending Critical Infrastructure Element Longevity Using Constellation-Based ID Verification. *J. Comput. Secur.* **2020**, *100*, 102073. [CrossRef]
6. Yang, K.; Forte, D.; Tehranipoor, M.M. CDTA: A Comprehensive Solution for Counterfeit Detection, Traceability, and Authentication in the IoT Supply Chain. *ACM Trans. Des. Autom. Electron. Syst.* **2017**, *22*, 42. [CrossRef]
7. Gutierrez del Arroyo, J.; Borghetti, B.; Temple, M. Consideration for Radio Frequency Fingerprinting Across Multiple Frequency Channels. *Sensors* **2022**, *22*, 2111. [CrossRef]
8. Maier, M.J.; Hayden, H.S.; Temple, M.A.; Fickus, M.C. Ensuring the Longevity of WirelessHART Devices in Industrial Automation and Control Systems Using Distinct Native Attribute Fingerprinting. *Int. J. Crit. Infrastruct. Prot.* **2022**. *Under Review*.

9. Mims, W.H.; Temple, M.A.; Mills, R.A. Active 2D-DNA Fingerprinting of WirelessHART Adapters to Ensure Operational Integrity in Industrial Systems, MDPI. *Sensors* **2022**, *22*, 4906. [CrossRef]
10. Rondeau, C.M.; Temple, M.A.; Schubert Kabban, C.M. TD-DNA Feature Selection for Discriminating WirelessHART IIoT Devices. In Proceedings of the 53rd Hawaii International Conference on System Sciences (HICSS), Maui, HI, USA, 7–10 January 2020. Available online: https://scholarspace.manoa.hawaii.edu/bitstreams/35252979-27c2-4ae0-b8fb-35529f731e5a/download (accessed on 7 February 2023).
11. Devan, P.A.M.; Hussin, F.A.; Ibrahim, R.; Bingi, K.; Khanday, F.A. A Survey on the Application of WirelessHART for Industrial Process Monitoring and Control. *Sensors* **2021**, *21*, 4951. [CrossRef] [PubMed]
12. FieldComm Group. *WirelessHART User Case Studies*; Technical Report; FieldComm Group: Austin, TX, USA, 2019. Available online: https://tinyurl.com/fcgwirelesscs (accessed on 7 February 2023).
13. Cyber Security and Infrastructure Agency (CISA). Cybersecurity and Physical Security Convergence. 2021. Available online: https://www.cisa.gov/cybersecurity-and-physical-security-convergence (accessed on 7 February 2023).
14. Society of Automobile Engineers (SAE). Counterfeit Electrical, Electronic, and Electromechanical (EEE) Parts; Avoidance, Detection, Mitigation, and Disposition, Issued: 4 April 2009. Available online: https://standards.globalspec.com/std/14217318/SAE%20AS6462 (accessed on 7 February 2023).
15. Society of Automobile Engineers (SAE). Counterfeit Electrical, Electronic, and Electromechanical (EEE) Parts; Avoidance, Detection, Mitigation, and Disposition Verification Criteria, Doc ID: SAE-AS6462, Quick Search, Last Update: 10 January 2023. Available online: https://quicksearch.dla.mil/qsDocDetails.aspx?ident_number=280435 (accessed on 7 February 2023).
16. Society of Automobile Engineers (SAE). Counterfeit Electrical, Electronic, and Electromechanical (EEE) Parts; Avoidance, Detection, Mitigation, and Disposition, Latest Revision: 14 April 2022. Available online: https://www.sae.org/standards/content/as5553d/ (accessed on 7 February 2023).
17. Raut, R.D.; Gotmare, A.; Narkhede, B.E.; Govindarajan, U.H.; Bokade, S.U. Enabling Technologies for Industry 4.0 Manufacturing and Supply Chain: Concepts, Current Status, and Adoption Challenges. *IEEE Eng. Manag. Rev.* **2020**, *48*, 83–102. [CrossRef]
18. Voetberg, B.; Carbino, T.; Temple, M.; Buskohl, P.; Denault, J.; Glavin, N. Evolution of DNA Fingerprinting for Discriminating Conductive Ink Specimens. In Proceedings of the Digest Abstract, 2019 Government Microcircuit Applications & Critical Technology Conference (GOMACTech), Albuquerque, NM, USA, 25–28 March 2019.
19. Lukacs, M.W.; Zeqolari, A.J.; Collins, P.J.; Temple, M.A. RF-DNA Fingerprinting for Antenna Classification. *IEEE Antennas Wirel. Propag. Lett.* **2015**, *14*, 1455–1458. [CrossRef]
20. Maier, M.J.; Temple, M.A.; Betances, J.A.; Fickus, M.C. Active Distinct Native Attribute (DNA) Fingerprinting to Improve Electrical, Electronic, and Electromechanical (EEE) Component Trust. In Proceedings of the Digest Abstract, 2022 Government Microcircuit Applications & Critical Technology Conference (GOMACTech), Maimi, FL, USA, 21–24 March 2022.
21. Paul, A.J.; Collins, P.J.; Temple, M.A. Enhancing Microwave System Health Assessment Using Artificial Neural Networks. *IEEE Antennas Wirel. Propag. Lett.* **2019**, *18*, 2230–2234. [CrossRef]
22. Siemens. *WirelessHART Adapter, SITRANS AW210, 7MP3111, User Manual*; Siemens: Munich, Germany, 2012; Available online: https://tinyurl.com/yyjbgybm (accessed on 7 February 2023).
23. Pepperl+Fuchs. WHA-BLT-F9D0-N-A0-*, WirelessHART Adapter, Manual. Available online: https://tinyurl.com/pepplusfucwirelesshart (accessed on 7 February 2023).
24. Soltanieh, N.; Norouzi, Y.; Yang, Y.; Karmakar, N.C. A Review of Radio Frequency Fingerprinting Techniques. *IEEE J. Radio Freq. Identif.* **2022**, *4*, 222–233. [CrossRef]
25. Chen, X.; Sobhy, E.A.; Yu, Z.; Hoyos, S.; Silva-Martinez, J.; Palermo, S.; Sadler, B.M. A Sub-Nyquist Rate Compressive Sensing Data Acquisition Front-End. *IEEE J. Emerg. Sel. Top. Circuits Syst.* **2012**, *2*, 542–551. [CrossRef]
26. Brunelli, D.M.; Caione, C. Sparse Recovery Optimization in Wireless Sensor Networks with a Sub-Nyquist Sampling Rate. *Sensors* **2015**, *15*, 16654–16673. [CrossRef]
27. Deng, W.; Jiang, M.; Dong, Y. Recovery of Undersampled Signals Based on Compressed Sensing. In Proceedings of the 2019 IEEE 4th International Conference on Signal and Image Processing (ICSIP), Wuxi, China, 19–21 July 2019; pp. 636–640. [CrossRef]
28. Fang, J.; Wang, B.; Li, H.; Liang, Y.C. Recent Advances on Sub-Nyquist Sampling-Based Wideband Spectrum Sensing. *IEEE Wirel. Commun. Mag.* **2021**, *28*, 115–121. [CrossRef]
29. Keysight Technologies. PNA Family Microwave Network Analyzer (N522x/3x/4xB), Configuration Guide, Doc ID: 5992-1465EN. 10 September 2021. Available online: https://www.keysight.com/us/en/assets/7018-05185/configuration-guides/5992-1465.pdf (accessed on 7 February 2023).
30. LeCroy. WaveMaster®8 Zi-A Series: 4 GHz-45GHz Doc ID: WM8Zi-A-DS-09May11. 2011. Available online: https://docs.rs-online.com/035e/0900766b8127e31c.pdf (accessed on 7 February 2023).
31. Reising, D.R.; Temple, M.A. WiMAX Mobile Subscriber Verification Using Gabor-Based RF-DNA Fingerprints. In Proceedings of the IEEE International Conference on Communications (ICC), Ottawa, ON, Canada, 10–15 June 2012. [CrossRef]
32. Talbot, C.M.; Temple, M.A.; Carbino, T.J.; Betances, J.A. Detecting Rogue Attacks on Commercial Wireless Insteon Home Automation Systems. *J. Comput. Secur.* **2018**, *74*, 296–307. [CrossRef]
33. Soberon, A.; Stute, W. Assessing Skewness, Kurtosis and Normality in Linear Mixed Models. *J. Multivar. Anal.* **2017**, *161*, 123–140. [CrossRef]
34. Tharwat, A. Classification Assessment Methods. *Appl. Comput. Inform.* **2020**, *17*, 168–192. [CrossRef]

35. Park, H.; Leemis, L.M. Ensemble Confidence Intervals for Binomial Proportions. *Stat. Med.* **2019**, *38*, 3460–3475. [CrossRef]
36. Memon, N.; Parikh, H.; Patel, S.; Patel, D.; Patel, V. Automatic Land Cover Classification of Multi-resolution Dualpol Data Using Convolutional Neural Network Remote Sensing Applications. *Soc. Environ.* **2021**, *22*, 100491.
37. Shi, F.; Zhao, C.; Zhao, X.; Zhou, X.; Li, X.; Zhu, J. Spatial Variability of the Groundwater Exploitation Potential in an Arid Alluvial-Diluvial Plain using GIS-based Dempster-Shafer Theory. *Quat. Int.* **2021**, *571*, 127–135. [CrossRef]
38. Tegegne, A.M. Applications of Convolutional Neural Network for Classification of Land Cover and Groundwater Potentiality Zones. *J. Eng.* **2022**, *2022*, 6372089. [CrossRef]
39. Rituraj, R.; Ecker, D. A Comprehensive Investigation into the Application of Convolutional Neural Networks (ConvNet/CNN) in Smart Grids, 17 November 2022. Available online: https://papers.ssrn.com/sol3/papers.cfm?abstract_id=4279873 (accessed on 7 February 2023).
40. Emmanuel, S.; Onuodu, F.E. Object Detection Using Convolutional Neural Network Transfer Learning. *Int. J. Innov. Res. Eng. Multidiscip. Phys. Sci.* **2022**, *10*. Available online: https://www.ijirmps.org/papers/2022/3/1371.pdf (accessed on 7 February 2023).
41. Nasiri, F.; Hamidouche, W.; Morin, L.; Dhollande, N.; Cocherel, G. Prediction-Aware Quality Enhancement of VVC Using CNN. In Proceedings of the IEEE International Conference on Visual Communications and Image Processing (VCIP), Macau, China, 1–4 December 2020. [CrossRef]
42. Huang, J.; Huang, S.; Zeng, Y.; Chen, H.; Chang, S.; Zhang, Y. Hierarchical Digital Modulation Classification Using Cascaded Convolutional Neural Network. *J. Commun. Inf. Netw.* **2021**, *6*, 72–81. [CrossRef]
43. Atik, I. Classification of Electronic Components Based on Convolutional Neural Network Architecture. *Energies* **2022**, *15*, 2347. [CrossRef]
44. Li, J.; Li, W.; Chen, Y.; Gu, J. A PCB Electronic Components Detection Network Design Based on Effective Receptive Field Size and Anchor Size Matching. *J. Comput. Intell. Neurosci.* **2021**, *2021*, 6682710. [CrossRef]
45. Rumelhart, D.E.; Hinton, G.; Williams, R.J. Learning Representations by Back-Propagation Errors. *Nature* **1986**, *323*, 533–536. [CrossRef]
46. Kiranyaz, S.; Avci, O.; Abdeljaber, O.; Ince, T.; Gabbouj, M.; Inman, D.J. 1D Convolutional Neural Networks and Applications: A survey. *Mech. Syst. Signal Process.* **2021**, *151*, 107398. [CrossRef]
47. Geron, A. *Hands-on Machine Learning with Scikit-Learn, Keras & TensorFlow*, 2nd ed.; O'Reilly: Sebastopol, CA, USA, 2019.
48. Shoelson, B. Deep Learning in Matlab: A Brief Overview. In Proceedings of the Mathworks Automotive Conference (MICHauto), Plymouth, MI, USA, 2 May 2018. Available online: https://tinyurl.com/3fy2ax5b (accessed on 7 February 2023).

electronics

MDPI

Article

Blockchain-Enabled IoT for Rural Healthcare: Hybrid-Channel Communication with Digital Twinning

Steve Kerrison [1], Jusak Jusak [1,*] and Tao Huang [2]

1 School of Science and Technology, James Cook University, Singapore 387380, Singapore; steve.kerrison@jcu.edu.au
2 College of Science and Engineering, James Cook University, Smithfield, Cairns, QLD 4878, Australia; tao.huang1@jcu.edu.au
* Correspondence: jusak.jusak@jcu.edu.au

Abstract: Internet of Things (IoT) and blockchains are enabling technologies for modern healthcare applications, offering the improved monitoring of patient health and higher data integrity guarantees. However, in rural settings, communication reliability can pose a challenge that constrains real-time data usage. Additionally, the limited computation and communication resources of IoT sensors also means that they may not participate directly in blockchain transactions, reducing trust. This paper proposes a solution to these challenges, enabling the use of blockchain-based IoT healthcare devices in low-bandwidth rural areas. This integrated system, named hybrid channel healthcare chain (HC^2), uses two communication channels: short-range communication for device authorisation and bulk data transfer, and long-range the radio for light-weight monitoring and event notifications. Both channels leverage the same cryptographic identity information, and through the use of a cloud-based digital twin, the IoT device is able to sign its own transactions, without disclosing the key to said twin. Patient data are encrypted end to end between the IoT device and data store, with the blockchain providing a reliable record of the data lifecycle. We contribute a model, analytic evaluation and proof of concept for the HC^2 system that demonstrates its suitability for the stated scenarios by reducing the number of long-range radio packets needed by 87× compared to a conventional approach.

Keywords: blockchain; digital twin; Internet of Things; healthcare; encryption; privacy; rural; LPWAN

Citation: Kerrison, S.; Jusak, J.; Huang, T. Blockchain-Enabled IoT for Rural Healthcare: Hybrid-Channel Communication with Digital Twinning. *Electronics* **2023**, *12*, 2128. https://doi.org/10.3390/electronics12092128

Academic Editor: Djuradj Budimir

Received: 29 March 2023
Revised: 28 April 2023
Accepted: 4 May 2023
Published: 6 May 2023

1. Introduction

Internet of Things (IoT) technology has given rise to many new and innovative applications. In manufacturing, organisations from small to large scale use IoT to improve the monitoring of production processes, respond immediately when process deviation occurs, and to provide better services to their customers [1]. Implementation of IoT in the healthcare domain is a focus area for many researchers, academics, and industry as well. Healthcare IoT (HIoT) devices equipped with sensors, computation capability, and radio communications collect and process a patient's health related data, such as body temperature, electrocardiograph (ECG), oxygen saturation, blood pressure, and others to be transmitted to a cloud storage system in the other parts of the world through the internet. The term *Healthcare 4.0*, analogous to *Industry 4.0*, has been used widely to mark the development of smart and connected healthcare offering a chance to shift from traditional patient treatment to technology-based solutions that allow remote monitoring and medication [2].

Healthcare IoT is expected to be widely adopted but primarily benefits those in city regions who are most likely enjoying more extensive communications capabilities compared to those living in remote areas. The deployment of HIoT-supporting infrastructure in rural areas may face several obstacles. Geographical features of remote areas may be dominated by mountains, forest, savanna, hills, and rivers. In such areas, due to impediments to signals and low population density, there is less incentive for telecommunication providers to invest in installing significant infrastructure. Therefore, in most rural environments, low-communication quality, such as low bandwidth and intermittent connections, is frequently experienced by IoT devices, which can pose a challenge for real-time data usage.

Several technologies have been introduced in an attempt to address these adverse impacts, such as low power wireless area network (LPWAN) solutions [3,4]. LPWAN networks were introduced to accommodate the need for long-range and energy-efficient communications IoT devices. An example of such a technology is the long range (LoRa) standard that has growing adoption and industry support [5,6].

In addition to rural communication issues, HIoT faces security and privacy challenges in managing massive amounts of collected data. Cloud-based electronic healthcare records (EHR) emerged as a widely adopted solution [7]. They have several advantages, including on-demand service, broad network access, resource sharing, rapid elasticity, and guaranteed quality of service from service providers. With these features, the implementation of the EHR contributes to reduced data storage and maintenance costs, improved speed and processing accuracy, and allows data exchange among parties within a particular EHR system [8,9]. However, the centralised nature of the EHR system creates a setback from the user's point of view, in that users are more concerned about security and privacy due to the loss of control over clinical data in cloud storage.

Alongside the advancement of cloud and IoT, blockchain technology, the engine behind the cryptocurrency hype, has led to many other applications leveraging its features. For example, an article by Pennino et al. in [10] outlined the use of blockchain to support secure economic transactions underlying the decentralised payment system independent, the work by Wang in [11] investigated the utilisation of blockchain to secure energy delivery in electric vehicles, and some works by Farooq and Marbouh documented in [12,13] highlighted blockchain-based frameworks to assist healthcare management to monitor, diagnose, and treat patients remotely by stressing its applications in the most current COVID-19 pandemic situation. With the blockchain, certain aspects of applications become decentralised, in which control and decision are now shifted from centralised organisations to a distributed network. Each member node in a blockchain network retains a duplicate of the exact same information represented in the form of a distributed ledger. In this distributed network, consensus must be reached in order to add or change data, and the integrity of such operations is cryptographically verifiable. Attempts to tamper with information in the ledger is almost impossible.

In this work, we propose an integrated IoT and a private blockchain system applied to rural healthcare monitoring, called hybrid channel healthcare chain (HC^2). We chose a private blockchain scheme to facilitate a controllable environment, which is more appropriate for the healthcare use case than a public blockchain. The system operates two communication channels: short-range communication via personal area networks (PANs) for device authorisation and bulk data transfer, and long-range radio via LPWAN for light-weight data transmission and event notifications. Both channels leverage the same cryptographic identity information, and through a form of cloud-based digital twin, the IoT device is able to sign its own transactions via templates, without disclosing the key to said twin. Patient data are encrypted end to end between the IoT device and data store, with the blockchain providing integrity and authority only, thus protecting privacy.

The main contributions of the paper can be summarised as follows:

1. We define an architecture and data model for HIoT data that connects rural patients' data with healthcare providers with integrity provided by a blockchain.
2. We introduce a hybrid-channel communication model, allowing HIoT devices to use two communication methods to accommodate healthcare data transmission suitable for rural areas.
3. To overcome the transmission limitations on one of the two transmission channels, we incorporate a digital twin to handle data transactions from both of the communication channels and assist with blockchain transaction message reconstruction without sharing private encryption keys.

4. We demonstrate the benefits of our approach over the state of the art with a performance analysis based on the real-world constraints of LoRaWAN, a widely used LPWAN technology.

The rest of the paper is organised in the following order. We begin by discussing related works in Section 2, and proceed to provide a detailed description of our proposed model in Section 3. In Section 4, we present an implementation of the model using LoRaWAN and Hyperledger Fabric, with an evaluation and discussion of limitations of our integrated system in Section 5. Finally, we draw conclusions and discuss potential future work in Section 6.

2. Related Work

Our examination of related work begins with the challenges of rural healthcare monitoring, details the current technologies used for long-range communication, then looks at the uses of blockchain within healthcare, before summarising the combined challenges that we seek to address.

2.1. Rural Healthcare Monitoring

Providing appropriate communication infrastructure for electronic rural healthcare monitoring has been one of the most challenging issues from both the technological and economics points of view [14]. The geographical structure and population of these areas are the main reasons for this. Rural areas are often dominated by hilly terrain for large distances. Therefore, investing in the telecommunication infrastructure, such as 4th or 5th generation networks, in such areas has a low return on investment due to low population density and the complexity of installation for adequate coverage.

Alternatively, it has been suggested to exercise LPWAN technology, which lends itself to such settings due to low power transmission, while offering long-range communications among IoT devices. There are various standards bodies that are extensively working on developing LPWAN systems, such as the Institute of Electrical and Electronics Engineers (IEEE), the European Telecommunications Standards Institute (ETSI), the 3rd Generation Partnership Project (3GPP), the Internet Engineering Task Force (IETF), and the LoRa Alliance [15].

For example, a study by Dimitrievski in [3] showed the use of LoRa to carry healthcare data from rural areas combined with fog computing and the low Earth orbit (LEO) satellite connectivity to provide real-time data transmission. This work also proposed techniques for energy conservation utilising the external ultra-low-power timers that allow the device to be powered down, and showed its advantage to extend battery life in the order of tens of times. The fog system is a computation machine that is usually located between the cloud and the end devices to enable computing, communications, storage, and data management within the close vicinity of IoT devices. Therefore, in this IoT setting, the fog computation gives advantages to any delay sensitive devices to accumulate and process their retrieved data quickly (i.e., to achieve its real-time mode operation) rather than pushing through all data into the cloud system. Furthermore, the edge computing can be used to alleviate computing, storage, and bandwidth burdens of the system by allowing data processing within the edge devices when the resources of the IoT devices can be exploited to support that purpose [16].

Another study highlights a healthcare IoT architecture integrating blockchain and LoRa network to monitor patient health data securely [4]. To achieve real-time data transmission, the proposed model employs edge and fog devices to run the LoRa communication protocol whereby the edge devices with sensors attached on them collect data from healthcare data sources and subsequently send those relevant patient data to the upper fog layer using LoRa. To guarantee security, the data are stored in the interplanetary file system (IPFS) combined with blockchain technology. Finally, data monitoring and analytics for patients' health status were performed through mobile or web applications.

The delivery of healthcare and the associated monitoring can be considered a complex system, with many changing variables that could change patient outcomes and affect decision making. The digital twin concept was first envisaged to aid the management of complex manufacturing systems, and the definition by NASA has become widely accepted [17]. Therein, a digital twin is considered a virtualisation of a physical system, maintained via the supply of data, for example, via IoT. With adequate data and modelling, scenarios can be simulated with a digital twin in order to predict outcomes for the physical system, allowing optimisations or corrections to be made. Unsurprisingly, this has been also been applied to healthcare settings [18]. In our work, we focus on the twinning aspects of HIoT sensors that allow the twin to facilitate blockchain-enabled activities that would not otherwise be possible over constrained network connections. As such, we assume that the wider benefits of digital twins (such as scenario simulation and physical/virtual linkages) can be realised elsewhere in the applications that make up the healthcare system as a whole. While we propose to use a twin to enable tighter integration between the HIoT device and the blockchain, a complementary (but not mutually exclusive) further example of their use can be in consensus-based decision making, such as that described for smart transportation, by Sahal et al. [19].

2.2. LPWAN and LoRaWAN

The term LPWAN, or low-power wide area network, refers to technologies that have the capability to reach long-range communications but at the same time maintain the minimum use of energy [6]. This communications model is particularly important to accommodate the need for various small devices which inherit features such as low computational power, low memory, and low battery capacity. However, contrasting these advantages of LPWAN, the nature of wireless signals dictates that most LPWANs have a low bit rate. Although there are many LPWAN architecture available on the market, LoRa has found its acceptance in both wider communities and broad industry support compared to other similar technologies in this scope, such as narrow band IoT (NB-IoT), LTE machine-type communication (LTE-M), and Sigfox [15].

Despite its long-range coverage and low-cost deployment, the most notable advantage of using LoRa is its reliance on a license-free operating frequency privilege operated on the industrial, scientific, and medical (ISM) frequency sub-band. The use of the chirp spread spectrum (CSS) modulation scheme on its bidirectional communications results in a signal transceiver with low noise levels yet high interference resilience. Utilising this modulation technique, the LoRa data rate varies from 250 bps to 50 kbps depending on the allocated spreading factor (SF) and channel bandwidth. For example, a lower spreading factor allows a higher data rate at the expense of a lower transmission range. The maximum payload length is 64–255 bytes, including its 13 bytes payload header, depending on the data rate chosen.

Alongside the growth of the LoRa adoption, LoRaWAN appeared as a protocol stack built on top of the LoRa physical layer. With its data link layer protocols support, this LoRaWAN shapes the LoRa network architecture into a typical gateway-nodes model that consists of a gateway that acts as a bridge between nodes, network servers and application servers over a backhaul interface [20]. In this structure, nodes can transmit messages to other LoRa devices or to a gateway. Hence, a gateway bears a task to gather data from all authorised sensor nodes (i.e., the end-devices) and pushes forward those data to the application server through the network servers.

The core of the LoRaWAN network resides in the network servers which maintain connectivity, routing, and security among devices. Therefore, gateways and network servers retain an important function in the LoRaWAN architecture to coordinate all nodes in its network, while at the same time synchronising data transmission to avoid collisions. This function was specifically defined in LoRaWAN as the medium access control (MAC) operation. Depending on how nodes should schedule their downlink traffic, users can alleviate the efficiency of LoRaWAN networks by properly selecting the class in which

LoRaWAN networks are deployed. The LoRaWAN allows operation in one of three different classes: A, B and C. In Class A (ALOHA) communications, an end device has the capacity to start transmitting data at any moment, whereas in Class B (Beacon), an end device can only open a receive window and transmit data between a periodic beacon signal duration according to the network-defined schedule. In Class C (Continue), an end device constantly listens to the downlink signal from the network unless the end-device is transmitting data.

Additionally, LoRaWAN enforces the duty cycle to limit the transmission of large amounts of data that may consume the whole bandwidth of a channel, which would cause congestion in the networks. The duty cycle defines how much of the total time a device is allowed to transmit data per hour on a particular sub-band. For example, a 1% duty cycle restricts the total amount of time a device spends transmitting data to 36 s per hour. Realistically, the amount of the duty cycle applied to a LoRaWAN is governed by regional regulatory authorities [21]. Furthermore, the things network (TTN), a service providing a public LoRaWAN network, applies a more rigid rule to lessen congestion by employing a fair access policy. This policy, applied to each end device, restricts the device's uplink airtime to 30 s per day (24 h) and downlink messages to only 10 in number per day [22].

2.3. Blockchain Systems for Healthcare

A considerable number of works have proposed IoT-based healthcare systems to provide a more timely and cost-efficient remote patient-care system [23,24]. Among other advantages, the IoT system might be identified as a substitute for the common in-hospital health monitoring with the remote one, where patients might stay at home or live in a rural area. While the traditional client–server and cloud computing paradigm offers significant improvement to the way patient data are stored, it also raises security and privacy concerns. For example, it suffers from the issues of single point of failure, data privacy, centralised data administration, and system vulnerability. The major threats to this cloud model may include spoofing identity, tampering with clinical data, and the data leaks [8].

Recently, the blockchain system has presented itself as a novel technology that could have a role in preserving healthcare data security and maintaining patient privacy. In a blockchain system, multiple data transactions, such as a patient's treatment and medical history, are grouped together in a structure called a block [25]. Each block is uniquely identified by its hash and timestamp and is chained to the previous block by incorporating the hash value of the previous block, thus creating a chain of blocks. The hash algorithm that is used acts as a one-way function, meaning it is computationally infeasible to produce a different block that would result in the same hash, effectively making the contents of the chain immutable. As such, validation of each block before they are chained in a blockchain network is paramount, as they typically cannot be removed or edited. Validation of transactions and blocks is performed by a consensus mechanism, whereby a shared ledger of blocks in the blockchain network can only be altered by the agreement or consensus of a majority of members [26].

Blockchain technology has a promising future in the healthcare domain, as it can solve some inherent issues facing modern health-management systems. It has advantages as a tamper-resistant distributed ledger for recording healthcare data and transactions, and its high availability and resiliency that will deter system failures and other cyber attacks [27–29]. However, the integration of blockchain into the IoT system in the healthcare rural area use cases may encounter several challenges to solve.

IoT devices may have difficulty to process and store even the smallest elements in the blockchain. Secondly, the geographical structure of rural areas and decreased availability of reliable transmission due to a sporadic communications infrastructure being in place are the other two notable problems faced by researchers to initiate such a secure healthcare monitoring system. As far as this study being carried out, we noticed there are only a few reported articles aiming to propose a solution in this domain. For example, the work by Munagala in [30] showed a blockchain-based traceable data sharing method to secure medical data transfer by incorporating software defined networking (SDN) technology to remove the clone nodes, and the work called Lorachaincare in [4] proposed a model of healthcare monitoring system which combines the blockchain, fog/edge computing, and the LoRa communications protocol. Besides focusing on its applications, there are also some blockchain-based frameworks proposed for managing secure healthcare systems, such as a framework for regulating mobile health apps and governing their safe use [13] and a framework for an asthma healthcare system that challenges its adoption during the COVID-19 pandemic [12]. All of these listed works use blockchain for medical data that have been collected in cloud storage, while the security of data transmission from IoT devices to the cloud is handled by encryption. However, the outlined works do not consider that transmitting medical data in rural settings is problematic, and several steps are required for user authentication in order to commit valid transactions in the blockchain system.

2.4. Use Case Definition

This paper seeks to further the state of the art by uniquely combining blockchain, LPWAN and HIoT technologies to deliver the possibility of improved healthcare services in rural areas. As such, we must address the following:

- HIoT data must be transmissible over an LPWAN technology that can be feasibly and cost-effectively deployed into rural settings.
- Integrity of data must be preserved through the use of blockchain, allowing lifecycle stages of the data (e.g., creation, storage, and granting of access) to be recorded.
- Patient confidentiality must be maintained, ensuring that persistent data such as those stored on the blockchain do not pose a privacy risk, nor are data transmitted over LPWAN a confidentiality or integrity risk if intercepted or manipulated.
- Mechanisms to provide the above security guarantees should be achieved alongside real-time transmission, avoiding the deferral of actions, such as the creation of transactions, wherever possible.

The following sections propose how to achieve these goals both architecturally and in implementation with the currently available technology, using the enhancements that we contribute.

3. Proposed Model

In this section, we describe our HC^2 model at a high level, and address healthcare entity participation, data flows, blockchain integration and security considerations. As an architecture, it does not dictate specific security, blockchain or communication technology selections, which we instead explore an example of in Section 4.

3.1. High-Level Architecture

Our HC^2 model uses Patel's framework for medical image sharing via blockchain [31] as a basis for its architecture. In the said work, image data are shared with the patient and physicians and forms the patient's health record (PHR), with access granted via transactions in the blockchain (as discussed in Section 2.4). First, we re-interpret this architecture to suit the HIoT use case, depicted in Figure 1.

The primary difference between this and the prior work is that the data provider is a healthcare IoT solution, rather than an imaging centre. The collection of data is not concentrated into a single location but rather streamed in real-time, or close to it, from a wide area, using many individual sensor devices. The HIoT data provider contributes sensor data to the patient's PHR and allows for any physician, authorised by the patient, to access them in order to provide them with healthcare services.

The sensor data are not stored on the blockchain, nor is any personally identifiable information regarding the patient. The access model for the blockchain is private, meaning that only authorised identities can view blockchain data and potentially transact on it. However, keeping personally identifiable information (PII) and sensor data off-chain provides additional protection of that data.

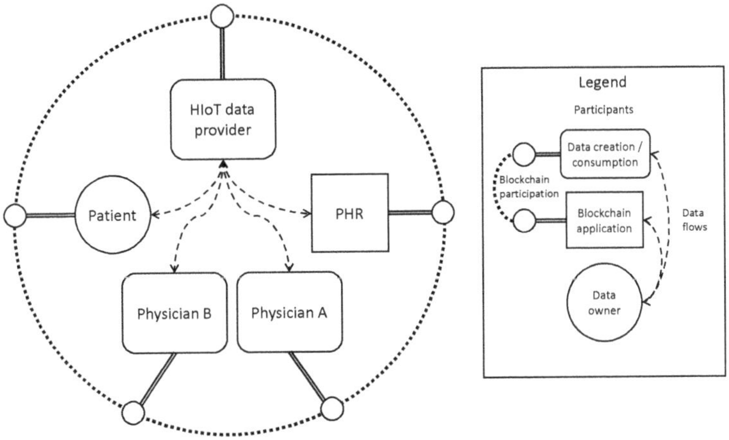

Figure 1. High-level architecture of HC^2.

3.2. HIoT Provider Entities and Data Flow

The HIoT provider component of the high-level architecture from Figure 1 comprises several entities that present unique challenges. We consider the following aspects:

- An HIoT sensor device which is paired with and attached to a patient for a duration of time. The device is expected to be portable and battery powered, for example, a health-monitoring watch or sensor pack.
- A "twin" of the HIoT sensor device used to represent the history and most-recent known state of the sensor device, regardless of connectivity status.
- Two communication methods between the device and its twin: one an LPWAN and one a PAN, where the LPWAN is low-bandwidth and possibly one-way, while the PAN is higher-bandwidth but intermittently available, for example, only when the patient visits a clinic.
- A data store for collected sensor and event data, obtained via the twin over either of the available communication methods.
- LPWAN connectivity is supported by base stations, uplinks to servers and subsequent internet connectivity to relay messages to the twin.
- PAN connectivity is achieved through short-range communication with an internet-connected bridging device, such as a phone over Bluetooth, or a physical docking station with USB or serial link.

The different types of participating components are represented with their own shapes. Potential data flows are represented by dashed lines, and the linkage to the blockchain, conceptually, is represented by the dotted circle around the diagram, to which the participants are all attached. Subsequent diagrams extend this concept further. For example, the participants responsible for maintaining the blockchain ledger and forming consensus are not represented at this stage.

The flow of data within this provision is visualised in Figure 2. The PAN is used for pairing, keying and detailed data transfers, whereas the LPWAN is used for small periodic data transmissions and events. For example, an HIoT device may monitor heart rate and ECG. After pairing, the device sends simple heart rate data over LPWAN every few minutes, along with an assessment of the patient's condition based on its own capabilities to analyse the heart rate and ECG data.

A healthcare provider may choose to act upon these data by calling the patient back to a clinic, or, under normal conditions, may await the next appointment. During the next visit, the detailed data logged by the HIoT device can be synchronised over PAN, via the twin, to the data store, and then immediately analysed for great insight.

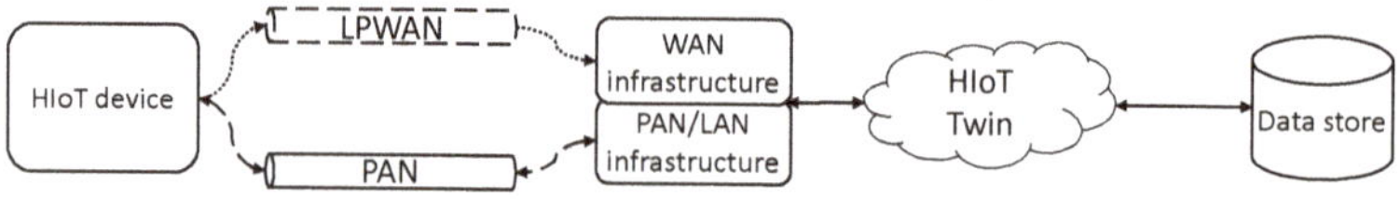

Figure 2. Hybrid-channel (LPWAN and PAN) architecture of HC^2.

3.3. Blockchain Integration

To integrate the HIoT provider into a decentralised blockchain, the following events must be recorded in the blockchain:

- Pairing between device and patient, whereby data that are generated by a device can be associated with the correct patient.
- Creation of data by the HIoT device to guarantee that a data record was produced by a legitimate, patient-paired source.
- Storage of data by an authorised data store to guarantee the retention of data that were generated by a device.
- Granting access to the data to additional entities to preserve a record of the management of permissions and, where necessary, encryption keys.

Pairing between the device and patient may be achieved through a transaction declaring the assignment of the device to the patient. The identifier for the device and patient must be sufficient to uniquely identify the relationship but does not need to be personally identifiable [32], and indeed this property may be necessitated by regulators now or in the future, who are advising on the best approaches to take [33].

While the HIoT provider may implement its own data store, this architecture does not preclude one or more external data stores being used, thus supporting a more decentralised approach to data handling. The data can be secured by a symmetric encryption key agreed between the sensor and store (discussed following subsection), and its creation, followed by its successful storage, recorded as transactions on the blockchain.

The events described above must be entered into the blockchain, and blockchain participants may refer to these in order to verify, authenticate, and progress to next steps in the process of providing healthcare. We focus mainly on the creation and storage of data in this paper (the middle two points), although all of these events can be considered blockchain transactions that must be recorded in a particular sequence in order for future actions to be allowed to proceed.

3.4. Security

The previous subsections alluded to several security considerations of the architecture, which we elaborate upon here. Firstly, sensor data are encrypted between the HIoT device and data store. To achieve this, a symmetric encryption method is used. If multiple data stores are used, then a key must be agreed between all of them and the device. To avoid overburdening the HIoT device, we assume that the data stores are responsible for coordinating key distribution among themselves.

This end-to-end encryption means that the device's twin cannot access the sensor data. It may store and forward the encrypted copy of the data, but will not possess the key needed to decrypt it. Data transferred over LPWAN or PAN are subject to this encryption, meaning the security of the WAN infrastructure or PAN link-layer poses no risk to the data's confidentiality.

The blockchain is largely responsible for protecting the integrity and availability of the data. Firstly, the creation of the data at the device is recorded as a transaction, verified by a signature that is cryptographically bound to the device's private key and associated identity. Similarly, the data store's acknowledgement of the receipt of the data has the same integrity assurances based on its own private key and identity information. Despite possessing the symmetric key used for data encryption, the data store cannot create data for itself, as it cannot sign a valid transaction representing the creation of data because its identity is not authorised to do so. Transformative processing of the data by other blockchain-enabled applications (for example, creating new data based upon analysis of the sensor data) remains possible and can be recorded as additional transactions, although the details of this are outside the scope of this work.

In terms of data availability, the loss of data over LPWAN can be established upon the synchronisation of data over PAN. At such a point in time, the device may verify that data it transmitted were correctly transacted, or the twin may observe the presence of records on the device that should have been received over LPWAN but were not. The cause may not be immediately knowable, but network outages, range issues or malicious interference can then be investigated. Finally, by agreeing on an expected data transmission interval, the twin may notify the HIoT provider system of missed data.

In summary, end-to-end encryption between device and data store provides confidentiality; blockchain transactions provide integrity and non-repudiation; the redundancy of communication channels (LPWAN + PAN) combined with the persistent presence/monitoring provided by the device's twin improves the detectability of availability issues; and the support for multiple external data stores improves the data's availability thereafter.

4. Implementation

To validate the architecture, we now discuss how it can be implemented in a realistic representation of our rural healthcare use case, under the constraints of contemporary HIoT devices, communication technologies, blockchain implementations and supporting software capabilities. First, we detail and justify our selections, then describe how the architecture can be realised within the technical constraints of the selections. Table 1 describes our selections, justifications for the choices, limitations/drawbacks and similar potential alternatives.

These technology selections pose challenges for how the components can fit the architecture of Figure 1 whilst achieving the requirements defined throughout in Section 3 in line with our use case. These are resolved in turn with the refinements detailed in this section, using proofs of concept where appropriate. Code for relevant proofs of concept, which are also used for data gathering used in Section 5.1, are collected into a group of repositories on GitLab [34].

Table 1. Technology selections made for HC^2 concept.

Component	Choice	Justification	Limitations	Alternatives
Sensor device	Micro-controller	Widely used for IoT-type devices. Relatively low cost. Capable of real-time sensor data acquisition.	Small amount of RAM and flash. Low processing power.	Smartphone or SBC with sensor attachments.
LPWAN	LoRaWAN	Multi-km range. Ability to create own infrastructure or use third party.	Limited or no downlinking. Very small uplink payloads and low duty cycles.	Narrow band IoT (NB-IoT), Weightless, Category M1 (Cat M1).
PAN	UART	Simplest communication method that can also be encapsulated within appropriate wireless protocols such as Bluetooth Serial Port Profile (SPP). Multi-kilobit to megabit transfer speeds are adequate for bulk data transfer.	Requires cable connection or dock to enable connection to twin.	Wi-Fi (LAN), ZigBee, Bluetooth low energy (BLE), serial peripheral interconnect (SPI), inter-integrated circuit (I2C).

Table 1. *Cont.*

Component	Choice	Justification	Limitations	Alternatives
Twin	Online deployment	Easier integration with LPWAN and connectivity to blockchain peers.	Link to device requires additional hardware with PAN + Internet capabilities. No offline capabilities.	Deployment onto LoRa gateways.
Blockchain	Hyperledger Fabric	Widely used for blockchain applications centring around business logic. Private access model. Certificate-based identities.	Transactions require round-trip communication with initiator.	Ethereum, Hyperledger Iroha.
Data store	MQTT historian	Commonly used protocol for IoT data simplifies collection of data. Multiple receivers can be implemented.	Blockchain application logic and data security must be additionally implemented and integrated.	Timescale, InfluxDB.

4.1. Blockchain Participation

We refer to the documentation for Hyperledger version 2.4, and, in particular, the "Key Concepts" topics, in describing the components relevant to this section [35]. Hyperledger Fabric uses public key infrastructure (PKI) to allow organisations to identify and enrol participants in the blockchain using certificates signed by certificate authorities (CAs). Fabric's blockchain comprises several types of participant:

- **Committing peers** are responsible for maintaining the ledger state.
- **Endorsing peers** execute chaincode or smart contracts (the state-changing code executed with a transaction's input arguments, described in [35]) to simulate a proposed transaction to determine if it would be valid.
- **Gateway peers** coordinate the dissemination of proposals to endorsing peers and the collection of endorsements on behalf of the proposer.
- **Orderer peers** construct blocks from endorsed transactions.
- **Clients** run applications that need to interact with Fabric peers to make transactions.
- **Admins** are able to perform privileged operations that change the configuration of Fabric and its peers, for example, by adding new organisations to a channel.

Identities for these participants are split into four groups: `client`, `peer`, `orderer` and `admin`. Most important to note is that not all participants maintain a copy of the blockchain or its current state. In the HC^2 model, this maps all participants of Figure 1 as *clients*, each possessing some form of application logic and an imperative to interact with Fabric to create, store and manage data.

Integrating the components of Fabric, alongside the other HC^2 components from Figure 2 into our architecture gives us a more detailed view, depicted in Figure 3. Here, we differentiate clients from peers. This creates a basis for visualising the sequence of transactions and flow of data in our rural HIoT use case.

The two biggest challenges from the limitations in Table 1 are the small packet size of LoRaWAN uplinks and the need for more than one round trip between client and Fabric peers to complete a transaction.

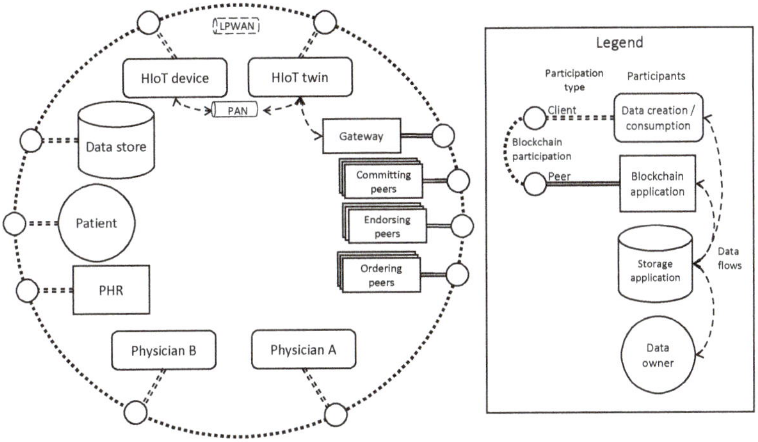

Figure 3. HC^2 architecture refined to accommodate selected technologies.

4.1.1. Payload Size

While various transmission profiles for LoRaWAN exist, the available payload size must accommodate the transmission of any data, in its encrypted form, along with a signature that might be usable in the blockchain. A 64-byte Elliptic Curve Digital Signature Algorithm (ECDSA) signature, as generated when using a P-256 (`secp256r1`) curve in Fabric, excludes most of the lower data rates from consideration. In the second Asia regulatory region (AS2 or AS923), which is of most interest to our research by virtue of locality, data rates providing 222 and 125 bytes of payload remain viable options [21].

Fabric uses `protobuf` to efficiently transfer messages between clients and peers. However, such messages still far exceed this payload limit, and when also faced with duty-cycle limitations as well, fragmentation is not practical.

We overcome this limitation by the pre-agreement of certain portions of Fabric messages, established between HIoT device and its twin over PAN (UART), prior to communication over LoRaWAN. Figure 4 shows agreements that take place between device, twin and data store. First, a template for Fabric messages is established between device and twin. Secondly, an encryption key is agreed between device and store, as discussed in Section 3.4, with the twin facilitating the transfer of the necessary key agreement messages. The fields and calculations that are relevant to the template agreement are detailed across Tables 2–4.

Table 2 lists the fields of a Fabric proposal that are agreed between the device and twin. This will be different for each device/twin pairing and each of their sessions but remain fixed between synchronisations.

The transmitted data are reduced down to that shown in Table 2, which is unique per transmission. Assuming messages may not be transmitted reliably and may not arrive in order necessitates the presence of a counter, C. These values are processed by the twin as indicated in Table 2 to complete the set of fields required to reconstruct the proposal message that was signed by the device.

Data transmission over LPWAN is reduced to three dynamic values: a counter C, encrypted data E, and a signature S. The latter two themselves are indexed by the counter value, and are all unique for each transmission. Within the 125-byte payload limit we selected for LoRaWAN, these values can be formed into a packet as shown in Table 5. The efficiency of this packet structure is discussed in Section 5.1.

From this packet, in combination with data prepared between the device and twin during PAN synchronisation, the twin is able to reconstruct the same message M that was signed by the device to produce its signature S. The twin can then submit this to the Fabric gateway on behalf of the device.

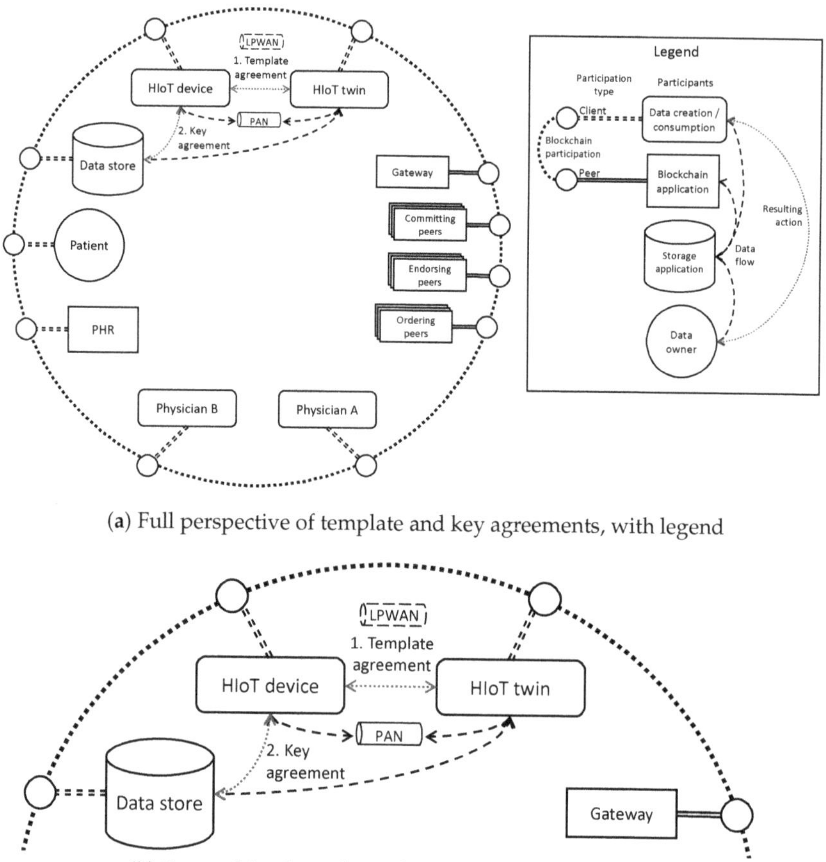

(a) Full perspective of template and key agreements, with legend

(b) Zoomed-in view of template and key agreements

Figure 4. Encryption key and proposal template agreement among device, twin and data store.

Table 2. Proposal fields agreed upon PAN synchronisation between device and twin.

Name	Symbol	Description
Header fields	H_x	Unchanging fields within the message header or headers of components within it.
Sync time	T_{sync}	Timestamp at synchronisation
Period	P	Time period stepped between transmissions
Identity	I_{dev}	Device's identity (certificate)
Seed	N_{seed}	Seed value used for per-transaction nonces
Args	$A_0 \dots A_n$	Unchanging chaincode arguments

Table 3. Dynamic data sent by device over LPWAN.

Name	Symbol	Description
Counter	C	Number of messages since last synchronisation
Data	E_c	Encrypted sensor data
Signature	S_c	Signature of proposal as computed device-side

Table 4. Re-computed data by twin based on dynamic data and pre-agreed field values.

Name	Symbol	Computation	Description
Nonce	N_c	$N_{\text{seed}} + C$	Proposal nonce based on seed and counter.
Timestamp	T_c	$T_{\text{sync}} + PC$	Timestamp at which data was sent based on counter value
Data hash	$A_{\texttt{hash}}$	$\texttt{Hash}(E_c)$	A chaincode argument dependent on received encrypted data
Transaction ID	X_c	$\texttt{Hash}(I_{\texttt{dev}}\|N_c)$	A unique ID for the proposed transaction based on nonce and creator.

Table 5. Payload format for HC^2 data over LoRaWAN.

Byte Position	0–1	2–65	66–124	***Total***
Length (bytes)	2	64	59	125
Purpose	Counter	Signature	Encrypted data	—

The precise construction of a transaction proposal is documented within Fabric's `protobuf` definitions [36]. However, referencing the values in Tables 2–4, we summarily describe the reconstructed proposal message in the partially abstracted form:

$$M = H_{\text{envelope}}|T_c|H_{\text{channel\_info}}|X_c|H_{\text{chaincode\_info}}| \\ |A_0|\ldots|A_n|A_{\text{hash}}|H_{\text{signature\_info}}|S_c \tag{1}$$

where the vertical bar symbol represents the concatenation of the values on either side of it, as an array of bytes. The exact ordering and encoding must respect that defined in the `protobuf` definitions [36].

4.1.2. Transaction Processing

Hyperledger Fabric ensures the integrity of transactions through endorsements. A client proposes a transaction, and the relevant chaincode is executed by several endorsing peers, and if valid, the peers sign and return endorsements to the client. The client can then combine these endorsements with the original proposal, signing them into a transaction which can be submitted, ordered and committed to the ledger, with the world state updated accordingly. In Fabric version 2.4 and above, the Fabric gateway can be used to distribute the client's proposal to necessary peers and collect the endorsement responses, prior to returning to the client for formation of the transaction submission.

This offloading is beneficial to the HIoT device, as it does not need to handle as much communication with Fabric. However, without refinement, transactions would only be proposed and endorsed but not committed, as the final endorsed transactions cannot be submitted to the orderer until the twin has an opportunity to return endorsements to the device, which we assume must happen over PAN. While data may be entered into the data store and the proposals/endorsements available in activity logs, this would delay the committing of any transactions to the blockchain.

To overcome this, we consider the chaincode for the data's early lifecycle in three parts:

1. The twin is responsible for submitting a transaction that *creates* the data.
2. The data store submits a transaction to register the *storage* of the data.
3. The device submits a transaction that *verifies* the data's origin.

It is counterintuitive to observe that the twin is responsible for the first transaction while the device is responsible for the last. Figure 5 shows the sequence of communications leading to transactions that achieve the desired outcome.

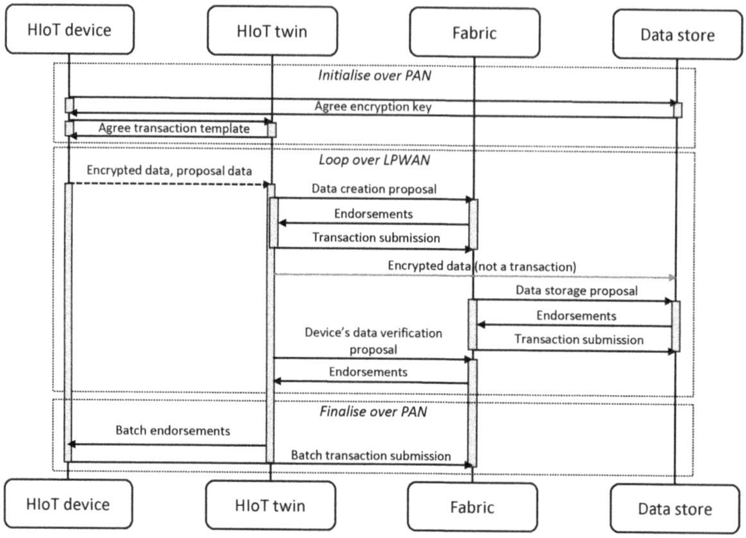

Figure 5. Sequence diagram of Fabric transactions representing HIoT sensor data early lifecycle.

The device and twin use the PAN to agree on a template for a proposal that verifies the data created by the device. Upon receiving a data packet over LPWAN, the twin can reconstruct this proposal for submission to the Fabric gateway. However, if it does so immediately, simulations of the proposal would fail, as it would refer to a non-existent data item.

Instead, the twin can propose its own transaction, using its own identity and private key, to execute chaincode that represents the creation of the data. The encrypted data are part of the payload they receive from the device, so they can produce a hash of it. The twin can also forward the same encrypted data to the data store, and the data store may have direct access to the data via the LPWAN's message queues (for example, an MQTT broker in the case of prominent LoRaWAN networks).

The data store, in possession of the encrypted data, should be able to decrypt them. It can also observe the twin's data creation transaction on the blockchain. Following this, it can submit a transaction that updates the status of this data item, indicating that it is intact and can be stored.

Observing the data store's transaction, the twin is now able to submit the device's proposal for endorsement. The endorsements can be collected, and once a PAN connection with the device is re-established, these can be relayed to the device for the creation of data validation transactions. The device's proposed transaction is created under the assumption that the other two transactions take place first. This is visible in the sequence diagram in the early activation of the device and twin at the start of the LPWAN loop portion, overlapping with two transactions by the twin and data store, before concluding with deactivation in the ending PAN communication portion. Multiple transactions may be batched together in this stage, as the LPWAN loop will have iterated many times between PAN-based synchronisations.

This approach more closely couples the existence of the data asset in the blockchain with the first transmission of it from the device, rather than deferring it until the next PAN connection. It also allows the data store to confirm the integrity and storage of the data before the device is finally able to confirm this also. It is a more fine-grained representation of the early stages of the data's lifecycle.

Figure 6 offers an alternative view of the same exchange. Data travel over LPWAN and PAN to the twin, which then interacts with both the Fabric gateway and the datastore, depicted by dashed arrows. In logical terms, this results in the device contributing data to the data store, along with the device, twin and data store contributing records to the blockchain that affect the PHR as depicted by dotted lines between said participants within Figure 6. The resulting data exchanges are numbered, with 1 being the sending of encrypted data from device to store, and 2, the record of the data's creation, which can be conceptualised as part of the patient's PHR on the blockchain. Subsequently, the data store can record its successful storage as item 3, and upon synchronisation over PAN, the final records of validity, 4, are entered into the blockchain to support the integrity of the PHR.

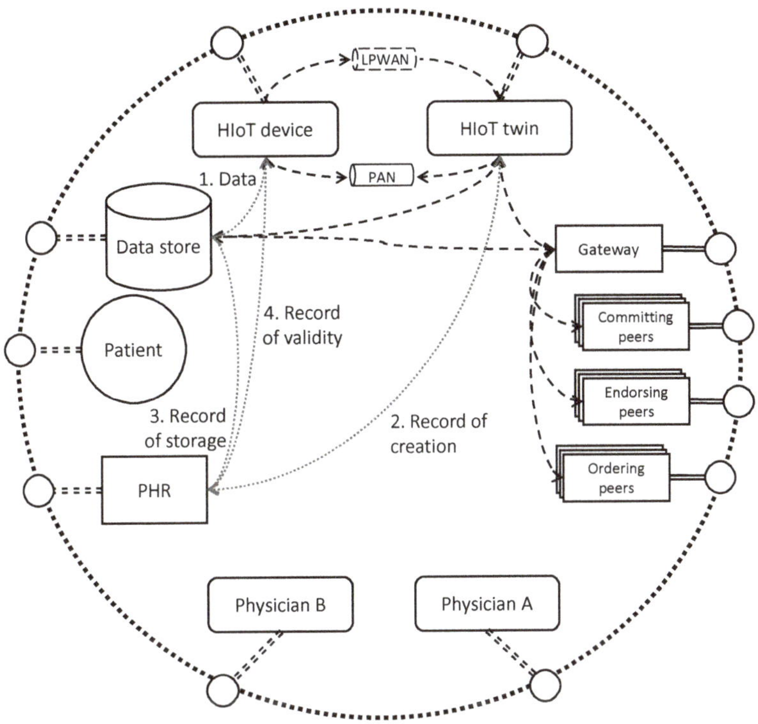

Figure 6. Paths of communication and resulting transactions in HIoT three-part early lifecycle. For legend, refer to Section 4.1.1.

4.2. Satisfying Security Requirements

In Section 3.4, we defined the security objectives sought by the HC^2 architecture in answer to our use case requirements from Section 2.4. Here, we explain how they are satisfied within the constraints of the technology choices made earlier within this section.

The primary security concerns and safeguards present in our system are summarised in Table 6 and explained in more detail in this section.

Table 6. CIA summary of HC^2. Some items discuss multiple security goals.

Confidentiality	Anonymity of data (Section 4.2.1), ownership of keys (Section 4.2.2), data keys (Section 4.2.3), off-chain data (Section 4.2.4), re-encryption (Section 4.2.5), forward secrecy (Section 4.2.7), post-quantum encryption (Section 4.2.8)
Integrity	Data keys (Section 4.2.3), post-quantum encryption (Section 4.2.8)
Availability	Missing data detection (Section 4.2.6)

4.2.1. Anonymity

No PII is transmitted by the device. Instead, the association between device and patient is maintained by reference in the blockchain. The identifiers used do not need to be directly attributable to a person; this can be resolved off-chain.

If device tracking is a concern, then additional countermeasures would be needed, such as changing device IDs and keys, for example, with each synchronisation. However, these are not considered further in this paper.

4.2.2. Device Key Ownership

The device can generate (or otherwise have injected) its own private key without being provided one by the Fabric CA, by leveraging the Fabric CA support for certificate signing requests (CSRs) during enrolment [37]. This precludes any possibility for impersonation of the device at the CA. If a TPM or a secure element is used on the device, the private key protection can be strengthened further [38].

Additionally, the twin does not share persistent key material with the device, so while both synchronise certain items of data (starting nonce, counter, and public keys), they cannot impersonate each other or tamper with signed messages. This remains the case despite the twin's ability to reconstruct signed messages from partial data transmitted from the device via LPWAN as described in Section 4.1.

4.2.3. Data Keys

During synchronisation between the device and twin, the twin also facilitates creating a secure session between the device and target data store. During this process, their respective identities are verified, and a symmetric encryption key is established for data transfer. The sensor data or events transmitted by the device are encrypted with this key, and thus the data store is the only other party able to decrypt it.

If the authentication encryption with associated data (AEAD) scheme, such as AES-GCM, is used, the encrypted data are accompanied by an authenticating tag that any party in possession of the symmetric key can use to verify the data integrity, independently of the message signature. In the case of 256-bit AES-GCM [39], the tag is 16 bytes, which must be included in the LPWAN transmission. Additionally, where AEAD is used, the integrity of the encrypted data is assured at this point, as well as later when the device verifies that it was stored.

4.2.4. Off-Chain Data

For privacy and efficiency, the sensor data are not stored on the blockchain. Instead, the hash of the encrypted data is stored. Any entity in possession of the encrypted data can verify that it is represented in the blockchain but can only decrypt them if in possession of the associated key.

Keeping data off-chain has the advantage of reducing the block sizes and growth rate of the blockchain by avoiding using the blockchain itself as a storage device. At significant scale, solutions such as IPFS may be used [4].

4.2.5. Re-Encryption of Data

The data store, or other accessors of the data, may re-encrypt the data to cease reliance on the key used between the device and data store. Provided the affected data assets can still be tracked, the integrity of the data in relation to the blockchain records can still be verified, provided the original encryption key is stored. This key should be stored with the equivalent protection as the re-encrypted data, for example, the original key could be stored encrypted by the new key.

4.2.6. Missing Data

Upon re-synchronisation, a device may additionally verify that the data it previously transmitted but also locally logged were indeed successfully stored. If they were not, a notification can be made. The data can then be provided during the synchronisation process instead. Although the benefits of real-time availability are lost, they will still eventually be available, and the evidence of their absence is provable.

4.2.7. Forward Secrecy

Forward secrecy is preserved through the use of ephemeral keys agreed between communicating parties. In the case of Fabric, this uses TLS. For the device and twin as well as device and data store, this may use DTLS [40] or EDHOC [41]. In all of these cases, the ephemeral keys used for data encryption are not related to the identifying keys of the participants. Thus, a successful attack on any of these ephemeral keys only affects data encrypted under that key. Each encrypted session must then be attacked independently.

In our use case, the session between the device and data store may last days or weeks, but the volume of data will not exceed a level that would pose a security risk through issues, such as initialisation vector reuse or exceeding data limits, which can affect AEAD ciphers, such as AES-GCM [39] (§8).

While we do not rely on the security of the LPWAN implementation for data or blockchain related activities, we remark that LoRaWAN agrees on a key during device activation [20] (pp. 62–63), [42] and that key management methods have been proposed or refined for it, too [43,44]. These could be more tightly integrated with blockchain identities and the Fabric CA/PKI if desired.

4.2.8. Post-Quantum Encryption

At the time of writing, post-quantum encryption (PQE) is a growing concern. Many of the cryptographic algorithms we use today are vulnerable to attack from the increased capabilities that quantum computers will eventually bring. New algorithms must be developed and adopted that are strong against conventional- and quantum-computing attacks but still feasible to run on conventional computers. For example, AES-256 encryption's security level is halved in the post-quantum area, and 256-bit EC-based key exchange and signing will be considered broken [45].

Institutions such as the USA's National Institute of Standards and Technology (NIST) continue to analyse and select candidate algorithms to address these concerns. However, these new algorithms are often more memory- and/or processor-intensive, which means they do not translate well to constrained IoT devices. Additionally, key and signature sizes in these PQE implementations can be significantly larger than those used today, making them unsuitable for use over LPWAN.

In early 2023 [46], NIST selected a family of lightweight cryptography algorithms targeting IoT and other constrained devices, named Ascon. These implement AEAD and hashing and so could substitute the existing algorithms that are part of the toolkits used in our demonstration codes. Additionally, one Ascon variant possesses some defences against quantum attacks; however, the NIST stance is that lightweight devices are less of a concern for PQE compared to systems responsible for long-term permanent storage.

In the case of our system, the data store may implement PQE and re-encrypt the data, using this to enhance protection. Forward secrecy remains in place on any data that were previously captured in transit for later decryption.

We do not explore the implications of PQE on algorithms used in the blockchain directly, as this is of interest to the community at large and not limited to HIoT. We do note, however, that the data are never stored in the blockchain, only a hash of the encrypted data (see Table 4).

5. Discussion and Limitations

In this section, we perform tests to analyse the efficiency of the HC^2 solution, discuss its performance and scaling properties, and consider the integration challenges faced when trying to develop an HC^2-enabled system.

5.1. Scaling and Integration Considerations

When deployed at scale, an HC^2 solution may encompass or interact with a variety of systems and many thousands of devices. In this section, we consider the efficiency of individual data packets, blockchain transactions, and the overall capacity of the blockchain, along with integration concerns. While we focus mainly on constraints relevant to regions that follow AS923 regulations, similar constraints must be considered in others, possibly with slightly differing results or optimal choices.

5.1.1. Data Payload Efficiency

At our proposed 125 byte LoRa payload (Table 5), 59 bytes are used by the encrypted sensor data, or 42.5% of the payload. If 16 bytes of that is also used for AEAD, 43 bytes of data remain, or 34.4%. In either case, more than half of the payload is used purely for blockchain-related data. If higher efficiency than this is required, then one must consider whether the benefits of the blockchain can be dispensed with, or substituted with a more lightweight alternative. Otherwise, a LoRa data rate that can accommodate a larger payload, or a different LPWAN technology altogether, may be preferred.

5.1.2. Fabric Payload Efficiency

This subsection examines the benefits brought by implementing the HC^2 scheme that we proposed when working within the transmission constraints of common LoRaWAN deployments. Our approach along with two alternatives are given as follows:

- **Hybrid Channel + Template**: The full HC^2 implementation, where we seek to send the bare-minimum non-templated data over LoRaWAN, relying on PAN, templates and the twin for data reconstruction and interactions with the Fabric gateway.
- **Hybrid Channel**: A simpler approach that still uses a PAN and twin to minimise LoRaWAN usage but uses a signed proposal generated and sent in full by the device.
- **Single Channel**: No PAN is used, and therefore messages for Fabric must be sent and received over LoRaWAN, even if a twin assists in transitioning between LoRaWAN and TCP/IP communication to the gateway.

An indicative set of data payloads is generated using our device and twin demonstration code [34] (Fabric samples: `scaling-data`), modified to output the length of the three messages that would be exchanged between the device and Fabric gateway (with or without twin assistance). They are the proposal, the signed endorsements for the proposal and finally the signed transaction. In the case of both hybrid channel variants, we assume that the endorsements and transaction are handled over PAN, which delays their processing but removes the need for LoRaWAN downlinks. However, for single channel, the downlinking of the endorsements would be required.

We chose a data size of 31 bytes, as this conveniently fit within our proof-of-concept use case whilst being a feasible length for the sensor data. It can be accommodated within a single LoRa packet in the Hybrid Channel + Template approach. When full Fabric messages are used, the message lengths vary due to the ASN.1 representation of signatures being 70–72 bytes long [47]. ECDSA signature values are represented as two signed integers, which may each need an additional octet to preserve their positive sign if the most-significant bit is set. This, combined with the headers in ASN.1, results in four possible lengths for an encoded signature. When determining the packet sizes in our demonstration code, we take the largest. This yields a proposal message of 1343 bytes, an endorsement message of 4709 bytes and a signed transaction of 4781 bytes.

Two LoRaWAN data rate profiles, DR5 and DR4, are used. In the higher data rate (DR5), a payload size of up to 222 bytes can be accommodated. However, HC^2 targets 125 byte payloads, which can be accommodated in both DR5 and DR4. We include both payload sizes and data rates in order to explore the effect that these choices have on the other transmission schemes that send full Fabric messages.

Figure 7 shows the number of LoRa packets needed when performing a single data transaction, that is, transmitting the encrypted sensor data, along with any signatures or other message components for Fabric, depending on the transmission scheme. HCT sets the baseline of using a single uplink packet, while HC uses more but benefits from the larger payload size available within DR5. The SC scheme, however, requires significantly more uplinks and also requires downlinks. Even with 222 byte payloads, the number of LoRa packets approaches fifty for each data transaction.

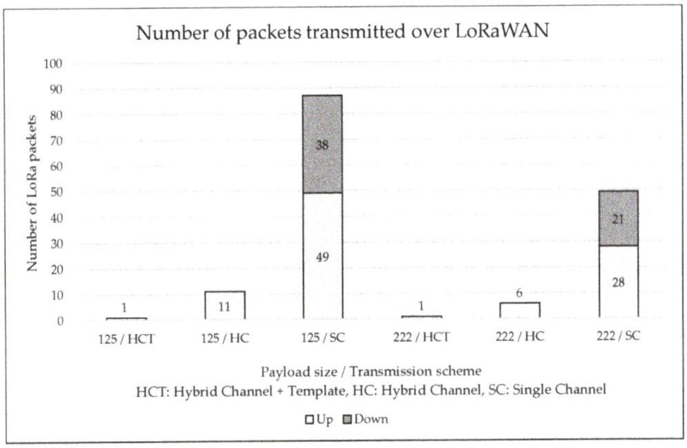

Figure 7. Packets transmitted in both directions (uplink and downlink) for each transmission scheme, using two available payload size limits over LoRa.

These data motivate the deferment of finalising transactions, as it significantly reduces LoRa utilisation (or allows data to be sent more frequently). If the three transactions proposed for HC^2 are used in the HCT and HC cases, this deferment is mitigated somewhat. With the assistance of templates, a further order of magnitude reduction in LoRa utilisation is achieved for 125 byte payloads. In the best case, HC^2 achieves an 87-times reduction in packets transmitted, thanks to combined message efficiencies and deferral.

Next, we examine the impact that duty cycle limits and fair access policies have on the amount of data that can be transmitted. As discussed in the literature view, regions impose limits on the amount of airtime that a LoRa device is able to occupy, in order to share the available bandwidth more fairly. Similarly, LoRaWAN providers may impose additional limits, such as even stricter duty cycles and limits on downlinks.

We take 1% as the duty cycle limit, which is applied in various global regions, including the AS923 region that is most relevant to the authors. To consider fair access policies, we use the things network (TTN), which limits airtime to 30 s per device per day and a maximum of ten downlinks. Figure 8 shows the results of applying these constraints to our selected payload sizes, data rates and transmission schemes.

To the left, Figure 8a shows that each of the three schemes are separated by an order of magnitude with respect to how many data transactions can be made per day. For HCT, the worst case is 2160 per day, or a data transaction every 40 s. HT is limited to every four minutes, whilst SC is limited to almost 33 min.

On the right, Figure 8b applies the TTN limits. Both HCT and HC are affected by the stricter airtime limits. SC incurs a much greater penalty due to the downlink limit, leading to two orders of magnitude, separating it from HC. In the 222-byte DR5 case, a small improvement is achieved due to more efficient use of the ten available downlinks per day because the larger transaction packets can be sent in fewer segments when the payload is larger; however, it does not substantially impact the results, as the transaction packets still exceed the payload size by several times.

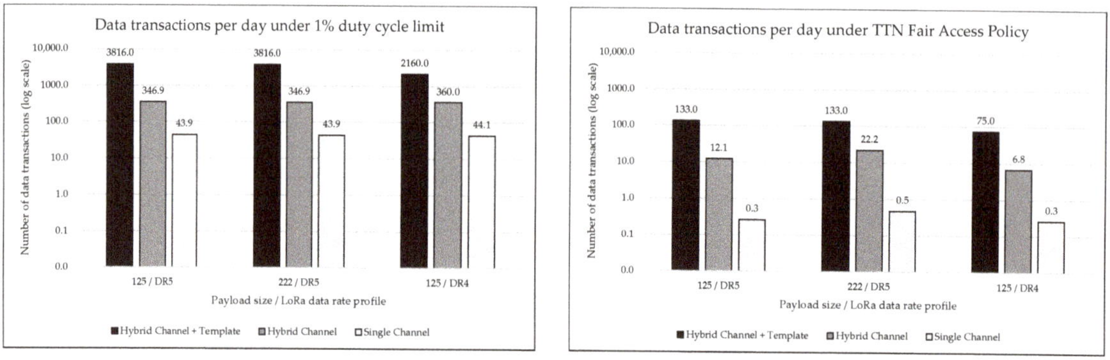

(**a**) With 1% duty cycle limit (**b**) Under TTN limits, including maximum 10 downlinks

Figure 8. Maximum data transactions per day under selected transmission schemes, payload limits and LoRa data rate profiles. Scales are consistent between sub-figures.

These data demonstrate the feasibility of conducting Fabric transactions over LoRaWAN, making a case for avoiding downlinks where possible. Additionally, by utilising the HC^2 scheme, a further improvement can be obtained. Looking beyond just the data transfer limitations, there are also likely to be significant energy savings to be had for the device. Assuming a data transmission period is chosen that is not close to the limit, the full HC^2 limitation will use far less radio airtime, preserving battery life and potentially allowing for more advanced computation to be performed on the device with the spare energy.

5.1.3. Blockchain Utilisation

Using 3000 transactions per second (tps) as the baseline performance of Hyperledger Fabric [48], and the DR5 data rate (spreading factor 7 with 125 kHz bandwidth) combined with the TTN usage policy yielding 5.5 messages per hour per device [21,22], we calculate that over 650 thousand devices could be supported by the solution in terms of blockchain throughput, assuming each data transfer produces three transactions. Hyperledger Fabric can be scaled to higher transaction throughputs than this [48], although we speculate that any particular healthcare ecosystem on a single blockchain is not likely to exceed one million active HIoT devices. Various scaling enhancements, such as side chains, can be employed [49] should they be necessary and can be implemented with existing frameworks, including Fabric.

Another scaling limitation is the number of devices supported by each LoRa concentrator. This is affected by the amount of airtime each device's transmissions will use as well as the number of available channels, which is governed by concentrator support and regional regulations. Continuing to use the DR5 data rate, eight uplink channels is a moderate selection that can be accommodated by most regions and concentrators.

Equation (2) is a simple equation to determine the number of devices D, that can be accommodated on a channel, given a packet airtime A and a periodicity of transmission P for each device:

$$D = \left\lfloor \frac{P}{A} \right\rfloor \tag{2}$$

In the AS923 region, the DR5 date rate requires $A = 225.5\,\text{ms} = 0.2255\,\text{s}$ of airtime for each uplink packet of the 125 byte payload size used for HC^2 throughout this section. If devices each transmit at a five-minute period, $P = 300\,\text{s}$, then applying Equation (2), we find an ideal upper limit of $D = 1330$ devices that can be accommodated by a single concentrator. However, in a rural setting, the device density is likely to be lower, negating this concern. A deployment of 488 concentrators at the full density of 1330 devices per concentrator (with no overlap in reception) would be needed to approach the Fabric blockchain transaction limit.

While the three-part early lifecycle approach (Section 4.1.2) increases transactions on the blockchain by 3× versus a single deferred transaction, the dispersal of endorsements helps to reduce bottlenecks in the system. At synchronisation time, endorsements have already been collected by the twin for the device to batch together, reducing the number of transaction submission messages to Fabric. Provided the synchronisation of all participating HIoT devices is not performed at the same time, excessive load should be avoidable.

5.2. Integration Challenges

Blockchain technology remains an active area of research and development, and so future changes to blockchain technologies may create new challenges for integration into architectures, such as HC^2. In the case of the selections made in Table 1, we note two integration challenges that are avoided based on the present state of Hyperledger Fabric.

The first such challenge is the introduction of the *epoch* value into transactions. This numerical value represents the height of the block into which the transaction will go (i.e., the number of blocks in blockchain). If a transaction's epoch value is lower than the current height, the orderer will not include it. While a field in proposals and transactions is defined for this, presently, it is set to the value zero and thus is not enforced. Hyperledger Fabric JIRA issue FAB-1430 https://jira.hyperledger.org/browse/FAB-1430 (accessed on 27 April 2023) proposes checking of this epoch value, however the status of the work is "won't do" and the issue is closed. Therefore, at the time of writing, this feature is not expected to be implemented by the orderer, but given that provision exists within the message framework, if that decision is reversed, it would have negative implications for the proposed HC^2 implementation, as the IoT device would have way to track the current block height.

Secondly, Fabric supports mutual TLS, where the secure connection between the client and peer (i.e., twin and gateway) verifies both participants' identities. While this provides additional security in some contexts, without addition considerations, its use may prevent the twin from submitting proposals on behalf of the device, as the identity bound to the TLS channel would not match the identity of the submitted message. Additional application logic in Fabric and/or extensions within the issued certificates that associate device and twin with each other could overcome this without compromising the intent of mutual TLS.

Looking at the security integration between device and data store, using AEAD over the symmetric encryption session affects the data payload efficiency as discussed in Section 5.1.1. Under some circumstances it may be desirable to remove this, relying instead upon verification of the data payload by checking the signatures of blockchain messages from the device and twin. However, doing so requires tighter integration between the logic used in the blockchain and the logic of the storage application, which may not be desirable. We see this as a trade-off for which the decision may vary depending on use case and constraints.

The transfer of data through the HIoT system needs to be compatible with the integration with HC^2. Section 3 does not define any strict underlying requirements in the HIoT system, and Section 4 provides an example implementation that is refined based on the combination of LoRaWAN and Hyperledger Fabric. The messaging patterns, both in the model and implementation, may benefit from representation in a clearer form, such as that proposed in [50]. For example, TTN provides an MQTT data API for its LoRaWAN network, which can already represented with the «`MQTT`» stereotype in [50], and a similar stereotype may be created for Fabric gateway interactions. The templating implemented in HC^2 between device and twin can be formalised with «`ContentEnricher`» and «`EnvelopeWrapper`» to represent the transformation of messages as they transit between the device/twin and, subsequently, Fabric gateway. The different messaging formats and delivery methods over WAN and PAN should also be well-defined. A full UML definition of these patterns, or suitable equivalent, is beyond the scope of this paper, however.

6. Conclusions

In this paper, we made the case for tightly integrating HIoT data with the blockchain. This benefits the patient and healthcare provider by ensuring data integrity, increasing trust between parties. We also showed how the data can be transacted and stored without undue risk to patient confidentiality, such as the disclosure of PII or unencrypted storage/transit of sensor data.

We focused on how this integration can be delivered in rural settings, where network connectivity may be limited for potential patients but for whom the benefits of HIoT devices are still sought. A combination of communication channels, LPWAN and PAN, into hybrid-channel connectivity, along with a novel use of a digital twin and transaction templates, enables blockchain participation without overburdening devices or networks.

Our results show that with an appropriate implementation of the HC^2 model, blockchain-backed data transactions become feasible where they would not otherwise be, such as using LoRaWAN in combination with the Hyperledger Fabric blockchain. An 87x reduction in the number of LoRa transmissions needed is shown, allowing two orders of magnitude more data to be transferred under normal LoRaWAN operating constraints. Proof-of-concept code is provided that can serve as the basis for a full implementation, or as a comparison point for alternative solution proposals.

Future Work

For future work, we envisage two main pursuits. Firstly, a full implementation of the platform described in Section 4 to validate its effectiveness and explore its performance under real-world usage. Secondly, the exploration of alternative implementations of the HC^2 model, such as by using a different LPWAN technology or another blockchain system. Both of these areas of work would help to increase the understanding of the practical applications of our model and technical decisions that can maximise its benefits. Additionally, formalising the messaging patterns present in the hybrid-channel approach of HC^2, both for models and any implementations, may aid in integration efforts.

Author Contributions: Conceptualization, J.J., S.K. and T.H.; Methodology, J.J., S.K. and T.H.; Formal analysis, J.J. and S.K.; Writing—original draft preparation, J.J. and S.K.; Writing—review and editing, J.J., S.K. and T.H.; Funding acquisition, J.J., S.K. and T.H. All authors have read and agreed to the published version of the manuscript.

Funding: This work was supported by the competitive James Cook University Singapore—Australia Cross-Campus Collaboration Grant Scheme 2022.

Data Availability Statement: The data analysed in this work are generated from modifications to the Hyperledger Fabric SDK and application source code. These modifications, including the means to generate the data, are made available via a group of GitLab projects [34]. No other data collection exercises were performed.

Acknowledgments: The authors give thanks to Belinda Lee and Kevin Wang for their reserach administration support. We also wish to thank the reviewers for their time and feedback.

Conflicts of Interest: The authors declare no conflict of interest.

References

1. Qadri, Y.A.; Nauman, A.; Zikria, Y.B.; Vasilakos, A.V.; Kim, S.W. The Future of Healthcare Internet of Things: A Survey of Emerging Technologies. *IEEE Commun. Surv. Tutor.* **2020**, *22*, 1121–1167. [CrossRef]
2. Li, J.; Carayon, P. Health Care 4.0: A vision for smart and connected health care. *IISE Trans. Healthc. Syst. Eng.* **2021**, *11*, 171–180. [CrossRef] [PubMed]
3. Dimitrievski, A.; Filiposka, S.; Melero, F.J.; Zdravevski, E.; Lameski, P.; Pires, I.M.; Garcia, N.M.; Lousado, J.P.; Trajkovik, V. Rural healthcare IoT architecture based on low-energy LoRa. *Int. J. Environ. Res. Public Health* **2021**, *18*, 7660. [CrossRef] [PubMed]
4. Dammak, B.; Turki, M.; Cheikhrouhou, S.; Baklouti, M.; Mars, R.; Dhahbi, A. LoRaChainCare: An IoT Architecture Integrating Blockchain and LoRa Network for Personal Health Care Data Monitoring. *Sensors* **2022**, *22*, 1497. [CrossRef] [PubMed]
5. Citoni, B.; Ansari, S.; Abbasi, Q.H.; Imran, M.A.; Hussain, S. Comparative Analysis of an Urban LoRaWAN Deployment: Real World Versus Simulation. *IEEE Sen. J.* **2022**, *22*, 17216–17223. [CrossRef]

6. Sun, Z.; Yang, H.; Liu, K.; Yin, Z.; Li, Z.; Xu, W. Recent Advances in LoRa: A Comprehensive Survey. *ACM Trans. Sens. Netw.* **2022**. [CrossRef]
7. Hathaliya, J.J.; Tanwar, S. An exhaustive survey on security and privacy issues in Healthcare 4.0. *Comput. Commun.* **2020**, *153*, 311–335. [CrossRef]
8. Chenthara, S.; Ahmed, K.; Wang, H.; Whittaker, F. Security and Privacy-Preserving Challenges of e-Health Solutions in Cloud Computing. *IEEE Access* **2019**, *7*, 74361–74382. [CrossRef]
9. Sun, P. Security and privacy protection in cloud computing: Discussions and challenges. *J. Netw. Comput. Appl.* **2020**, *160*, 102642. [CrossRef]
10. Pennino, D.; Pizzonia, M.; Vitaletti, A.; Zecchini, M. Blockchain as IoT Economy Enabler: A Review of Architectural Aspects. *J. Sens. Actuator Netw.* **2022**, *11*, 20. [CrossRef]
11. Wang, Y.; Su, Z.; Zhang, N. BSIS: Blockchain-Based Secure Incentive Scheme for Energy Delivery in Vehicular Energy Network. *IEEE Trans. Ind. Inform.* **2019**, *15*, 3620–3631. [CrossRef]
12. Farooq, M.S.; Suhail, M.; Qureshi, J.N.; Rustam, F.; de la Torre Díez, I.; Mazón, J.L.V.; Rodríguez, C.L.; Ashraf, I. Consortium Framework Using Blockchain for Asthma Healthcare in Pandemics. *Sensors* **2022**, *22*, 8582. [CrossRef]
13. Marbouh, D.; Simsekler, M.C.E.; Salah, K.; Jayaraman, R.; Ellahham, S. A Blockchain-Based Regulatory Framework for mHealth. *Data* **2022**, *7*, 177. [CrossRef]
14. Semwal, N.; Mukherjee, M.; Raj, C.; Arif, W. An IoT based smart e-health care system. *J. Inf. Optim. Sci.* **2019**, *40*, 1787–1800. [CrossRef]
15. Mekki, K.; Bajic, E.; Chaxel, F.; Meyer, F. A comparative study of LPWAN technologies for large-scale IoT deployment. *ICT Express* **2019**, *5*, 1–7. [CrossRef]
16. Xue, H.; Chen, D.; Zhang, N.; Dai, H.N.; Yu, K. Integration of blockchain and edge computing in internet of things: A survey. *Future Gener. Comput. Syst.* **2023**, *144*, 307–326. [CrossRef]
17. Glaessgen, E.; Stargel, D. The digital twin paradigm for future NASA and US Air Force vehicles. In Proceedings of the 53rd AIAA/ASME/ASCE/AHS/ASC Structures, Structural Dynamics and Materials Conference 20th AIAA/ASME/AHS Adaptive Structures Conference 14th AIAA, Honolulu, HI, USA, 23–26 April 2012; p. 1818. [CrossRef]
18. Liu, Y.; Zhang, L.; Yang, Y.; Zhou, L.; Ren, L.; Wang, F.; Liu, R.; Pang, Z.; Deen, M.J. A Novel Cloud-Based Framework for the Elderly Healthcare Services Using Digital Twin. *IEEE Access* **2019**, *7*, 49088–49101. [CrossRef]
19. Sahal, R.; Alsamhi, S.H.; Brown, K.N.; O'Shea, D.; McCarthy, C.; Guizani, M. Blockchain-Empowered Digital Twins Collaboration: Smart Transportation Use Case. *Machines* **2021**, *9*, 193. [CrossRef]
20. Sornin, N.; Yegin, A. *LoRaWAN™ 1.1 Specification 2*; Technical Report; LoRa Alliance, Inc.: Fremont, CA, USA, 2017.
21. van Bentem, A. Airtime Calculator for LoRaWAN. Available online: https://avbentem.github.io/airtime-calculator/ttn/as923/125 (accessed on 17 March 2023).
22. The Things Network. LoRaWAN Duty Cycle-Fair Use Policy. Available online: https://www.thethingsnetwork.org/docs/lorawan/duty-cycle/#fair-use-policy (accessed on 17 March 2023).
23. Baker, S.B.; Xiang, W.; Atkinson, I. Internet of Things for Smart Healthcare: Technologies, Challenges, and Opportunities. *IEEE Access* **2017**, *5*, 26521–26544. [CrossRef]
24. YIN, Y.; Zeng, Y.; Chen, X.; Fan, Y. The internet of things in healthcare: An overview. *J. Ind. Inf. Integr.* **2016**, *1*, 3–13. [CrossRef]
25. Gordon, W.J.; Catalini, C. Blockchain Technology for Healthcare: Facilitating the Transition to Patient-Driven Interoperability. *Comput. Struct. Biotechnol. J.* **2018**, *16*, 224–230. [CrossRef] [PubMed]
26. Haleem, A.; Javaid, M.; Singh, R.P.; Suman, R.; Rab, S. Blockchain technology applications in healthcare: An overview. *Int. J. Intell. Netw.* **2021**, *2*, 130–139. [CrossRef]
27. Maftei, A.A.; Mutescu, P.M.; Popa, V.; Petrariu, A.I.; Lavric, A. Internet of Things Healthcare Application: A Blockchain and LoRa Approach. In Proceedings of the 2021 International Conference on e-Health and Bioengineering (EHB), Iasi, Romania, 18–19 November 2021; pp. 1–4. [CrossRef]
28. Ahmad, R.W.; Salah, K.; Jayaraman, R.; Yaqoob, I.; Ellahham, S.; Omar, M. The role of blockchain technology in telehealth and telemedicine. *Int. J. Med. Inform.* **2021**, *148*, 104399. [CrossRef] [PubMed]
29. Adere, E.M. Blockchain in healthcare and IoT: A systematic literature review. *Array* **2022**, *14*, 100139. [CrossRef]
30. Munagala, N.V.L.M.K.; Rani, A.D.; Reddy, D.V.R.K. Blockchain-Based Internet-of-Things for Secure Transmission of Medical Data in Rural Areas. *Comput. J.* **2022**. [CrossRef]
31. Patel, V. A framework for secure and decentralized sharing of medical imaging data via blockchain consensus. *Health Inform. J.* **2019**, *25*, 1398–1411. [CrossRef]
32. Li, C.T.; Weng, C.Y.; Lee, C.C.; Wang, C.C. A Hash Based Remote User Authentication and Authenticated Key Agreement Scheme for the Integrated EPR Information System. *J. Med. Syst.* **2015**, *39*, 144. [CrossRef]
33. Agencia española de protección de datos. *Introduction to the Hash Function as a Personal Data Pseudonymisation Technique*; Technical Report; European Data Protection Supervisor: Brussels, Belgium, 2019.
34. Kerrison, S. HIoT Blockchain. Available online: https://gitlab.com/hiot-blockchain (accessed on 23 March 2023).
35. Hyperledger. Hyperleder Fabric: Key Concepts. Available online: https://hyperledger-fabric.readthedocs.io/en/release-2.4/key_concepts.html (accessed on 20 March 2023).

36. Sykes, M.; Yellick, J.; Enyeart, D. Hyperledger Fabric gRPC Service Definitions-Proposal. Available online: https://github.com/hyperledger/fabric-protos/blob/f0d57a53cb997351d8066fd6ab24cb48da1155b2/peer/proposal.proto (accessed on 17 March 2023).
37. Hyperledger. Hyperledger Fabric SDK for node.js. Available online: https://hyperledger.github.io/fabric-sdk-node/release-1.4/FabricCAClient.html#enroll__anchor (accessed on 20 March 2023).
38. Bouazzouni, M.A.; Conchon, E.; Peyrard, F. Trusted mobile computing: An overview of existing solutions. *Future Gener. Comput. Syst.* **2018**, *80*, 596–612. [CrossRef]
39. Dworkin, M.J. *SP 800-38D. Recommendation for Block Cipher Modes of Operation: Galois/Counter Mode (GCM) and GMAC*; Technical Report; National Institute of Standards & Technology: Gaithersburg, MD, USA, 2007. [CrossRef]
40. Rescorla, E.; Tschofenig, H.; Modadugu, N. The Datagram Transport Layer Security (DTLS) Protocol Version 1.3. Technical Report; RFC Editor. April 2022. Available online: https://datatracker.ietf.org/doc/html/draft-carpenter-rfc-citation-recs-01#section-5.2 (accessed on 27 April 2023).
41. Selander, G.; Mattsson, J.P.; Palombini, F. *Ephemeral Diffie-Hellman Over COSE (EDHOC)*; Internet-Draft draft-ietf-lake-edhoc-19; Internet Engineering Task Force: Fremont, CA, USA, 2023; *Work in Progress*.
42. The Things Network. LoRaWAN Security. Available online: https://www.thethingsnetwork.org/docs/lorawan/security/ (accessed on 17 March 2023).
43. Sanchez-Iborra, R.; Sánchez-Gómez, J.; Pérez, S.; Fernández, P.J.; Santa, J.; Hernández-Ramos, J.L.; Skarmeta, A.F. Enhancing LoRaWAN Security through a Lightweight and Authenticated Key Management Approach. *Sensors* **2018**, *18*, 1833. [CrossRef]
44. Chen, X.; Lech, M.; Wang, L. A Complete Key Management Scheme for LoRaWAN v1.1. *Sensors* **2021**, *21*, 2962. [CrossRef]
45. Bernstein, D.; Lange, T. Post-quantum cryptography. *Nature* **2017**, *549*, 188–194. [CrossRef]
46. Boutin, C. NIST Selects 'Lightweight Cryptography' Algorithms to Protect Small Devices. Available online: https://www.nist.gov/news-events/news/2023/02/nist-selects-lightweight-cryptography-algorithms-protect-small-devices (accessed on 20 March 2023).
47. Housley, R.; Polk, T.; Bassham, L. Algorithms and Identifiers for the Internet X.509 Public Key Infrastructure Certificate and Certificate Revocation List (CRL) Profile; Technical Report; RFC Editor. April 2002. Available online: https://datatracker.ietf.org/doc/html/draft-carpenter-rfc-citation-recs-01#section-5.2 (accessed on 27 April 2023).
48. Gorenflo, C.; Lee, S.; Golab, L.; Keshav, S. FastFabric: Scaling Hyperledger Fabric to 20,000 Transactions per Second. In Proceedings of the 2019 IEEE International Conference on Blockchain and Cryptocurrency (ICBC), Seoul, Republic of Korea, 14–17 May 2019; pp. 455–463. [CrossRef]
49. Kim, S.; Kwon, Y.; Cho, S. A Survey of Scalability Solutions on Blockchain. In Proceedings of the 2018 International Conference on Information and Communication Technology Convergence (ICTC), Jeju, Republic of Korea, 17–19 October 2018; pp. 1204–1207. [CrossRef]
50. Górski, T. UML Profile for Messaging Patterns in Service-Oriented Architecture, Microservices, and Internet of Things. *Appl. Sci.* **2022**, *12*, 12790. [CrossRef]

electronics MDPI

Article

A Distance Vector Hop-Based Secure and Robust Localization Algorithm for Wireless Sensor Networks

Rameez Asif [1,*], Muhammad Farooq-i-Azam [2], Muhammad Hasanain Chaudary [3], Arif Husen [3,4] and Syed Raheel Hassan [1]

1 School of Computing Sciences, University of East Anglia, Norwich NR4 7TJ, UK
2 Department of Electrical and Computer Engineering, COMSATS University Islamabad, Lahore Campus, Lahore 54000, Pakistan
3 Department of Computer Science, COMSATS University Islamabad, Lahore Campus, Lahore 54000, Pakistan
4 Department of Computer Science and Information Technology, Virtual University of Pakistan, Lahore 54000, Pakistan
* Correspondence: rameez.asif@uea.ac.uk

Abstract: Location information of sensor nodes in a wireless sensor network is important. The sensor nodes are usually required to ascertain their positions so that the data collected by these nodes can be labeled with this information. On the other hand, certain attacks on wireless sensor networks lead to the incorrect estimation of sensor node positions. In such situations, when the location information is not correct, the data may be labeled with wrong location information that may subvert the desired operation of the wireless sensor network. In this work, we formulate and propose a distance vector hop-based algorithm to provide secure and robust localization in the presence of malicious sensor nodes that result in incorrect position estimation and jeopardize the wireless sensor network operation. The algorithm uses cryptography to ensure secure and robust operation in the presence of adversaries in the sensor network. As a result of the countermeasures, the attacks are neutralized and the sensor nodes are able to estimate their positions as desired. Our secure localization algorithm provides a defense against various types of security attacks, such as selective forwarding, wormhole, Sybil, tampering, and traffic replay, compared with other algorithms which provide security against only one or two types. Simulation experiments are performed to evaluate the performance of the proposed method, and the results indicate that our secure localization algorithm achieves the design objectives successfully. Performance of the proposed method is also compared with the performance of basic distance vector hop algorithm and two secure algorithms based on distance vector hop localization. The results reveal that our proposed secure localization algorithm outperforms the compared algorithms in the presence of multiple attacks by malicious nodes.

Keywords: secure localization; positioning; distance vector hop; DV-Hop; security attacks; wireless sensor network

Citation: Asif, R.; Farooq-i-Azam, M.; Chaudary, M.H.; Husen, A.; Hassan, S.R. A Distance Vector Hop-Based Secure and Robust Localization Algorithm for Wireless Sensor Networks. *Electronics* **2023**, *12*, 2237. https://doi.org/10.3390/electronics12102237

Academic Editors: Hirokazu Kobayashi and Pingyi Fan

Received: 11 February 2023
Revised: 28 March 2023
Accepted: 10 May 2023
Published: 15 May 2023

1. Introduction

Location information of sensor nodes in a wireless sensor network (WSN) is considered important due to several factors. For example, the data gathered by the sensor nodes must be labeled with the coordinates of the geographic location from where these are collected. Without location information, the data may not make much sense [1]. Examples of such applications where position information is significant include area surveillance [2], habitat monitoring [3], agricultural monitoring [4], and rescue operations [5]. Position information also enables the WSN to make route decisions in the case of certain routing protocols. Using such routing decisions, the data may be routed, for example, to the closest sink [6]. Transmission and communication costs are reduced in this way and the network is energy efficient. Location information also enables the sensor nodes to self organize and form an optimized WSN [7].

Due to aforementioned significance of location information, unknown sensor nodes, i.e., the sensor nodes which do not know their positions, employ a localization algorithm to estimate their position coordinates in the sensor network [8]. By using a localization algorithm, the unknown sensor nodes usually estimate their positions with the help of a few beacon nodes [9]. The beacon nodes, also called anchor nodes, reference nodes, or landmark nodes, know their position coordinates a priori either because these are deployed at known positions or are equipped with a location finding device, such as a global navigation satellite system (GNSS) receiver. A number of localization algorithms for WSNs have been proposed in the literature. A localization scheme for WSN proposed in [10] relies on Voronoi diagram-based grouping tests. This approach involves dividing the sensor nodes in a WSN into several groups and utilizing the closest corresponding Voronoi cells to determine location information. A localization method for WSN which does not need anchor nodes and instead uses cross technology for communication has been proposed in [11]. Instead of using anchor nodes, the method exploits the position information of wireless fidelity (Wi-Fi) access points (APs) for range estimation. Once an unknown node has ascertained its position, it helps other unknown nodes to estimate their positions. A localization algorithm based upon a selection strategy of appropriate beacon nodes has been proposed in [12]. The algorithm uses the signal strength information between the nodes for the selection strategy. With the help of signal strength information topology diagram of a set of nodes is formed. This diagram is then further exploited for position estimation. Localization in WSN is an active area of research and many other location estimation algorithms have also been proposed, such as [13–19].

The majority of these localization algorithms do not take security into consideration. Therefore, these algorithms are prone to various types of security attacks. As a result of these attacks, different types of problems may arise in the localization process. The positions estimated by some of the sensor nodes may have large errors. It is also possible that some nodes are not able to estimate their positions at all due to a security attack. To counter these problems, security measures and secure localization algorithms are being proposed. Two secure localization algorithm against different types of security attacks have been presented in [20]. The first algorithm, named improved randomized consistency position algorithm, exploits position information of beacon nodes and particle swarm optimization (PSO) for localization of unknown nodes. The second algorithm, referred to as the enhanced attack-resistant secure localization algorithm, utilizes a combination of methods, including a voting system, location optimization, and PSO, to estimate the positions of sensor nodes whose locations are unknown. The method proposed in [21] utilizes a blockchain based trust management model to combat malicious nodes in a sensor network. The trust evaluation is composite and involves behavior and data for this purpose. Different parameters, such as honesty, closeness, frequency of interaction, and intimacy, are used for the evaluation of behavior-based trust of the beacon nodes. Honesty is measured using the number of successful and unsuccessful interactions among sensor nodes. The number of sensor nodes covered by a beacon node in one hop neighborhood determines the closeness factor. The frequency of interaction is dependent upon total number of interactions between beacon nodes. Intimacy is quantified by the time of interaction. The beacon nodes with the least trust values are discarded to ensure localization reliability. A received signal strength-based localization algorithm for a WSN with malicious nodes has been proposed in [22]. The algorithm uses different localization techniques, i.e., weighted least square, secure weighted least square, and two norm-based techniques. The different techniques are meant to counter different types of security attacks.

Traditionally, cryptography is used to counter different types of security attacks in various categories of networks. However, conventional cryptography may not be used in resource constrained networks, such as WSN. Therefore, lightweight cryptography techniques have been proposed for such networks. A lightweight public key infrastructure

(PKI) has been proposed in [23] for networks with limited resources, such as the Internet of things (IoT) and WSN. PKI is a security system that uses encryption to authenticate the identity of devices and secure the communication between them. However, the PKI was not designed for devices with constrained resources. Therefore, the conventional PKI system may also not be deployed in networks, such as WSN and IoT, where the devices have small energy resource in the form of a battery, limited memory and storage, and small processing power. The work in [23] has developed a lightweight public key infrastructure (PKI) for registration and distribution of digital certificates in networks with highly constrained devices. The proposed lightweight PKI can be used to secure IoT and WSN devices in a variety of industries, such as healthcare, industrial, transportation, and smart cities. An aggregate signature technique based on a linearly homomorphic signature for resource constrained electronic healthcare system has been proposed in [24]. By combining the advantages of aggregate signature and linearly homomorphic signature, this method offers benefits from both. Under the security model, an aggregate signature is considered valid only if each individual signature utilized to construct the aggregate signature is also valid. Lightweight security algorithms have been used in [25] for reliable data collection from healthcare WSN and to improve security efficiency. The scheme uses elliptic curve digital signature algorithm with BLAKE2bp for the security. Privacy of the patients is ensured by masking the sensor identifications with pseudonyms. Similar works, such as [26–30] have proposed lightweight cryptography techniques for WSN and IoT.

In our work, we propose a secure and robust localization algorithm for WSN. The proposed algorithm is based on distance vector hop (DV-Hop) localization [31,32], which is a popular technique for position estimation in WSN. The traditional DV-Hop method is prone to different types of security attacks. We employ cryptography techniques to provide a secure localization algorithm, which we call the Secure DV-Hop. Compared to other secure algorithms based on DV-Hop which provide protection only against a single type of attack, our proposed secure localization algorithm provides security against multiple types of attacks. The performance of the proposed algorithm is evaluated and compared with the benchmark traditional DV-Hop algorithm and two other secure algorithms based on DV-Hop using simulation experiments. Results show that our proposed algorithm provides a secure, robust, and consistent performance in the presence of malicious nodes.

The rest of this paper is organized as follows. We discuss previously published research related to our work in Section 2. In the next Section 3, we delineate the network model. The DV-Hop localization algorithm is described in Section 4. We present our secure localization algorithm in Section 5. Performance evaluation of the algorithm is reported in Section 6. We finally conclude with Section 7.

2. Related Work

Previously, work has been performed to investigate different types of attacks in wireless sensor networks and their impact on localization and positioning accuracy. In this section, we describe and discuss the related work that has been completed to develop secure localization algorithms for wireless sensor networks.

The work in [33] proposed to secure the DV-Hop localization algorithm against wormhole attacks. The wormhole attack is usually carried out by more than one node in the network. One of the malicious nodes collects and forwards data from the compromised nodes through a tunnel to another malicious node located somewhere else in the network. The secondary malicious node then may transmit the data to the destination while masquerading the identity of the compromised nodes. In this way, the receiving node may be lead to believe that the sender is located at a different hop count other than the actual value. As a result, the localization process may be severely disrupted and the reported positions may have large errors.

Chen et al. analyzed the impact of the wormhole attack and thereby proposed a label-based secure DV-Hop scheme to mitigate this attack in [33]. The proposed method consists of three phases. In the first phase, the beacon nodes are labeled according to their geographic locations. Next, in the second phase, the sensor nodes are differentiated and labeled according to beacon node labeling results. By exploiting these labels, malicious wormhole communication links between the nodes can be prevented. In the final and third phase, the localization process may be completed by using the DV-Hop. This scheme, however, does not take packet loss into consideration. Moreover, it assumes that all nodes have the same transmission radii and does not consider the scenario where different nodes may have different transmission coverage.

Another secure localization algorithm, which is based on DV-Hop and provides defense against the wormhole attack was presented in [34]. This work considers the default wormhole attack with an out of band hidden channel and without data modification. All the nodes in the network are aware of their identification numbers except for the attack nodes. The proposed scheme comprises three stages, i.e., detection of the wormhole attack, resistance against the wormhole attack, and error sources analysis. At first, the proposed scheme establishes a neighbor node relationship list through a broadcast mechanism. The suspect nodes are then identified by comparing the actual number with the theoretical number of nodes. Further, to isolate the actually attacked beacon nodes, the suspect nodes estimate distances from other nodes in their neighbor node relationship list. After the victim nodes have been identified, the attacked nodes mark themselves as either type 1 or type 2 depending upon the attacker node and assuming that there are only two types of attacker nodes in the network. Next, the unknown nodes also mark themselves as either type 1 or type 2 according to their neighbor nodes relationship list. Finally, the nodes marked as type 1 disconnect from nodes marked as type 2 and vice versa to mitigate the wormhole attack. After the attack has been mitigated the localization can be performed. The main limitation of this proposed scheme is that the attack model considers only two attacker nodes. Information modification is also not considered in the attack model.

Prashar et al. proposed a secure localization algorithm for WSN using digital signatures in [35]. At first, the private and public key pair for each node are created. Next, digital signatures for the nodes are generated so that the nodes can authenticate each other. After this, secure localization is performed based upon a procedure derived from DV-Hop. In the DV-Hop localization algorithm, the essential steps for node localization are, hop count determination, average hop size calculation, distance estimation and position determination using trilateration. However, the method proposed in [35], uses a scheme called hyperbolic and mid-perpendicular with centroid to estimate the node positions. If the unknown node is an immediate neighbor of an anchor node, then the mid perpendicular with centroid method is used. Otherwise, hyperbolic scheme is leveraged for position determination.

Another secure localization algorithm for WSN was presented in [36]. The work proposes a malicious node detection algorithm and also presents its extended version. The proposed algorithm, which is range-based, has four stages. In the first stage, the location data of an unknown are obtained using trilateration. In the second stage, the location data are divided into normal and abnormal clusters using self-adaptive density-based spatial clustering of applications with noise. Next, in the third stage, the reference error interval is calculated for the difference between two separate distance measurements based on time of arrival and received signal strength of the reference node. In the final fourth stage, a sequential probability ratio test is performed to test the difference between two measured distances of the suspected malicious node. After all these four stages have been completed, the malicious nodes are detected and the information provided by these malicious nodes can be discarded and the locations of the unknown nodes can be estimated through multilateration.

A secure localization algorithm against the Sybil attack was proposed in [37]. In the Sybil attack, a malicious node may monitor, listen, capture, and modify the data in a network. As a result, the malicious node is able to forge and present multiple identities to the other nodes in the network. This is accomplished by either generating false identities or by simply stealing and spoofing identities of other legitimate nodes on the network. The nodes with forged identities are usually referred to as the Sybil nodes [38]. The

Sybil nodes communicate with other nodes in the network using the forged identities and propagate false information. As a result, the integrity of the data in the network is compromised and network functions based upon this false information are severely damaged. The work in [37] proposed a defense against the Sybil attack which is based upon number allocation and neighbor nodes guarantee. Each node in the network is allotted a number by guaranteed nodes. The number acts as the identity of the node and is verified by its guaranteed node. As a result, any malicious nodes which are not able to present a valid number can be identified and isolated thereby securing the network and the localization process.

Another work in [39] has proposed secure localization using DV-Hop against the Sybil attack. In this proposed method, the beacon nodes broadcast test information. The replies from the neighbor nodes are monitored and a neighbor list is established. If a node has a different neighbor list, then it is concluded that the node is under Sybil attack. If the node has the same neighbor list, then the hop difference between the nodes in the neighbor list is determined. If the hop difference is zero, then it is concluded that the node is under Sybil attack. All the nodes which are found to be under the attack are added to a black list. All the remaining nodes then estimate their positions using the DV-Hop localization algorithm. This proposed method provides protection against only Sybil attack and does not provide defense against other types of attacks on the confidentiality, integrity, and availability of information.

3. Network Model

We consider a WSN deployed in a two-dimensional unconstrained sensor field. The sensor field has finite geographic boundaries. Two types of nodes are deployed in the WSN. The beacon nodes, also known as anchor, landmark or reference nodes, are fixed nodes which know their exact position coordinates. This is possible because these beacon nodes are equipped with navigation devices, such as a global positioning system (GPS), which is a type of global navigation satellite system (GNSS) or because the beacon nodes are deployed at known position coordinates. The other type of nodes in the sensor field are the sensor nodes which perform the sensing and collect the required data. These nodes are not aware of their location. Therefore, these nodes are usually termed as unknown nodes. Alternatively, some literature may refer to these nodes with less plausible names, such as dumb nodes or blind nodes. The unknown nodes estimate their positions with the help of the beacon nodes using a localization algorithm.

In our present work, the localization algorithm to be used by the unknown nodes is DV-Hop ad hoc positioning system. An assumption is made that all nodes in the network have the same radio range. However, the radio range of the unknown nodes is greater than their sensing range. This results in a higher sensing granularity of the WSN, allowing the transmission of sensed data over longer distances. Additionally, all nodes are outfitted with omnidirectional antennas, enabling them to communicate equally well in all directions. We represent a beacon node as B_i where $B_i \in \mathcal{B} = \{B_1, B_2, B_3, \cdots, B_L\}$. So, B_i is a member of $\mathcal{B}$, where the number of beacon nodes in the set is L. The position of a beacon node B_i is given by (x_{Bi}, y_{Bi}). Similarly, we represent an arbitrary unknown sensor node as U_i, where $U_i \in \mathcal{U} = \{U_1, U_2, U_3, \cdots, U_N\}$. Therefore, there are N unknown sensor nodes in the set $\mathcal{U}$ which are deployed in the sensor field. The actual position of an unknown node U_i is represented using (x_{Ui}, y_{Ui}), whereas the estimated position is denoted by $(\hat{x}_{Ui}, \hat{y}_{Ui})$. Each node in the network is pre-installed with a secret key K for encryption and decryption using secret key cryptography. Each node also generates a public and private key pair using an asymmetric encryption algorithm. The network also operates a lightweight public key infrastructure (PKI) for secure management and distribution of the public keys. Secret key encryption is used to ensure confidentiality whereas public key encryption is employed for authentication of hash values only as the latter encryption technique is computationally expensive [40]. The cryptographic keys are stored using a secure storage mechanism [41–46], such as a hardware security module.

We consider that the sensor network is deployed in a hostile environment where malicious nodes are present. The malicious nodes can launch one or a combination of security attacks to disrupt the network operations and localization system. It is considered that the malicious nodes are able to use different types of attacks, including wormhole, tampering, Sybil, traffic replay, and selective forwarding attacks. In the wormhole attack, the malicious nodes create a tunnel between two points in the network. Packets are captured at one point and tunneled to the other point. In the tampering attack, a malicious node modifies the contents of the intercepted packets, such as changing of beacon node position coordinates in the beacon message. Consequently, the position estimated by the unknown nodes is not correct. In the Sybil attack, a malicious nodes uses forged identities to spread false information and disrupt localization system and network operations. A malicious node can intercept and capture packets in a network communication and then later replay the packets to impersonate the identity of one of the nodes involved in the original communication. This type of attack falls in the category of traffic replay attack. In the selective forwarding attack, a malicious node selectively forwards some of the packets while dropping the other packets.

4. Distance Vector Hop Localization

In this section, we briefly describe and discuss the DV-Hop ad hoc positioning system [31,32] for wireless sensor networks.

The DV-Hop algorithm uses distributed processing. To estimate its location, each unknown node calculates its distance from three or more beacon nodes and then uses multilateration to calculate position coordinates. In a multi-hop sensor network, an unknown node may not have direct communication link with three beacon nodes. In other words, the unknown node may be more than one hop away from the beacon nodes. To address this problem, the DV-Hop localization algorithm leverages the connectivity information and the hop count to estimate the distance of an unknown node which may be at a multi-hop distance from the beacon node. Similar to the nature of operation of distance vector (DV) routing protocols, the DV-Hop localization algorithm uses flooding to propagate information in the multi-hop sensor network [47]. Beginning with the beacon nodes, each of the nodes propagates information only to its immediate first hop neighbors. Leaving out the next hop nodes saves bandwidth and power making the approach suitable for WSNs with limited resources. The signaling complexity of this scheme depends upon the number of beacon nodes in the sensor field and average degree of each node, i.e., the number of single hop neighbors of a node.

All the unknown and the beacon nodes in the WSN maintain a table with an entry corresponding to each of the beacon nodes from which it receives messages. The entry is of the form $\{x_{Bi}, y_{Bi}, h_i\}$, where (x_{Bi}, y_{Bi}) are the position coordinates of the beacon node B_i and h_i is the hop count of the node maintaining the table from the beacon node B_i. To obtain the hop count, the hop count field in the beacon message is incremented as the message is transmitted from the beacon node to its nearest neighbor nodes and so on. The beacon nodes in the WSN also maintain this table. After a beacon node B_i has obtained position information and hop count of all other beacon nodes B_j from which it receives messages, it proceeds to ascertain the average size of a hop [31] as follows,

$$c_i = \frac{\sum \sqrt{(x_{Bi} - x_{Bj})^2 + (y_{Bi} - y_{Bj})^2}}{\sum h_j}, \tag{1}$$

for all beacon nodes B_j and $B_j \neq B_i$. The numerator of Equation (1) is the sum of the distances between a beacon node B_i and other beacon nodes B_j. The denominator is the sum of hop counts between the beacon node B_i and other beacon nodes B_j. Therefore, Equation (1) gives average size of a hop as the sum of distances divided by the sum of hop counts. The DV-Hop algorithm terms this average size of the hop c_i, calculated by the beacon node B_i, as the correction factor. Using controlled flooding, this correction factor is propagated through the network as described earlier. After receiving the correction factor and with the knowledge of position coordinates of at least three beacon nodes, an unknown node performs multilateration to estimate its own position information. The steps involved in the position estimation using the DV-Hop ad hoc positioning system are summarized as follows:

- At the beginning of the algorithm, the beacon nodes transmit their location data to their nearest neighbor nodes in the first hop.
- All the other nodes in the network receive and propagate the beacon node position coordinates using the same method as the distance vector routing protocol. The intermediate nodes increment the hop count field as they propagate the information to the next hop neighbor. Eventually, all the nodes in the network obtain the position coordinates of all the beacon nodes along with the hop count to these beacon nodes.
- After a beacon node has obtained position coordinates of other beacon nodes and the hop count to them, it computes the average hop size using Equation (1).
- A beacon node propagates its computed average hop size as correction factor throughout the network using the controlled flooding approach of the DV protocol.
- Once an unknown node receives the correction factor, it calculates the distance to the beacon node from which it received the correction factor. The distance is calculated by multiplying the hop count to the correction factor, i.e., $h_i \times c_i$.
- After knowing the position coordinates of at least three beacon nodes and distance estimates to them, the unknown node performs multilateration to estimate its position.

It should be noted that the correction factor calculated by one beacon node may differ from the correction factor computed by another beacon node. Moreover, each unknown node will receive different correction factors from different beacon nodes. The DV-Hop ad hoc positioning system [31,32] suggests that, for position estimation, an unknown node should store and utilize the initial correction factor it receives and disregard any other correction factors received subsequently.

5. Secure Localization

In this section, we describe our secure localization algorithm based on DV-Hop using cryptography. In the ensuing description of our proposed secure localization algorithm, concatenation of two items, a and b, is denoted by $a||b$. We denote the encryption operation of a message M using key K to obtain ciphertext C by $C = E(K, M)$. The decryption operation of the ciphertext C using the key K to obtain the message M is denoted by $M = D(K, C)$. When A sends a message M to B, we represent this as follows.

$$A \xrightarrow{M} B \tag{2}$$

At the time of first deployment, an unknown node U_i sends a registration request to the nearest beacon node B_i from which it receives messages. The registration message M_{Ui} is prepared as follows,

$$M_{Ui} \leftarrow ID_{Ui}||REG||N_{Ui}||S_{Ui}||T_{Ui}, \tag{3}$$

where ID_{Ui} is the unique public identification of the unknown node U_i, REG represents registration request, N_{Ui} is a cryptographic nonce, S_{Ui} is the sequence number, and T_{Ui} is the time stamp by the unknown node U_i. The unknown node U_i computes one way cryptographic hash of the message M_{Ui} using an agreed upon hash function h. The computed hash of the message M_{Ui} is encrypted using the private key PR_{Ui} of the unknown node U_i to obtain $C_{Uih} = E(PR_{Ui}, h(M_{Ui}))$. The message M_{Ui} and the hash value are then encrypted using the secret key K as $C_{Ui} = E(K, M_{Ui}||E(PR_{Ui}, h(M_{Ui})))$ and transmitted to the beacon node B_i as follows,

$$U_i \xrightarrow{E(K, M_{Ui}||E(PR_{Ui}, h(M_{Ui})))} B_i. \tag{4}$$

The localization algorithm identifies a sensor node with the help of the application layer identifier ID_{Ui}. Legitimate nodes are able to decrypt the encrypted application layer messages and, hence, are able to retrieve the application layer identifier ID_{Ui}. The cryptographic nonce N_{Ui} serves the purpose of authentication as a legitimate receiver should be able to retrieve it from the encrypted message and send it back. The time stamp T_{Ui} serves as a defense against the traffic replay and other man in the middle attacks. Sequence numbers prevent disruption of traffic by an attacker by reordering the packets. The sequence numbers are unpredictable and are generated according to Algorithm 1. The length of the message is added to the sequence number to ensure that each message has a unique sequence number and that any messages received out of order can be identified. When a receiver receives a message with a sequence number that is not an expected number, it knows that some data have been lost or delivered out of order. This also helps thwart selective forwarding attack.

Algorithm 1 Sequence number generator.

```
procedure SEQUENCENUMBER(SeqNumber)
    if (SessionStart==true) then
        SeqNumber ← Secure.Random(value)
    else
        SeqNumber ← SeqNumber + Length(Message)
        if (SeqNumber > MAXSEQNUM) then
            SeqNumber ← 0
        end if
    end if
    return SeqNumber
end procedure
```

Upon receiving the encrypted message, the beacon node B_i decrypts it as $D(K, C_{Ui})$. The beacon node B_i is able to determine the length of the message after this decryption process. The encrypted hash value is further retrieved using the public key PU_{Ui} of the unknown node U_i as $D(PU_{Ui}, C_{Uih})$. If the beacon node B_i is able to successfully decrypt the encrypted hash using the public key PU_{Ui}, it is confirmed that the message was indeed sent by the sensor node U_i as no other node could have encrypted the hash using the private key PR_{Ui} of the sensor node U_i. The beacon node B_i also computes the hash value of the message using the hash function h. If the computed and the retrieved hash values do not match, the message is discarded. However, if the two values match, then the beacon node B_i prepares the following message M_{Bi} for the unknown node U_i.

$$M_{Bi} \leftarrow ID_{Bi}||N_{Ui}||N_{Bi}||S_{Bi}||T_{Bi}, \tag{5}$$

where ID_{Bi} is the unique identification of the beacon node B_i, N_{Ui} is the cryptographic nonce which was sent by the unknown node U_i, N_{Bi} is the cryptographic nonce prepared by the beacon node B_i, S_{Bi} is the sequence number generated according to Algorithm 1, and T_{Bi} is the time stamp by the beacon node B_i. The hash of this message is encrypted using the private key PR_{Bi} of the beacon node B_i and concatenated with the message M_{Bi}. This is then encrypted using the secret key K to produce ciphertext $C_{Bi} = E(K, M_{Bi}||E(PR_{Bi}, h(M_{Bi})))$, which is transmitted to the unknown node U_i.

$$B_i \xrightarrow{E(K,M_{Bi}||E(PR_{Bi},h(M_{Bi})))} U_i \tag{6}$$

The unknown node U_i decrypts this message as $D(K, C_{Bi})$. If it is unable to retrieve the cryptographic nonce N_{Ui}, the message is discarded and is not processed further. However, if the nonce is retrieved successfully, it generates a code word W_{Ui} and sends it to the beacon node B_i using the following message.

$$M_{Ui} \leftarrow ID_{Ui}||N_{Bi}||W_{Ui}||S_{Ui}||T_{Ui}. \tag{7}$$

Similar to the previous messages, ciphertext $C_{Ui} = E(K, M_{Ui}||E(PR_{Ui}, h(M_Ui)))$ is prepared and sent to the beacon node B_i as follows,

$$U_i \xrightarrow{E(K,M_{Ui}||E(PR_{Ui},h(M_Ui)))} B_i. \tag{8}$$

The beacon node B_i decrypts the received message as $D(K, C_{Ui})$. If it cannot find the nonce N_{Bi} in the message, it discards the message and does not process it further. However, if the nonce N_{Bi} is retrieved successfully, it proceeds to process the received code word W_{Ui}. The beacon node B_i adds a salt to the code word W_{Ui}, computes the hash of the salted code word. The hash is stored along with the salt and the ID_{Ui} of the unknown node U_i. This process is performed as depicted in Algorithm 2. The registration procedure of an unknown node with a beacon node is illustrated in Figure 1.

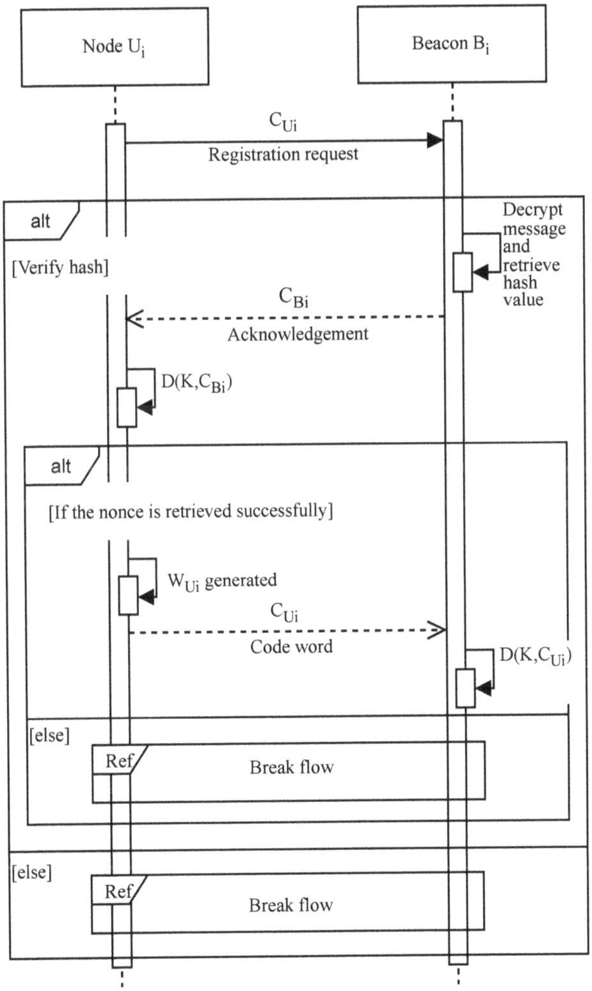

Figure 1. Registration procedure of an unknown node with a beacon node.

Algorithm 2 Code Word Storage.

```
procedure UIDSTORAGE(ID_Ui, CodeWord, SaltLength)
    AllowedChars ← "abcdefghijklmnopqrstuvwxyz"
    AllowedChars ← AllowedChars + "ABCDEFGHIJKLMNOPQRSTUVWXYZ"
    AllowedChars ← AllowedChars + "0123456789"
    AllowedChars ← AllowedChars + "!@#$%^&*"
    MAX ← Length(AllowedChars)
    Salt ← ""
    Count ← 0
    while (Count < SaltLength) do
        RandNum ← Secure.Random(0, MAX)
        Salt ← Salt + AllowedChars[RandNum]
        Count ← Count + 1
    end while
    WordHash ← HashAlgo(CodeWord + Salt)
    Handle ← Open(SecureFile)
    Write(Handle, ID_Ui, WordHash, SaltLength)
    Close(Handle)
end procedure
```

Subsequently, if the unknown node U_i wants to communicate with another unknown node U_j, the latter asks U_i to provide its surety. The unknown node U_j prepares the following message for this purpose.

$$M_{Uj} \leftarrow ID_{Uj}||SURETY||N_{Uj}||S_{Uj}||T_{Uj} \tag{9}$$

Next, this message and its hash are encrypted as $C_{Uj} = E(K, M_{Uj}||E(PR_{Uj}, h(M_{Uj})))$ and sent to the node U_i, as follows,

$$U_j \xrightarrow{E(K,M_{Uj}||E(PR_{Uj},h(M_{Uj})))} U_i. \tag{10}$$

After decrypting this message and confirming its validity with the help of the hash, the unknown node responds with the following message.

$$C_{WUi} \leftarrow E(PU_{Bi}, W_{Ui}||E(PR_{Ui}, h(W_{Ui}))) \tag{11}$$

$$M_{Ui} \leftarrow ID_{Ui}||N_{Uj}||N_{Ui}||ID_{Bi}||C_{WUi}||S_{Ui}||T_{Ui} \tag{12}$$

$$U_i \xrightarrow{E(K,M_{Ui}||E(PR_{Ui},h(M_{Ui})))} U_j. \tag{13}$$

The unknown node U_j decrypts and verifies this message using the hash function. It also retrieves the encrypted code word C_{WUi} and sends it to the beacon node B_i for verification.

$$M_{Uj} \leftarrow ID_{Uj}||N_{Uj}||C_{WUi}||S_{Uj}||T_{Uj} \tag{14}$$

The encrypted text $C_{Uj} = E(K, M_{Uj}||E(PR_{Uj}, h(M_{Uj})))$ is prepared and sent to the beacon node B_i as follows.

$$U_j \xrightarrow{E(K,M_{Uj}||E(PR_{Uj},h(M_{Uj})))} B_i. \tag{15}$$

The beacon node B_i decrypts and checks the validity of the message as described previously. It then decrypts C_{WUi} and retrieves the code word of the unknown node U_i. It confirms its validity by computing its hash using the stored salt and then comparing with the stored value of the hash. It then communicates the result back to the unknown node U_j.

$$M_{Bi} \leftarrow ID_{Bi}||N_{Uj}||N_{Bi}||RESULT||S_{Bi}||T_{Bi} \tag{16}$$

$$B_i \xrightarrow{E(K,M_{Bi}||E(PR_{Bi},h(M_{Bi})))} U_j, \quad (17)$$

where the variable $RESULT$ contains OK if the code word is verified or NOK otherwise. The node U_j proceeds with its data exchange with the node U_i in the former case and drops the communication in the latter instance. The procedure to store the code word and to establish trust between two nodes takes place only once. The authentication process of two unknown nodes with the help of a beacon node is illustrated in Figure 2.

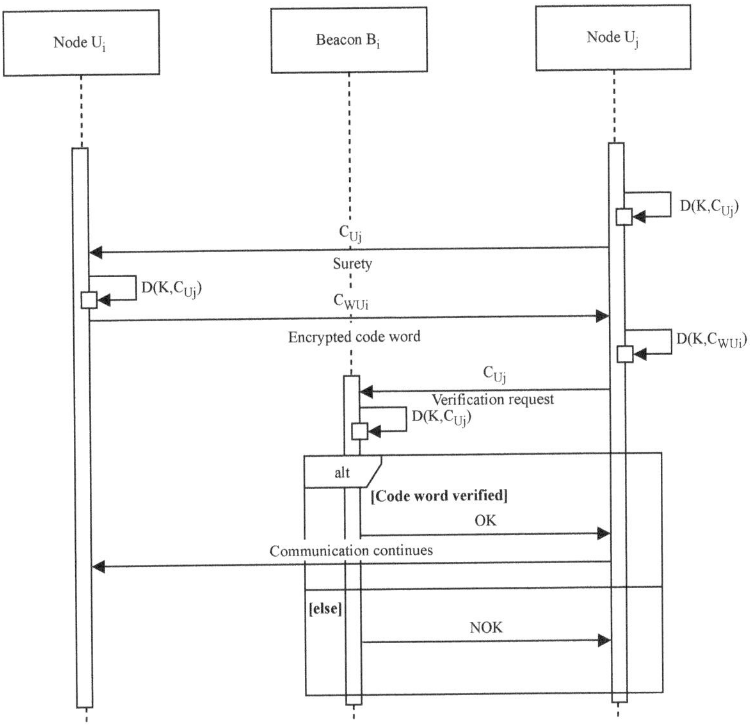

Figure 2. Authentication process between two unknown nodes.

The beacon nodes broadcast their positions using beacon messages at regular intervals. The message may contain the identification of the beacon node and a time stamp. For example, a typical beacon message M_{Bi} of a beacon node B_i is as follows.

$$M_{Bi} \leftarrow ID_{Bi}||(x_{Bi}, y_{Bi})||hc_{Bi}||T_{Bi}, \quad (18)$$

where ID_{Bi} is the unique identification of the beacon node B_i, (x_{Bi}, y_{Bi}) is its position information, hc_{Bi} is a variable to store the hop count and is initialized to zero, and T_{Bi} is the time stamp by the beacon node B_i. The beacon node B_i computes one way cryptographic hash of the message using a hash function h. The computed cryptographic hash of the message M_{Bi} is encrypted using the private key PR_{Bi} of the beacon node B_i to obtain $C_{Bih} = E(PR_{Bi}, h(M_{Bi}))$. The message M_{Bi} and its encrypted hash are then broadcast. The broadcast message is $M_{Bi}||E(PR_{Bi}, h(M_{Bi}))$, and is depicted as below.

$$B_i \xrightarrow{M_{Bi}||E(PR_{Bi},h(M_{Bi}))} \text{All} \quad (19)$$

When an unknown node U_i receives this message, it decrypts the encrypted hash using the public key PU_{Bi} of the beacon node B_i using $D(PU_{Bi}, C_{Bih})$. It also computes the one way cryptographic hash of the received message M_{Bi} using the same hash function which was used by the beacon node B_i. If $h(M_{Bi}) \neq D(PU_{Bi}, C_{Bih})$, that is, the hash computed by the unknown node U_i does not match the received hash value, then the unknown node U_i discards the message. However, if the computed hash and the received hash values match each other, i.e., $h(M_{Bi}) = D(PU_{Bi}, C_{Bih})$, then the message is considered legitimate. The unknown node U_i stores the position (x_{Bi}, y_{Bi}) of the beacon node B_i. Moreover, the unknown node U_i increments the hop count variable hc_{Bi}, and constructs a message M_{Ui} for the next hop neighbor as follows.

$$M_{Ui} \leftarrow ID_{Ui}||M_{Bi}||hc_{Bi}||T_{Ui}, \tag{20}$$

where ID_{Ui} is the unique identification of the unknown node U_i, hc_{Bi} is the hop count variable with incremented value, and T_{Ui} is the time stamp by the unknown node U_i. The unknown node U_i also computes one way cryptographic hash of the message using the hash function h. The cryptographic hash value of the message M_{Ui} is then encrypted using the private key PR_{Ui} of the unknown node U_i to obtain $C_{Uih} = E(PR_{Ui}, h(M_{Ui}))$. The message M_{Ui} containing the new hop count and the encrypted hash value are then sent to the next hop neighbor U_j as follows.

$$U_i \xrightarrow{M_{Ui}||E(PR_{Ui},h(M_{Ui}))} U_j \tag{21}$$

Upon receiving this message, the node U_j performs a procedure similar to the procedure performed by the node U_i when it received the beacon message. It decrypts the encrypted hash as $D(PU_{Ui}, C_{Uih})$ and also computes the hash value $h(M_{Ui})$. The message is processed if the two hash values match and is discarded otherwise. The message is propagated further until it reaches another beacon node.

After a beacon node B_i has obtained position coordinates of other beacon nodes and the hop count to them, it computes the average hop size or the correction factor c_{Bi} using Equation (1), as stated earlier. The beacon node B_i, then prepares a message M_{Bi} to propagate this correction factor as follows.

$$M_{Bi} \leftarrow ID_{Bi}||(x_{Bi}, y_{Bi})||c_{Bi}||T_{Bi} \tag{22}$$

This message and its cryptographic hash encrypted using the private key of the beacon node B_i are concatenated as $M_{Bi}||E(PR_{Bi}, h(M_{Bi}))$. This is then propagated through the next hop neighbors U_i as follows.

$$B_i \xrightarrow{M_{Bi}||E(PR_{Bi},h(M_{Bi}))} U_i \tag{23}$$

When an unknown node U_i receives the correction factor c_{Bi}, it then computes the distance to the beacon node B_i from which it received the correction factor. The distance is calculated by multiplying the hop count hc_{Bi} to the correction factor c_{Bi}, i.e., $hc_{Bi} \times c_{Bi}$. After an unknown node U_i has received the position coordinates of at least three beacon nodes and estimated distance to them, the unknown node performs multilateration to estimate its position as already described in Section 4. Propagation of beacon messages and correction factor is illustrated in Figure 3.

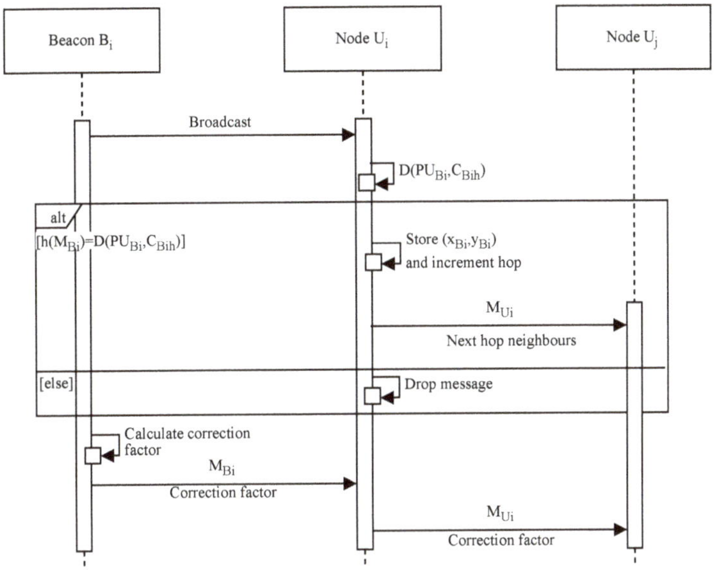

Figure 3. Propagation of beacon messages and correction factor.

6. Simulation Results

We evaluate the performance of our proposed Secure DV-Hop localization algorithm using simulation experiments. A sensor field with dimensions of 100 m × 100 m is considered for the experiments. The number of sensor nodes is 100. The number of beacon nodes and the number of malicious nodes are varied for the performance evaluation. Moreover, the performance is also evaluated both in the absence and presence of the malicious nodes. The malicious nodes use different types of security attacks which include wormhole, Sybil, tampering, traffic replay, and selective forwarding attacks. In addition, performance of the proposed Secure DV-Hop algorithm is compared with those of basic DV-Hop [31], label-based DV-Hop (LBDV-Hop) [33] and Security Positioning DV-Hop (SPDV-Hop) [39] localization algorithms.

Localization error of a single node is the distance between the actual position and the estimated position. Therefore, if the actual position of a sensor node U_i is (x_{Ui}, y_{Ui}) and the estimated position is $(\hat{x}_{Ui}, \hat{y}_{Ui})$, then the localization error, e_L, of an unknown node U_i is given by,

$$e_L = \sqrt{(\hat{x}_{Ui} - x_{Ui})^2 - (\hat{y}_{Ui} - y_{Ui})^2}. \tag{24}$$

The average normalized localization error, e_{LN}, of the sensor network is given by,

$$e_{LN} = \frac{\sum_{i=1}^{N} \sqrt{(\hat{x}_{Ui} - x_{Ui})^2 - (\hat{y}_{Ui} - y_{Ui})^2}}{NR}, \tag{25}$$

where N is the total number of unknown nodes and R is the radio range of a sensor node.

The localization efficiency η_L is the ratio of the number of unknown sensor nodes which are able to estimate their positions to the total number of unknown sensor nodes [48]. The unknown sensor nodes which are able to ascertain their positions may be termed as settled nodes. If the total number of settled nodes is represented by N_s, then the localization efficiency, η_L, is given by

$$\eta_L = \frac{N_s}{N} \times 100. \tag{26}$$

In Figure 4, we plot the average normalized localization error for the basic DV-Hop, SPDV-Hop, LBDV-Hop, and the Secure DV-Hop localization algorithms as the number of beacon nodes is varied in the sensor field in the absence of any attack. The localization efficiency of these algorithms against the varying number of beacon nodes in the absence of any attack is plotted in Figure 5. From both Figures 4 and 5, it can be observed that all the compared localization algorithms perform as good as the basic DV-Hop methods in the absence of any attack. Therefore, these algorithms work in the same fashion under normal circumstances. This also validates the localization performance of the proposed algorithm.

However, all these algorithms perform differently when malicious nodes are introduced in the sensor network. This can be observed from Figure 6 where average normalized localization error of each of the proposed Secure DV-Hop and three compared algorithms is plotted for varying number of malicious nodes and 20% beacon nodes. It is evident that, when the malicious nodes are present, the basic DV-Hop, SPDV-Hop, and LBDV-Hop localization algorithms do not perform the way as they do in the absence of any attack. The average normalized localization error for each of these algorithms increases significantly as the count of the malicious nodes is increased while keeping the number of beacon nodes fixed at 20%. However, our proposed Secure DV-Hop localization algorithm remains unaffected and shows consistent results in the presence of any number of malicious nodes.

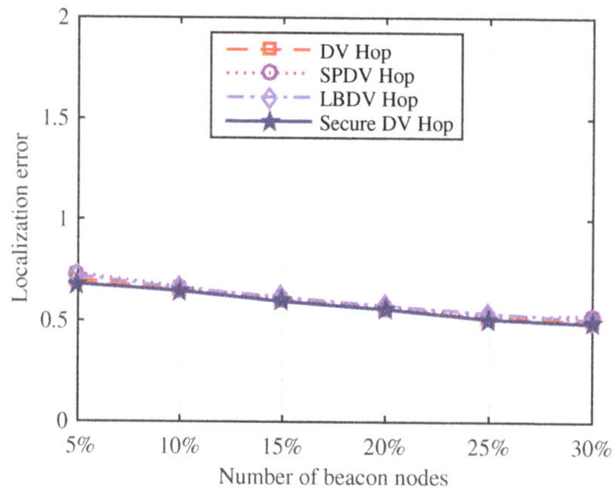

Figure 4. Average normalized localization error as the number of beacon nodes is varied in the absence of an attack.

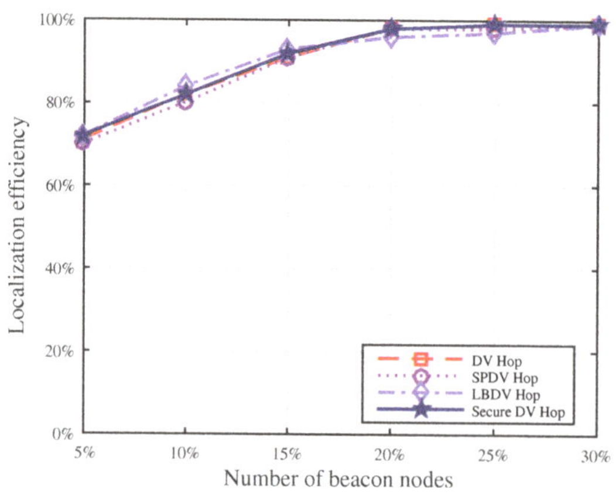

Figure 5. Localization efficiency as the number of beacon nodes is varied in the absence of an attack.

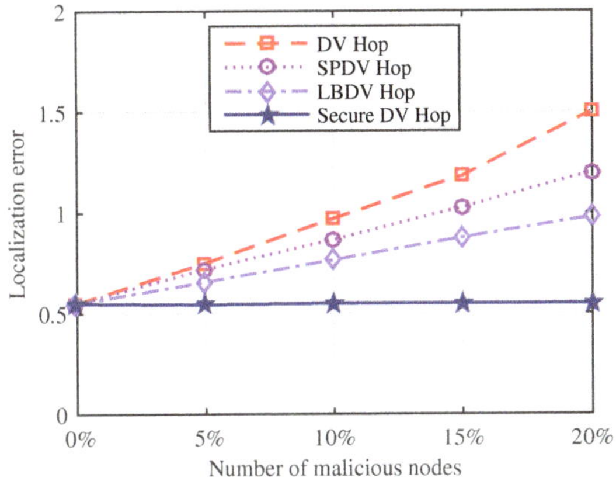

Figure 6. Average normalized localization error as the number of malicious nodes is varied with fixed 20% beacon nodes in the sensor network.

We plot average normalized localization error with a varying number of beacon nodes and a fixed 10% number of malicious nodes in Figure 7. It can be observed that the localization error decreases for all the compared localization algorithms as the number of beacon nodes is increased in the sensor field. However, in the case of DV-Hop, SPDV-Hop, and LBDV-Hop, when we compare their performance in Figure 4 in the absence of attack to their performance in Figure 7 when attacked by 10% hostile nodes, a degradation in the performance is clearly observed. For example, from Figure 4, in the absence of attack, the localization error resulting from DV-Hop with 20% beacon nodes is almost 0.5. However, in the case of Figure 7, the localization error with 20% beacon nodes in the presence of 10% malicious nodes is almost 1. This is twice as high as the error in Figure 4 for the same number of beacon nodes. Similar observation can be made for SPDV-Hop and LBDV-Hop as well. On the other hand, the Secure DV-Hop algorithm remains robust and its localization results remain unaffected by the malicious nodes. If we compare the performance of our proposed Secure DV-Hop localization algorithm in Figures 4 and 7, we see that it provides similar performance in the presence or absence of the malicious nodes.

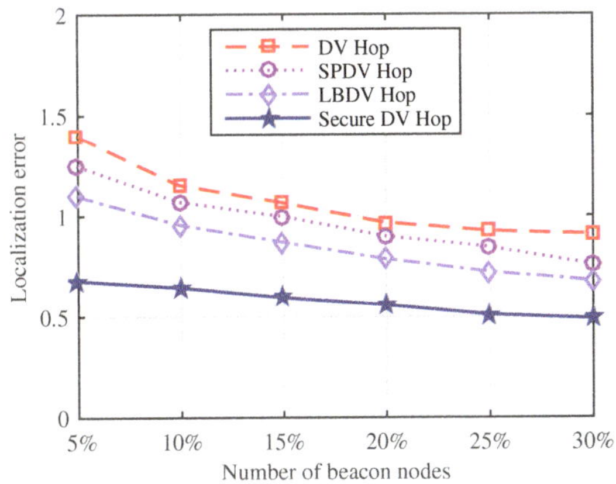

Figure 7. Average normalized localization error with varying number of beacon nodes when the number of malicious nodes is fixed at 10%.

Localization efficiency of each of the compared algorithms is plotted in Figure 8 against a varying number of malicious nodes when the number of beacon nodes is 20%. The localization efficiencies of the DV-Hop, SPDV-Hop, and LBDV-Hop algorithms decrease as the number of malicious nodes in the sensor field increases. This implies that lesser and lesser number of unknown sensor nodes are able to estimate their positions as the number of malicious nodes increases. On the other hand, the Secure DV-Hop localization algorithm is not affected and its localization efficiency does not change with the varying number of malicious nodes in the sensor field.

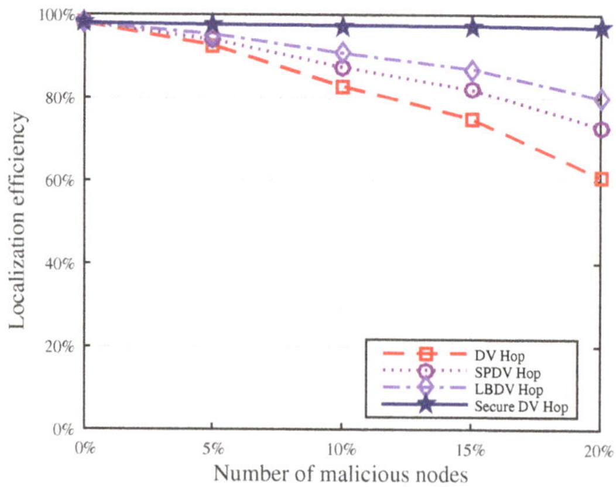

Figure 8. Localization efficiency when the number of malicious nodes is varied and the number of beacon nodes is fixed at 20%.

In Figure 9, we plot localization efficiencies of DV-Hop, SPDV-Hop, LBDV-Hop, and Secure DV-Hop algorithms against a varying number of beacon nodes in the presence of 10% malicious nodes. Results in Figure 9 corroborate previous findings. Although localization efficiencies of DV-Hop, SPDV-Hop, and LBDV-Hop increase with an increase in the number of beacon nodes in the sensor field, these do not attain the same values as they do in the absence of malicious nodes. However, Secure DV-Hop localization algorithm once again shows consistent and robust performance even in the presence of adverse conditions. From Figures 5 and 9, it can be readily observed that the localization efficiency of the Secure DV-Hop algorithm remains unaffected in the presence or absence of the malicious nodes.

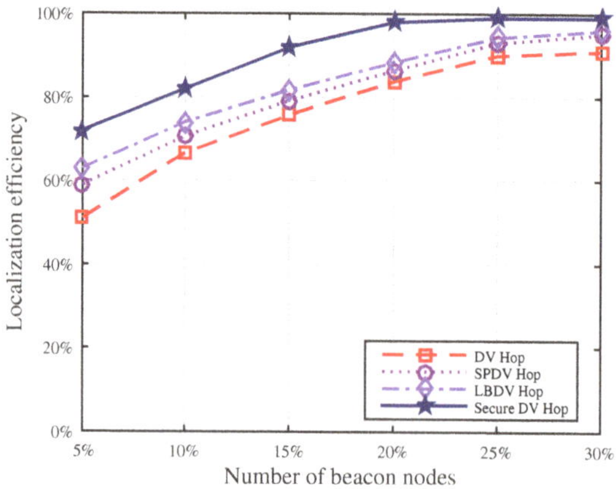

Figure 9. Localization efficiency with varying number of beacon nodes when the number of malicious nodes is fixed at 10%.

This robust performance of Secure DV-Hop localization algorithm can be attributed to effective authentication and communication implemented through encryption. SPDV-Hop provides protection against Sybil attack only and LBDV-Hop is designed for security against wormhole attack alone. However, both these algorithms do not provide effective protection against other types of attacks, such as tampering, selective forwarding, and traffic replay. On the other hand, robust authentication and communication implemented by the Secure DV-Hop localization algorithm protect against all these types of attacks. Hence, it remains unaffected by these security attacks.

7. Conclusions

In this work, we have proposed distance vector hop-based secure and robust localization algorithm for wireless sensor networks. The algorithm uses secret and public key cryptography to secure the localization process against different types of security attacks. These attacks include wormhole, Sybil, selective forwarding, traffic replay and tampering attacks. A number of simulation experiments were performed to evaluate the performance of the proposed algorithm both in the presence and the absence of malicious nodes using these attacks. The results were compared with the basic distance vector hop method and two secure algorithms based on distance vector hop localization. The average normalized localization error and the localization efficiency were measured in the presence, as well as in the absence of malicious nodes. The results revealed that the performance of the compared algorithms was severely affected in the presence of malicious nodes. However, the proposed secure localization algorithm provided secure and robust performance in either scenario. As a result of the countermeasures, the algorithm provided similar performance in the presence of adversaries as it did in the absence of any attacks. The secure localization algorithm can be implemented in a wireless sensor network which is deployed in a hostile environment and where unknown sensor nodes have to estimate their position coordinates. Future work includes improvement of localization performance of the algorithm and its implementation and practical evaluation in a real wireless sensor network.

Author Contributions: Conceptualization, M.F.-i.-A., M.H.C. and A.H.; methodology, M.F.-i.-A.; software, R.A.; validation, M.H.C. and A.H.; formal analysis, M.F.-i.-A.; investigation, M.F.-i.-A. and S.R.H.; resources, R.A.; data curation, M.H.C. and A.H.; writing—original draft preparation, M.F.-i.-A., M.H.C. and A.H.; writing—review and editing, M.F.-i.-A., R.A. and S.R.H.; visualization, M.H.C. and A.H.; supervision, M.F.-i.-A.; project administration, M.F.-i.-A., R.A. and S.R.H. All authors have read and agreed to the published version of the manuscript.

Funding: The authors acknowledge the internal research start-up fund, reference: 1012606FA1, from University of East Anglia (UEA), Norwich, UK.

Data Availability Statement: Data sharing not applicable.

Conflicts of Interest: The authors declare no conflict of interest.

References

1. Najarro, L.A.C.; Song, I.; Kim, K. Fundamental limitations and State-of-the-art Solutions for Target Node Localization in WSNs: A Review. *IEEE Sens. J.* **2022**, 22, 23661–23682. [CrossRef]
2. Benahmed, T.; Benahmed, K. Optimal barrier coverage for critical area surveillance using wireless sensor networks. *Int. J. Commun. Syst.* **2019**, *32*, e3955. [CrossRef]
3. Lahoz-Monfort, J.J.; Magrath, M.J. A comprehensive overview of technologies for species and habitat monitoring and conservation. *BioScience* **2021**, *71*, 1038–1062. [CrossRef] [PubMed]
4. Abdollahi, A.; Rejeb, K.; Rejeb, A.; Mostafa, M.M.; Zailani, S. Wireless sensor networks in agriculture: Insights from bibliometric analysis. *Sustainability* **2021**, *13*, 12011. [CrossRef]
5. Bravo-Arrabal, J.; Zambrana, P.; Fernandez-Lozano, J.; Gomez-Ruiz, J.A.; Barba, J.S.; García-Cerezo, A. Realistic deployment of hybrid wireless sensor networks based on ZigBee and LoRa for search and Rescue applications. *IEEE Access* **2022**, *10*, 64618–64637. [CrossRef]
6. Nakas, C.; Kandris, D.; Visvardis, G. Energy efficient routing in wireless sensor networks: A comprehensive survey. *Algorithms* **2020**, *13*, 72. [CrossRef]
7. Ullah, I.; Youn, H.Y. A novel data aggregation scheme based on self-organized map for WSN. *J. Supercomput.* **2019**, *75*, 3975–3996. [CrossRef]
8. Huang, X.; Han, D.; Cui, M.; Lin, G.; Yin, X. Three-Dimensional Localization Algorithm Based on Improved A* and DV-Hop Algorithms in Wireless Sensor Network. *Sensors* **2021**, *21*, 448. [CrossRef]

9. Messous, S.; Liouane, H.; Cheikhrouhou, O.; Hamam, H. Improved Recursive DV-Hop Localization Algorithm with RSSI Measurement for Wireless Sensor Networks. *Sensors* **2021**, *21*, 4152. [CrossRef]
10. Li, G.; Xu, M.; Teng, G.; Yang, W.; Mak, S.L.; Li, C.Y.; Lee, C.C. A Voronoi Diagram-Based Grouping Test Localization Scheme in Wireless Sensor Networks. *Electronics* **2022**, *11*, 2961. [CrossRef]
11. Jing, N.; Zhang, B.; Wang, L. A Novel Anchor-Free Localization Method Using Cross-Technology Communication for Wireless Sensor Network. *Electronics* **2022**, *11*, 4025. [CrossRef]
12. Liu, W.; Luo, X.; Wei, G.; Liu, H. Node localization algorithm for wireless sensor networks based on static anchor node location selection strategy. *Comput. Commun.* **2022**, *192*, 289–298. [CrossRef]
13. Kaur, A.; Gupta, G.P.; Mittal, S. Comparative study of the different variants of the dv-hop based node localization algorithms for wireless sensor networks. *Wirel. Pers. Commun.* **2022**, *123*, 1625–1667. [CrossRef]
14. Yuvarasu, M.; Balaram, A.; Chandramohan, S.; Sharma, D.K. A Performance Analysis of an Enhanced Graded Precision Localization Algorithm for Wireless Sensor Networks. *Cybern. Syst.* **2023**, 1–16. [CrossRef]
15. Liouane, H.; Messous, S.; Cheikhrouhou, O. Regularized least square multi-hops localization algorithm based on DV-Hop for wireless sensor networks. *Telecommun. Syst.* **2022**, *80*, 349–358. [CrossRef]
16. Luomala, J.; Hakala, I. Adaptive range-based localization algorithm based on trilateration and reference node selection for outdoor wireless sensor networks. *Comput. Netw.* **2022**, *210*, 108865. [CrossRef]
17. Liu, J.; Liu, M.; Du, X.; Stanimirovi, P.S.; Jin, L. An improved DV-Hop algorithm for wireless sensor networks based on neural dynamics. *Neurocomputing* **2022**, *491*, 172–185. [CrossRef]
18. Du, J.; Yuan, C.; Yue, M.; Ma, T. A novel localization algorithm based on RSSI and multilateration for indoor environments. *Electronics* **2022**, *11*, 289. [CrossRef]
19. Zhang, H.; Yang, J.; Qin, T.; Fan, Y.; Li, Z.; Wei, W. A Multi-Strategy Improved Sparrow Search Algorithm for Solving the Node Localization Problem in Heterogeneous Wireless Sensor Networks. *Appl. Sci.* **2022**, *12*, 5080. [CrossRef]
20. Nguyen, T.N.; Le, V.V.; Chu, S.I.; Liu, B.H.; Hsu, Y.C. Secure Localization Algorithms Against Localization Attacks in Wireless Sensor Networks. *Wirel. Pers. Commun.* **2022**, *127*, 767–792. [CrossRef]
21. Kim, T.H.; Goyat, R.; Rai, M.K.; Kumar, G.; Buchanan, W.J.; Saha, R.; Thomas, R. A Novel Trust Evaluation Process for Secure Localization Using a Decentralized Blockchain in Wireless Sensor Networks. *IEEE Access* **2019**, *7*, 184133–184144. [CrossRef]
22. Mukhopadhyay, B.; Srirangarajan, S.; Kar, S. RSS-Based Localization in the Presence of Malicious Nodes in Sensor Networks. *IEEE Trans. Instrum. Meas.* **2021**, *70*, 1–16. [CrossRef]
23. Höglund, J.; Lindemer, S.; Furuhed, M.; Raza, S. PKI4IoT: Towards public key infrastructure for the Internet of Things. *Comput. Secur.* **2020**, *89*, 101658. [CrossRef]
24. Gu, Y.; Shen, L.; Zhang, F.; Xiong, J. Provably Secure Linearly Homomorphic Aggregate Signature Scheme for Electronic Healthcare System. *Mathematics* **2022**, *10*, 2588. [CrossRef]
25. Al-Zubaidie, M.; Zhang, Z.; Zhang, J. REISCH: Incorporating Lightweight and Reliable Algorithms into Healthcare Applications of WSNs. *Appl. Sci.* **2020**, *10*, 2007. [CrossRef]
26. Revanesh, M.; Acken, J.M.; Sridhar, V. DAG block: Trust aware load balanced routing and lightweight authentication encryption in WSN. *Future Gener. Comput. Syst.* **2022**, *140*, 402–421.
27. Nagarajan, M.; Rajappa, M.; Teekaraman, Y.; Kuppusamy, R.; Thelkar, A.R. Renovated XTEA encoder architecture-based lightweight mutual authentication protocol for RFID and green wireless sensor network applications. *Wirel. Commun. Mob. Comput.* **2022**, *2022*, 8876096. [CrossRef]
28. Mezrag, F.; Bitam, S.; Mellouk, A. An efficient and lightweight identity-based scheme for secure communication in clustered wireless sensor networks. *J. Netw. Comput. Appl.* **2022**, *200*, 103282. [CrossRef]
29. Hussein, S.M.; López Ramos, J.A.; Ashir, A.M. A Secure and Efficient Method to Protect Communications and Energy Consumption in IoT Wireless Sensor Networks. *Electronics* **2022**, *11*, 2721. [CrossRef]
30. Chen, C.M.; Chen, Z.; Kumari, S.; Lin, M.C. LAP-IoHT: A lightweight authentication protocol for the internet of health things. *Sensors* **2022**, *22*, 5401. [CrossRef] [PubMed]
31. Niculescu, D.; Nath, B. DV Based Positioning in Ad Hoc Networks. *Telecommun. Syst.* **2003**, *22*, 267–280. [CrossRef]
32. Niculescu, D.; Nath, B. Ad hoc positioning system (APS) using AOA. In Proceedings of the IEEE INFOCOM 2003. Twenty-Second Annual Joint Conference of the IEEE Computer and Communications Societies (IEEE Cat. No. 03CH37428), San Francisco, CA, USA, 30 March–3 April 2003; Volume 3, pp. 1734–1743. [CrossRef]
33. Chen, H.; Lou, W.; Wang, Z.; Wu, J.; Wang, Z.; Xia, A. Securing DV-Hop localization against wormhole attacks in wireless sensor networks. *Pervasive Mob. Comput.* **2015**, *16*, 22–35. [CrossRef]
34. Li, J.; Wang, D.; Wang, Y. Security DV-hop localisation algorithm against wormhole attack in wireless sensor network. *IET Wirel. Sens. Syst.* **2018**, *8*, 68–75. [CrossRef]
35. Prashar, D.; Rashid, M.; Siddiqui, S.T.; Kumar, D.; Nagpal, A.; AlGhamdi, A.S.; Alshamrani, S.S. SDSWSN—A Secure Approach for a Hop-Based Localization Algorithm Using a Digital Signature in the Wireless Sensor Network. *Electronics* **2021**, *10*, 3074. [CrossRef]
36. Liu, X.; Su, S.; Han, F.; Liu, Y.; Pan, Z. A Range-Based Secure Localization Algorithm for Wireless Sensor Networks. *IEEE Sens. J.* **2019**, *19*, 785–796. [CrossRef]

37. Tang, Q.; Wang, J. A secure positioning algorithm against Sybil attack in wireless sensor networks based on number allocating. In Proceedings of the 2017 IEEE 17th International Conference on Communication Technology (ICCT), Chengdu, China, 27–30 October 2017; pp. 932–936. [CrossRef]
38. Douceur, J.R. The Sybil Attack. In *Peer-to-Peer Systems*; Druschel, P., Kaashoek, F., Rowstron, A., Eds.; Springer: Berlin/Heidelberg, Germany, 2002; pp. 251–260.
39. Dong, S.; Zhang, X.G.; Zhou, W.G. A Security Localization Algorithm Based on DV-Hop Against Sybil Attack in Wireless Sensor Networks. *J. Electr. Eng. Technol.* **2020**, *15*, 919–926. [CrossRef]
40. Council, N.R. *Trust in Cyberspace*; The National Academies Press: Washington, DC, USA, 1999; p. 126. [CrossRef]
41. Sidhu, S.; Mohd, B.J.; Hayajneh, T. Hardware security in IoT devices with emphasis on hardware Trojans. *J. Sens. Actuator Netw.* **2019**, *8*, 42. [CrossRef]
42. Mehrabi, M.A.; Doche, C.; Jolfaei, A. Elliptic curve cryptography point multiplication core for hardware security module. *IEEE Trans. Comput.* **2020**, *69*, 1707–1718. [CrossRef]
43. Khan, A.A.; Laghari, A.A.; Shaikh, A.A.; Dootio, M.A.; Estrela, V.V.; Lopes, R.T. A blockchain security module for brain–computer interface (BCI) with multimedia life cycle framework (MLCF). *Neurosci. Inform.* **2022**, *2*, 100030. [CrossRef]
44. Butun, I.; Sari, A.; Österberg, P. Hardware security of fog end-devices for the internet of things. *Sensors* **2020**, *20*, 5729. [CrossRef]
45. Pearson, B.; Luo, L.; Zhang, Y.; Dey, R.; Ling, Z.; Bassiouni, M.; Fu, X. On misconception of hardware and cost in IoT security and privacy. In Proceedings of the ICC 2019-2019 IEEE International Conference on Communications (ICC), Shanghai, China, 20–24 May 2019; pp. 1–7.
46. Kim, T.; Park, J.; Woo, J.; Jeon, S.; Huh, J. Shieldstore: Shielded in-memory key-value storage with sgx. In Proceedings of the Fourteenth EuroSys Conference 2019, Dresden, Germany, 25–28 March 2019; pp. 1–15.
47. Farooq-i-Azam, M.; Ayyaz, M.N. Location and position estimation in wireless sensor networks. In *Wireless Sensor Networks: Current Status and Future Trends*; CRC: Boca Raton, FL, USA, 2016; pp. 179–214.
48. Farooq-I-Azam, M.; Ni, Q.; Ansari, E.A. Intelligent Energy Efficient Localization Using Variable Range Beacons in Industrial Wireless Sensor Networks. *IEEE Trans. Ind. Inform.* **2016**, *12*, 2206–2216. [CrossRef]

electronics

MDPI

Article

Covert Channel Based on Quasi-Orthogonal Coding

Krystian Grzesiak *, Zbigniew Piotrowski and Jan M. Kelner

Institute of Communications Systems, Faculty of Electronics, Military University of Technology, 00-908 Warsaw, Poland; zbigniew.piotrowski@wat.edu.pl (Z.P.); jan.kelner@wat.edu.pl (J.M.K.)
* Correspondence: krystian.grzesiak@wat.edu.pl; Tel.: +48-261-885-509

Abstract: The paper presents a method of creating a hidden channel using a signals' superposition. According to this idea, a transmitter simultaneously sends overt and covert (secret) signals, whereby the overt signal is a carrier for the covert one. Due to the need to ensure a low probability of detection for covert communication, the covert signal should have low power. This implies a number of problems relating to its correct reception. This is similar to non-orthogonal multiple access (NOMA) systems, where the collective signal is a superposition of signals with different powers dedicated to different users. In this case, the successive interference cancellation (SIC) process is used in the receiver for the separation of the component signals. SIC requires accurate channel estimation. Even a small channel estimation error causes a significant increase in bit error rate (BER), performance degradation, or connection loss for covert transmission. This is due to the residual signal, i.e., the remnant of the cover signal after an imperfect SIC operation. The paper proposes a method of transforming (i.e., encoding) the applied hidden signal in such a way that the residual signal in the receiver is quasi-orthogonal to the hidden signal. The proposed model is based on appropriate sorting and, compared to methods with fixed constellation points, provides the covert channel with a low BER while maintaining high protection against detection as measured by the Kolmogorov–Smirnov distance. The proposed solution was tested using the USRP-2920 software-defined radio platform.

Keywords: security; steganalysis; covert channel; steganography; undetectability

Citation: Grzesiak, K.; Piotrowski, Z.; Kelner, J.M. Covert Channel Based on Quasi-Orthogonal Coding. *Electronics* **2023**, *12*, 2249. https://doi.org/10.3390/electronics12102249

Academic Editor: Athanasios D. Panagopoulos

Received: 30 March 2023
Revised: 5 May 2023
Accepted: 11 May 2023
Published: 15 May 2023

1. Introduction

Wireless transmission, in its physical layer, is susceptible to all kinds of manipulation, which can be used to create covert channels [1]. It is currently assumed that any method of communication used to illegally transmit information, which violates the system security policy, is a covert channel. Steganography of the physical layer of a radio signal essentially boils down to subtle changes of the parameters of the modulated cover signal. Such parameters can be the carrier frequency and, in the case of an orthogonal frequency-division multiplexing (OFDM) signal, the training sequence of the cover signal [2]. The vast majority of physical layer steganography methods consist of manipulating the position of the constellation points of the in-phase (I) and quadrature (Q) components of the cover signal. Small changes in position correspond to transmitted classified (covert) information, and an uninformed receiver (with no knowledge or ability to ascertain additional transmission) treats them as (channel or hardware) noise.

Examples of such solutions are presented in the literature:

(a) In [3], quadrature amplitude modulation (QAM) covert information constellation points are distributed around the QAM cover constellation points, forming the so-called dirty constellation. Thus, an additional constellation is formed based on the cover constellation.

(b) The authors [4] propose a similar solution, with the dirty constellation being formed using phase drift [5]. The solution can be applied to phase-shift keying (PSK) and QAM modulation.

(c) Hiding the data by moving the constellation points by a given angle (right or left) is shown in [1]. In this case, binary PSK (BPSK) modulation was used for practical implementation.
(d) In [6], the use of 8 frequency-shift keying (8FSK) modulation to embed information in QAM constellation points is presented. In this case, artificial neural networks were used to extract hidden data.
(e) The authors [7] noted that PSK modulations do not use the channel fully in terms of Shannon capacity. Therefore, the so-called residual capacity can be used to hide information. In order to hide this emission from potential observers, pseudo-noise asymmetric shift keying (PN-ASK) modulation is proposed.
(f) An extension to the pseudo randomness element of the [7] method is presented in [8]. The solution, called SteaLTE or Stegano LTE, is a steganographic technique for transmitting hidden data over Long-Term Evolution (LTE) radio networks. The developed method is resistant to steganalysis.
(g) The approach described in [9] is based on the [3] method with elements of additional randomness, which involves an additional shift in the phase of the constellation points of the covert signal. In addition, the authors [9] propose using polar codes to reduce bit error rate (BER).
(h) The transmission of stealth information in the form of noise on a QAM basis is presented in [10]. In this case, the cover's signal is not used to carry information.

The aforementioned literature on the covert channel ignores the issue of channel estimation error [11] and, directly related to this, non-perfect (non-ideal) successive interference cancellation (SIC) [12–15]. The assumption of being able to easily separate the cover signal from the very low-power covert signal is fundamentally difficult to implement. Hence, in this paper, the authors propose an original and novel approach, which is to transmit the covert signal in such a way that it is quasi-orthogonal to the cover signal at the receiver. This is accomplished by sorting, that is, by appropriately ranking the IQ samples of the covert signal against the cover signal over time. In the receiver this results in mutual orthogonalization, thus, easier frequency separation of the signals. The method used can be used for both amplitude–phase and frequency modulations. In the paper, however, FSK modulation is indicated as the optimal solution in terms of transmission capabilities (calculated by transmission speed) as well as protection against steganalysis. The proposed approach and the considerations presented here are a continuation of the work presented in [16].

The work is organized as follows. In Chapter 2, the basics of creating a covert channel in a radio channel are presented. Chapter three describes how the proposed transceiver with sorting circuit works. The results of the computer simulations are included in Chapter 4. The rationale for using FSK modulation to create a covert channel is placed in Chapter 5. Chapter 6 contains the results of the tests conducted based on the universal software radio peripheral (USRP). A summary is included at the end.

2. Radio Physical Layer Steganography

2.1. Creating a Covert Channel

In mathematical terms, the process of creating a covert channel in the physical layer of a radio channel can be described as a superposition of a cover signal (cover) and a covert signal, which can be represented by the formula:

$$s(t) = \sqrt{P_1}x_1(t) + \sqrt{P_2}x_2(t) \tag{1}$$

where x_1 and x_2 are the cover signal and the covert signal, respectively, P ($P = P_1 + P_2$) is the transmitter power.

In order to reduce the probability of detecting a covert channel, the following conditions should be met:

(a) the covert signal power should be significantly less than the cover signal ($P_1 \gg P_2$).

(b) constellation points of the covert signal should have a pseudo-random (noise) characteristic.

The receiver input signal is represented by the following simplified formula:

$$y(t) = h\sum_{i=1}^{2} \sqrt{P_i}x_i(t) + w(t) \tag{2}$$

where h ($h \sim CN(0,1)$) is a channel gain and w is a Gaussian noise.

In an ideal case, when $w(t) = 0$ and the value of the parameter h is known in the receiver, the recovery process of the signal $x_2(t)$ would proceed as follows (ideal SIC):

$$x_2(t) = \frac{y(t) - \sqrt{P_1}x_1(t)}{\sqrt{P_2}} + w(t) \tag{3}$$

Transmission and reception of the cover and covert signal was presented in Figure 1.

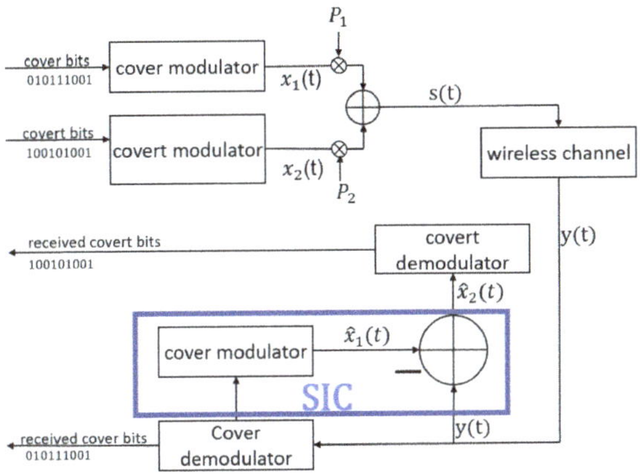

Figure 1. Transmission and reception of a covert signal as a superposition of cover and covert signals.

2.2. Channel Estimation Error

In a real-world situation, the receiver does not have full information about the wireless channel (2), but only estimates the parameter h with a certain error value ε ($\varepsilon \sim CN(0,\sigma_\varepsilon^2)$):

$$\hat{h} = h + \varepsilon, \tag{4}$$

where $\hat{h}$ is a channel gain estimation.

Taking into account the channel estimation error (4), the recovered covert signal $\hat{x}_2(t)$ is distorted by the residual cover signal according to the following formula (non-ideal SIC):

$$\hat{x}_2(t) = h\sqrt{P_2}x_2(t) + \varepsilon\sqrt{P_1}\hat{x}_1(t) + w(t) \tag{5}$$

The error introduced by the SIC propagates and affects the demodulation of the covert signal, i.e., the interference that arose during the recovery of the cover signal (stronger signal) propagates during the recovery of the covert signal (weaker signal) [17–19]. Errors arising during SIC mostly depend on the channel coefficients and the power allocation coefficient [20].

According to Formula (5), the channel estimation error variance value σ_ε^2 has a fundamental effect in the recovery of the covert signal. In this paper, we consider a situation where $\sigma_\varepsilon^2 \neq 0$ as a non-ideal SIC. In such a case, the best way to distinguish signals $x_1(t)$ from $x_2(t)$ is to ensure their mutual orthogonality or (if this is not possible) quasi-orthogonality.

2.3. Quasi-Orthogonality

We speak of orthogonality when the inner product of the $x_1(t)$ and $x_2(t)$ signals is zero, according to the formula:

$$\int_{-\infty}^{\infty} x_1(t) \cdot x_2^*(t)dt = 0 \tag{6}$$

The correlation of two time-limited signals defined over the time interval $0 < t < \tau$ is defined as:

$$r(t) = \int_{-\infty}^{\infty} x_1(\rho) \cdot x_2^*(\rho - t)d\rho, \tag{7}$$

where ρ is a dummy variable.

For two orthogonal signals for each t shift, a zero correlation is obtained if these signals are disjoint in frequency domain. Hence, separated band-limited signals in frequency are the main method of obtaining orthogonal signals. For signals that occupy the same frequency band, there is no possibility of zero cross-correlation for any t time shift.

Quasi-orthogonality [21] refers to signals (waveforms) that exhibit low cross-correlation. Two waveforms are quasi-orthogonal if:

$$\left| \int_{-\infty}^{\infty} x_1(\rho) \cdot x_2^*(\rho - t)d\rho \right| < \varepsilon \tag{8}$$

where $\varepsilon \ll 1$ for any t, and x_1 and x_2 are normalized to a unit energy value.

3. Proposed Model

3.1. Basic Assumptions

In the presented model, a single covert symbol is superimposed on several cover signal symbols. Such a solution stems from the need to reduce the distortion/interference of the cover signal (in order to preserve the energy per bit, variance of the covert signal is reduced), and thus the detection by outsiders of the fact that a covert channel exists. In the analyzed model, as in previous solutions [2,3,7,10], we assume that the cover is QAM amplitude–phase modulation in the form of IQ samples. In order to facilitate the process of demodulation of covert information from Figure 1, it is proposed to cross-orthogonalize the cover and covert signals by sorting. It is assumed that the coherence time of the radio channel, and therefore the channel estimation error, remains constant at least for a single data block for which the sorting operation is performed. Knowing that a typical channel coherence time is between approximately 10 ms to 200 ms [22–24], this assumption will usually remain.

Sorting occurs in both the transmitter and receiver as shown in Figure 2, whereby:

- At the transmitter, samples of the covert signal $x_2(t)$ are sorted based on a given sequence of the cover (QAM modulation) signal $x_1(t)$. In this way, the sorted signal x_{2_p} has a pseudo-random (noise) form.
- At the receiver, the $\hat{x}_2(t)$ signal re-sorting is performed after the SIC operation. Sorting in the receiver aims to:
    - (a) restore the original sample order of the covert signal after the SIC operation $\hat{x}_{2_p}$ to the original order (in an ideal case $\hat{x}_{2_i} = x_2$)
    - (b) restoring the original order of $\hat{x}_{2_p}$ is followed by a simultaneous change in the sample order of the residual signal associated with x_1 Thus, the residual signal becomes orthogonal (quasi-orthogonal) to the covert signal.
    - (c) The covert signal $\hat{x}_{2_i}$ is fed to the input of the covert channel demodulator

Sorting involves dividing covert information into blocks. The block length depends on the covert information modulation used, the value of the cover modulation and the number of IQ samples per signal. Every covert symbol consists of several IQ points imposed on the several cover symbols. In Section 4 are presented simulation results for blocks equalling 16 or 64 cover symbols with (imposed) 1 to 4 covert symbols, respectively. From the proposed principle of sorting, in order to correctly reproduce the covert signal, it is necessary to correctly (without error) reproduce the unclassified (cover) signal (the equation $\hat{x}_1 = x_1$ must be true) because based on the cover signal, the reverse operation of sorting in the transmitter is reproduced (in the covert samples reorder system). Any error in the reception of the cover data in the block results in an error in the covert signal. That means, for example, that if the block has a length of 16 cover symbols with one imposed covert symbol, we lose one covert symbol in case of any cover error. And similarly, for a larger number of covert symbols per block, we lose all covert symbols. Therefore, the block length and the number of covert symbols in the block should be taken carefully.

As shown in Figure 2, the proposed solution, compared with traditional SIC (Figure 1), is based on two additional sorting operations: one in the transmitter and one in the receiver. It can be assumed that the complexity of every sorting operation has complexity O(n^2).

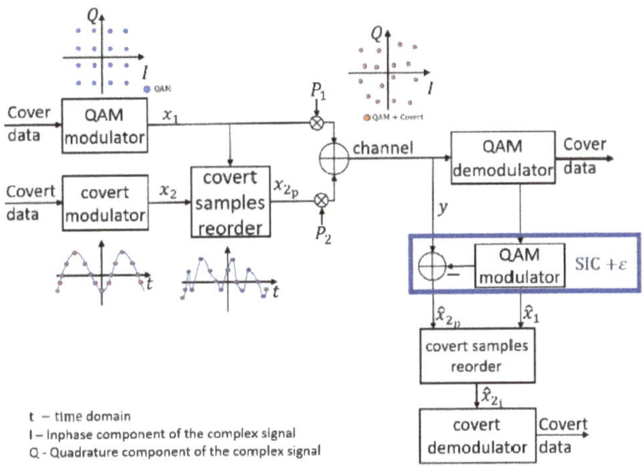

Figure 2. Covert signal transceiver system.

The selection of the optimal modulation of covert information was preceded by simulation tests included in the following chapters. The easiest way to explain orthogonalization (the creation of quasi-orthogonal signals) is to assume that the covert signal is 8FSK modulation. By orthogonalization we mean the mutual transformation of signals in such a way that their spectra are disjointed, i.e., they do not overlap. The 8FSK signal spectrum has eight peaks corresponding to each symbol. The averaged spectrum of the QAM signal is flat over the bandwidth except roll-off, but the instantaneous spectrum calculated for a packet/block of data is characterized by high variability. Hence the concept of using such sorting that will reduce the instantaneous spectrum variation for random IQ values representing QAM symbols. It is reasonable to assume that sorting can, by decoupling the instantaneous spectra of the cover and covert information, reduce the impact of channel estimation error during the recovery of covert information.

3.2. Cover Signal Sorting

In this section, it will be shown on a selected example that by sorting the IQ samples of the cover signal in an appropriate way, its spectral characteristics can be changed, so that the influence of the residual signal on the covert signal x_2 is reduced (Figure 2).

The impact of cover IQ sample sorting will be tested in the frequency domain using fast Fourier transform (FFT) analysis on the example of QAM amplitude–phase modulation. Sorting is performed for a sequence (block) of random IQ samples of signal x_1 consisting of $N = 64$ samples of 64 order QAM modulation ($M = 64$). The probability of each symbol (from 0 to 63) is equally likely. The averaged FFT spectrum of such a signal does not have a clear main peak. We then sort the samples according to the phase increment defined as the angle $angle(x_1)$. The signal sorted in this way is denoted x_{1_sorted}. Since the phases of the QAM constellation points vary from $-\pi$ to π, a x_{1_sorted} signal close to a sine wave (Figure 3a) with a period equal to N samples will be obtained. The FFT spectrum of the signal thus formed has one strong spectral line for the fundamental frequency (f_0), the normalized value of which is $f_0 = 2/N$. The same will be true if N is a multiple of the modulation order M. By reordering the x_{1_sorted} samples, we can obtain shifted spectral lines on the frequency $f_n = nf_0$, where f_0 is a fundamental frequency, n is a power of 2 and satisfies the condition $-0.5 \leq f_n \leq 0.5$. The above actions can be written using matrix operations. If we have a set of sorted (according to phase increase) samples of the N-QAM x_{1_sorted} signal with indices [1:N], the i indices of the sorted signal with the fundamental frequency $\mathrm{k}\cdot f_0$ ($\mathrm{k} = 1, 2, 4 \ldots \frac{\mathrm{N}}{2}$) are obtained according to the formula:

$$i = reordered\_index\_for\ x_{1\_sorted} = \text{parallel\_to\_serial}\left\{\begin{bmatrix} 1:\mathrm{k}:\mathrm{N} \\ 2:\mathrm{k}:\mathrm{N} \\ \cdot \\ \cdot \\ \cdot \\ \mathrm{k}:\mathrm{k}:\mathrm{N} \end{bmatrix}^{\mathrm{T}}\right\} \tag{9}$$

For example, if we have a sorted (according to the phase increment) set of samples of a 64QAM signal labelled x_{1_sorted} with base frequency f_0 and corresponding indexes from 1 to 64, then the indices of the signal with frequency $16\cdot f_0$ and $32\cdot f_0$ (which corresponds to the normalized frequency equal to $^1/_4$ and $^1/_2$) are as follows:

$$i = \text{parallel\_to\_serial}\left\{\begin{bmatrix} 1,\ 17, 33, 49 \\ 2,\ 18,\ 34,\ 50 \\ 3,\ 19,\ 35,\ 51 \\ \cdot \\ \cdot \\ \cdot \\ 16,\ 32,\ 48,\ 64 \end{bmatrix}^{\mathrm{T}}\right\} \Rightarrow \tag{10}$$

and for $k = 32$ ($32\cdot f_0$)

$$i = \text{parallel\_to\_serial}\left\{\begin{bmatrix} 1,\ 33 \\ 2,\ 34 \\ \cdot \\ \cdot \\ \cdot \\ 32, 64 \end{bmatrix}^{\mathrm{T}}\right\} = [1,\ 33,\ 2,\ 34,\ \ldots,\ 32,\ 64] \tag{11}$$
$$\Rightarrow reordered\left(x_{1_{sorted}}\right) = x_{1_i} = \left[x_{1_1},\ x_{1_{33}}, x_{1_2}, \ldots x_{1_{64}}\right]$$

In Figure 3, for 64-QAM modulation, for a sequence of 64 random IQ samples, the time and frequency spectrum waveforms are shown for sequences sorted with fundamental frequencies of f_0, $16\cdot f_0$ and $32\cdot f_0$, respectively.

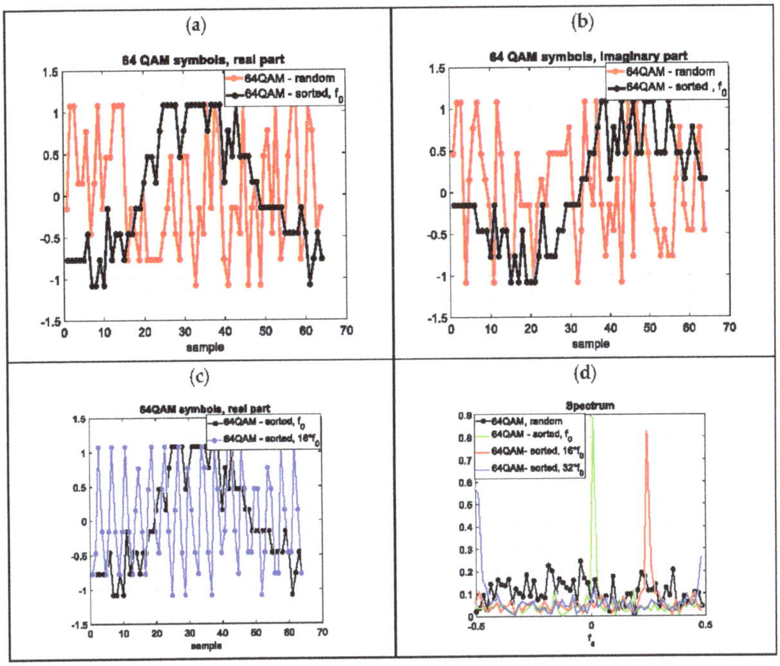

Figure 3. 64QAM signal. (**a**) Random and sorted (x_{1_sorted}, f_0) 64QAM signal-real part of the signal, (**b**) Random and sorted 64QAM signal-imaginary part of the signal (c) Sorted 64QAM signal with frequencies of f_0 and $16{\cdot}f_0$ (d) Instantaneous FFT spectrum for 64 samples of random and sorted signals.

According to Figure 3, for 64 random 64QAM symbols we can, as a result of sorting, control (to a certain extent) the position of the spectral lines. This enables the parameters of the covert signal to be selected for optimal reception. On the basis of the presented method, the reception of the covert signal in the form of FSK and PSK modulation and the different modulation values of the cover and covert signal will be presented.

4. Simulation Tests

A preliminary evaluation of the feasibility of using the system in Figure 2 to receive a covert signal in the presence of a channel estimation error resulting in a non-perfect SIC was carried out in the MATLAB environment. The following graphs are shown in the figures to better illustrate the phenomena taking place:

(a) Averaged FFT spectrum of cover and covert signal (before and after sorting)
(b) Averaged value and variance of cross-correlation of signals
(c) Probability distribution of cross-correlation of signals estimated using histograms

In order to provide an understanding of the phenomena occurring during the sorting process, the signals were assumed to be unnoisy and of equal power to produce charts containing spectra and cross-correlations (Figure 5a–f). This corresponds to the situation when the $\hat{x}$ signal in Figure 2 consists of a cover and covert signal of equal strengths. This is to show how distinguishable the signals are in the time and frequency domain.

Detection capabilities Figure 5f were investigated in accordance with the diagram in Figure 1 (system without sorting) and Figure 2 (system with sorting), assuming that the signal-to-noise ratio (SNR) of the channel (calculated for the aggregate signal) is 45 dB, and the estimation error of channel ε has a variance σ_{ε}^2. It was assumed that the covert symbol is transmitted with l IQ samples that are submultiples of the number of cover symbols for which the sorting operation is performed. The simulation was performed for 10,000 (for each value of σ_ε^2) executions of a random sequence of cover and covert data. It was assumed that the value $\sqrt{P_1} = 1$, $\sqrt{P_2} = 0.005$.

In the case of frequency modulation, it was assumed that the M-FSK covert signal [25–29] in the baseband is defined as:

$$g(t) = \exp(i \cdot \pi \cdot k \cdot \Delta f \cdot t + \theta) \tag{12}$$

$k = \pm 1,\ \pm 3,\ \ldots \pm M/2$, and Δf is a frequency deviation, θ random initial phase (random for each symbol).

4.1. Simulation No. 1

The elementary signal from the quadrature transmitter is a composite (superposition) of one symbol of 8FSK (each symbol as 16 samples) and 16 symbols (samples) of 16QAM (Figure 4). The covert signal was sorted in such a way that, at the receiver, the cover fundamental frequency f_0 was increased four times (the normalized frequency of the cover is $4 \cdot f_0 = 0.25$).

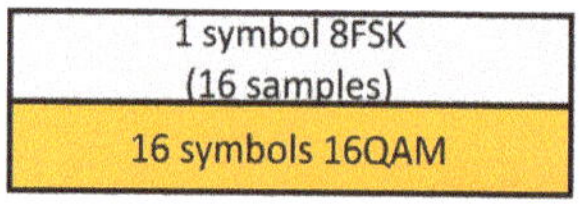

Figure 4. Block of samples subject to sorting.

Figure 4 shows the impact of sorting on the cover spectrum and the cross-correlation between the cover and the covert signal (8FSK) (cross-correlation between two signals). As a result of sorting, the value of the averaged cross-correlation for the 8FSK modulation symbol equal to "5" has increased (Figure 5c), but its variation is eight times smaller (Figure 5e). For the other symbols, the cross-correlation is lower, as is its variance. As a result, the SIC operation runs with a lower probability of BER error.

(a) (b)

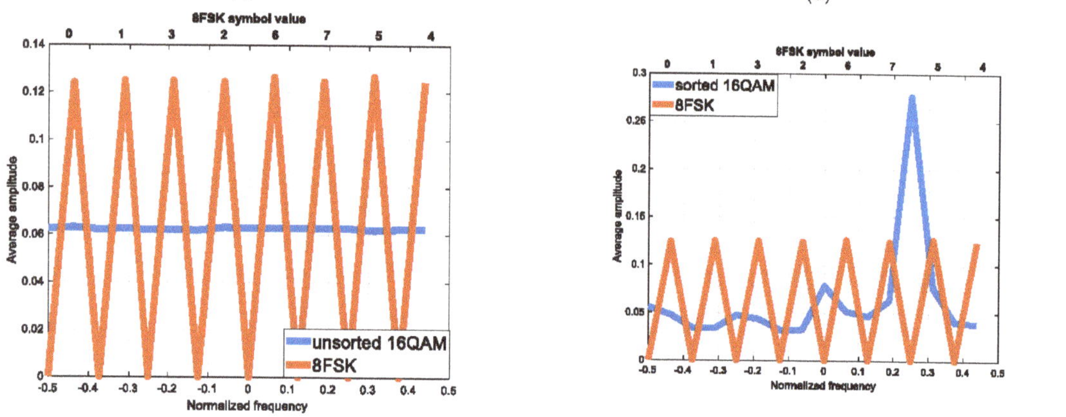

Figure 5. *Cont.*

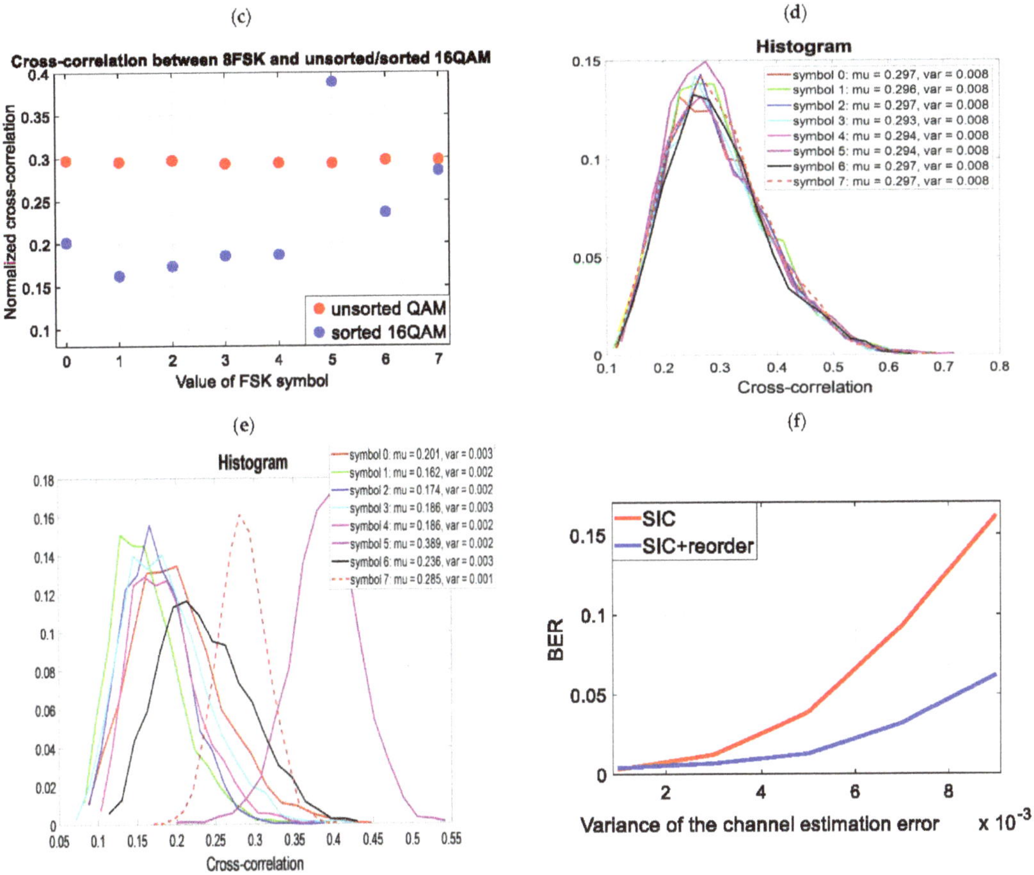

Figure 5. Testing the impact of cover orthogonalization on the non-ideal SIC process (**a**) Averaged FFT spectrum of unsorted cover and 8FSK (**b**) Averaged FFT spectrum of sorted cover and 8FSK (**c**) Averaged cross-correlation value of unsorted cover and 8FSK (**d**) Histogram of the cross-correlation value of the unsorted cover and 8FSK (**e**) Histogram of the cross-correlation value of the sorted cover and 8FSK (**f**) Reception of covert information (8FSK) for the unsorted and sorted cover (non-ideal SIC).

4.2. Simulation No. 2

In order to reduce cross-correlations (from Simulation No. 1), the sorting was changed (the number of covert samples for the block remains the same as in Figure 5). The spectrum in Figure 6b) was obtained by means of two successive repetitions of sorting (according to Formulas (13) and (14)) of the originally sorted x_{1_sorted}.

$$i_1 = \text{paralel\_to\_serial}\left\{ \begin{bmatrix} 1:2:\mathrm{N} \\ 2:2:\mathrm{N} \end{bmatrix}^{\mathrm{T}} \right\} \tag{13}$$

$$i_2 = f(i_1) = \left[1:\mathrm{k}:\frac{\mathrm{N}}{2}, \ \mathrm{N}:-1:\frac{\mathrm{N}}{2}+1 \right] \tag{14}$$

As a result of the sorting, the cross-correlation for the symbol "7" is the same before and after the orthogonalization process. Nevertheless, the variance of the cross-correlation is twice as small. Hence, a much lower BER was obtained for the sorted cover than in the previous case (Figure 5f) vs. (Figure 6f).

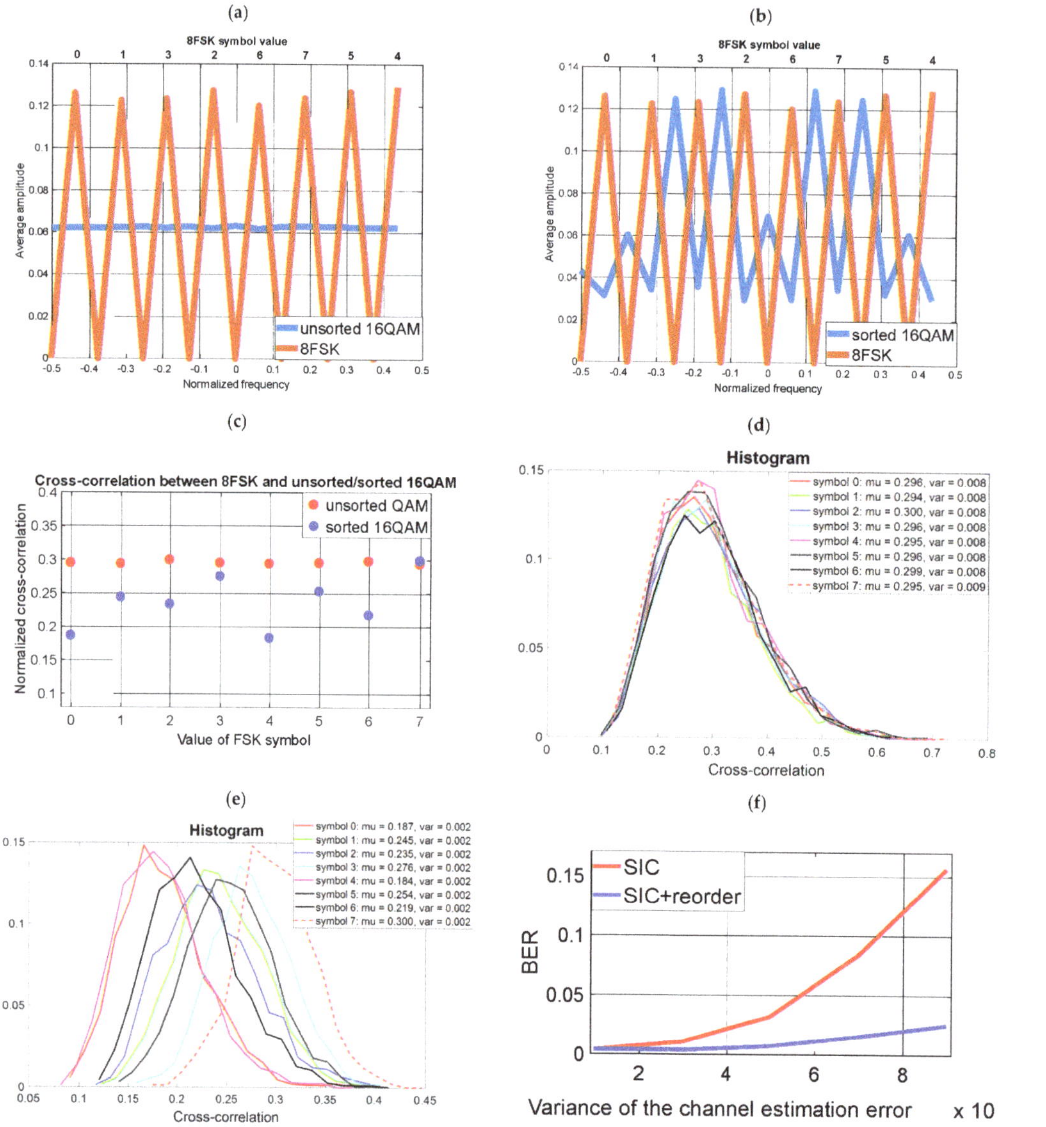

Figure 6. Testing the impact of cover orthogonalization on the non-ideal SIC process (**a**) Averaged FFT spectrum of unsorted cover and 8FSK (**b**) Averaged FFT spectrum of sorted cover and 8FSK (**c**) Averaged cross-correlation value of unsorted cover and 8FSK (**d**) Histogram of the cross-correlation value of the unsorted cover and 8FSK (**e**) Histogram of the cross-correlation value of the sorted cover and 8FSK (**f**) Reception of covert information (8FSK) for the unsorted and sorted cover (non-ideal SIC).

4.3. Simulation No. 3

The cover is 64QAM. For every 64 samples of the cover signal, there are 4 symbols (16 samples each) of the covert signal. To correctly decode the covert data, the sorting operation must be performed sequentially for each block of data shown in Figure 7. The data in the transmitter and receiver are sorted according to the cover signal. The normalized frequency of the sorted cover is $4 \cdot f_0 = 0.5$. The results are presented in Figure 8.

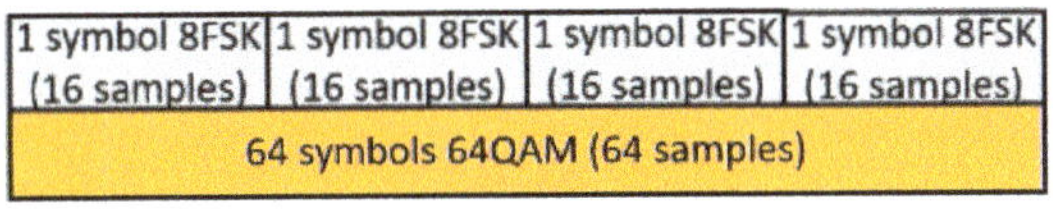

Figure 7. Block of samples subject to sorting.

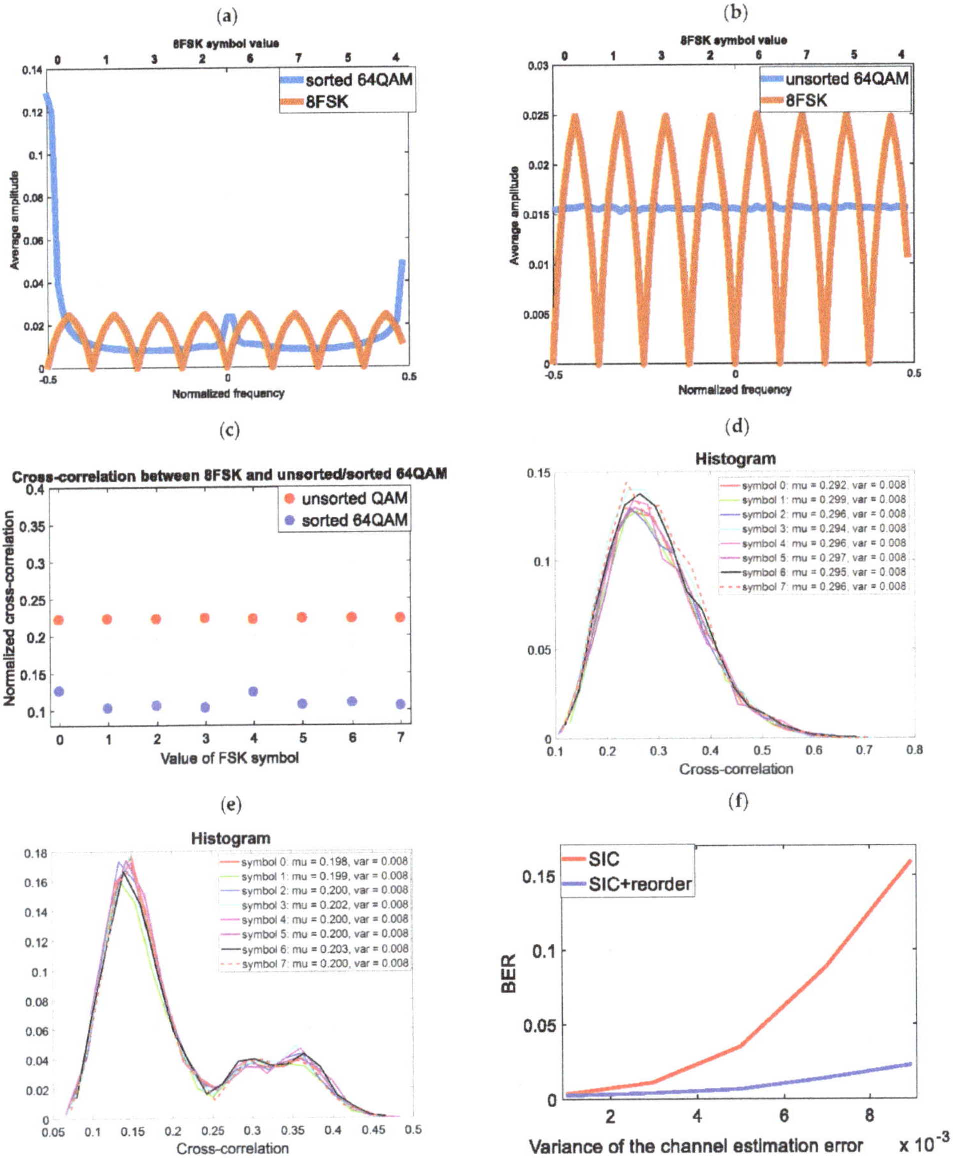

Figure 8. Testing the impact of cover orthogonalization on the non-ideal SIC process (**a**) Averaged FFT spectrum of unsorted cover and 8FSK (**b**) Averaged FFT spectrum of sorted cover and 8FSK (**c**) Averaged cross-correlation value of unsorted cover and 8FSK (**d**) Histogram of the cross-correlation value of the unsorted cover and 8FSK (**e**) Histogram of the cross-correlation value of the sorted cover and 8FSK (**f**) Reception of covert information (8FSK) for the unsorted and sorted cover (non-ideal SIC).

As a result of the sorting, the cross-correlation of the cover and the covert signal was reduced. This has reduced the BER for a non-perfect SIC. The results are compared with Figure 6f).

4.4. Simulation No. 4

In previous simulations (simulation 1 to 3), the modulation of the covert information was 8FSK. However, it should be examined what effect sorting has when the covert signal is the amplitude–phase modulation used in [1,3,7,10]. Let's assume, as in simulations 1 and 2, that one 2PSK covert symbol is transmitted using 16 16QAM cover symbols. The FFT analysis (Figure 9a) of such a signal will show that each covert symbol is a constant value (on the frequency scale it has a non-zero value only for $f = 0$). Orthogonalization will not provide any benefit (Figure 9f) because the block for which we perform orthogonalization is equal to the modulation value of the cover and, at the same time, the number of samples per covert symbol. This is because the average value of the random sorted and unsorted cover in such a case remains constant.

Note that we obtained a relatively low BER for both the sorted and unsorted signal, even for a large estimation error. This is due to the high energy per bit (16 samples represent one bit of data) and, unfortunately, this comes at the expense of reduced resistance to steganalysis (as will be demonstrated in Section 5).

4.5. Simulation No. 5

Simulation conditions are the same as in the previous example, except that we increase the value of modulation of the covert information to 8PSK. The results are presented in Figure 10.

Multi-valued 8PSK modulation requires a higher ratio of energy per bit of information. For this reason (with an assumed SNR = 45 dB), even for zero estimation error, the BER is different from zero (Figure 10f) and there is less immunity to channel estimation errors.

4.6. Simulation No. 6

Sorting was carried out for the parameters as for case no. 3 (sorting a block of data equal to 64, and a covert symbol with a length of 16 samples). The modulation for covert data is 4PSK. The results are presented in Figure 11. The benefits of the sorting are noticeable, although the transmission rate compared with 8FSK is twice smaller.

4.7. Simulation No. 7

Sorting was carried out for the parameters as for case no. 6 (sorting a block of data equal to 64, and a covert symbol with a length of 16 samples). The modulation for covert data is 8PSK. The results are presented in Figure 12.

As expected, sorting yields a lower BER. However, comparing Figure 7 with Figure 12 graphs, it is clear that for the given bit rates and power levels of the covert signal, better results are obtained (regardless of the sorting process) for 8FSK modulation. 8PSK modulation relative to 8FSK requires more energy per bit.

All simulations presented above aimed to show that, for the given waveform of the covert channel, it is possible to find an optimal sorting pattern to minimize imperfect SIC operation in the covert signal demodulator. This seems to be easier for FSK modulation and longer frames. However, it is necessary to keep in mind that the longer the frame, the lower the probability that channel gain is constant, which is the main assumption of this method.

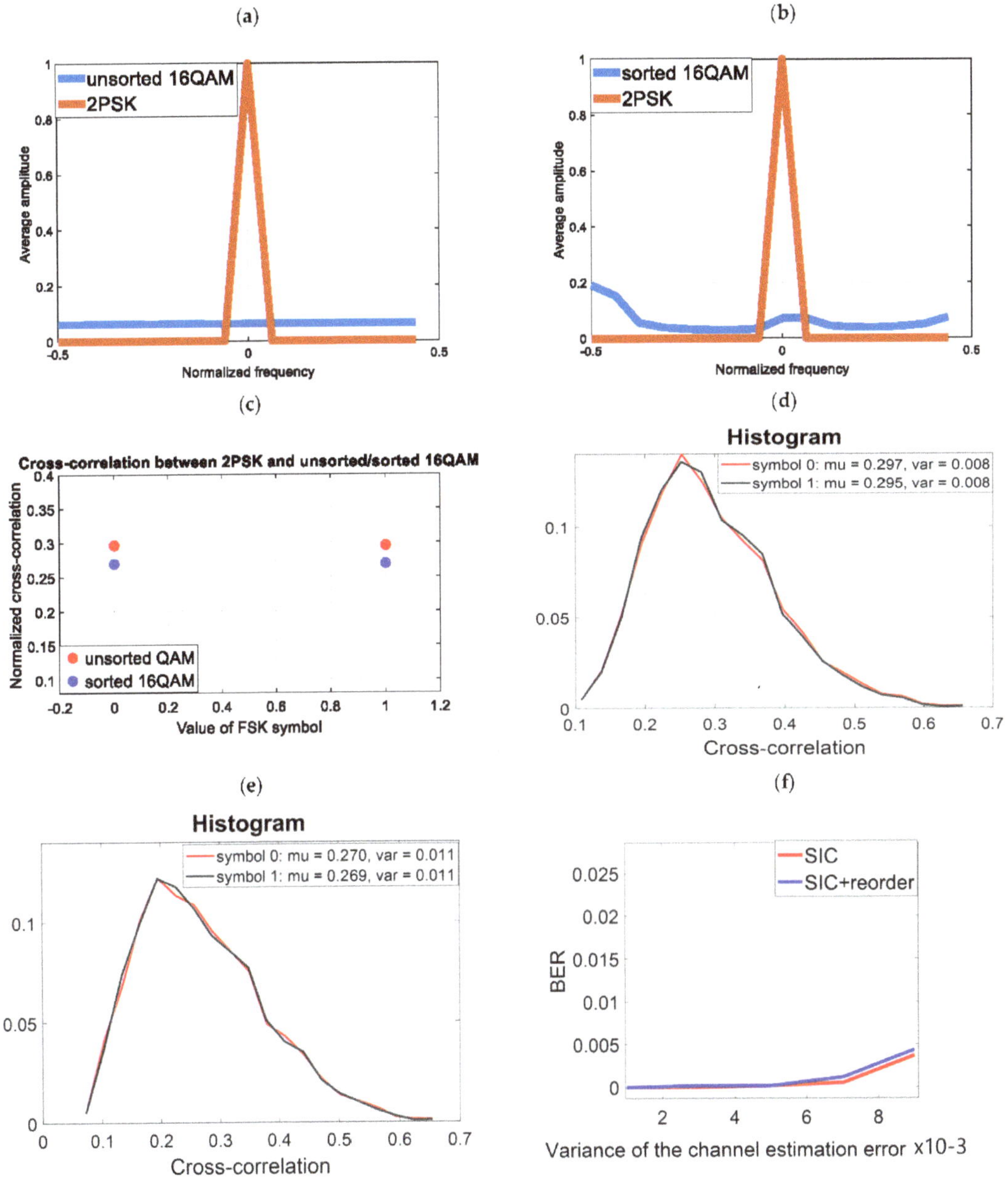

Figure 9. Use of 2-PSK modulation for covert transmission (**a**) Averaged FFT spectrum of unsorted cover and 2PSK (**b**) Averaged FFT spectrum of sorted cover and 2PSK (**c**) Averaged cross-correlation value of unsorted cover and 2PSK (**d**) Histogram of the cross-correlation value of the unsorted cover and 2PSK (**e**) Histogram of the cross-correlation value of the sorted cover and 2PSK (**f**) Reception of covert information (2PSK) for the unsorted and sorted cover (non-ideal SIC).

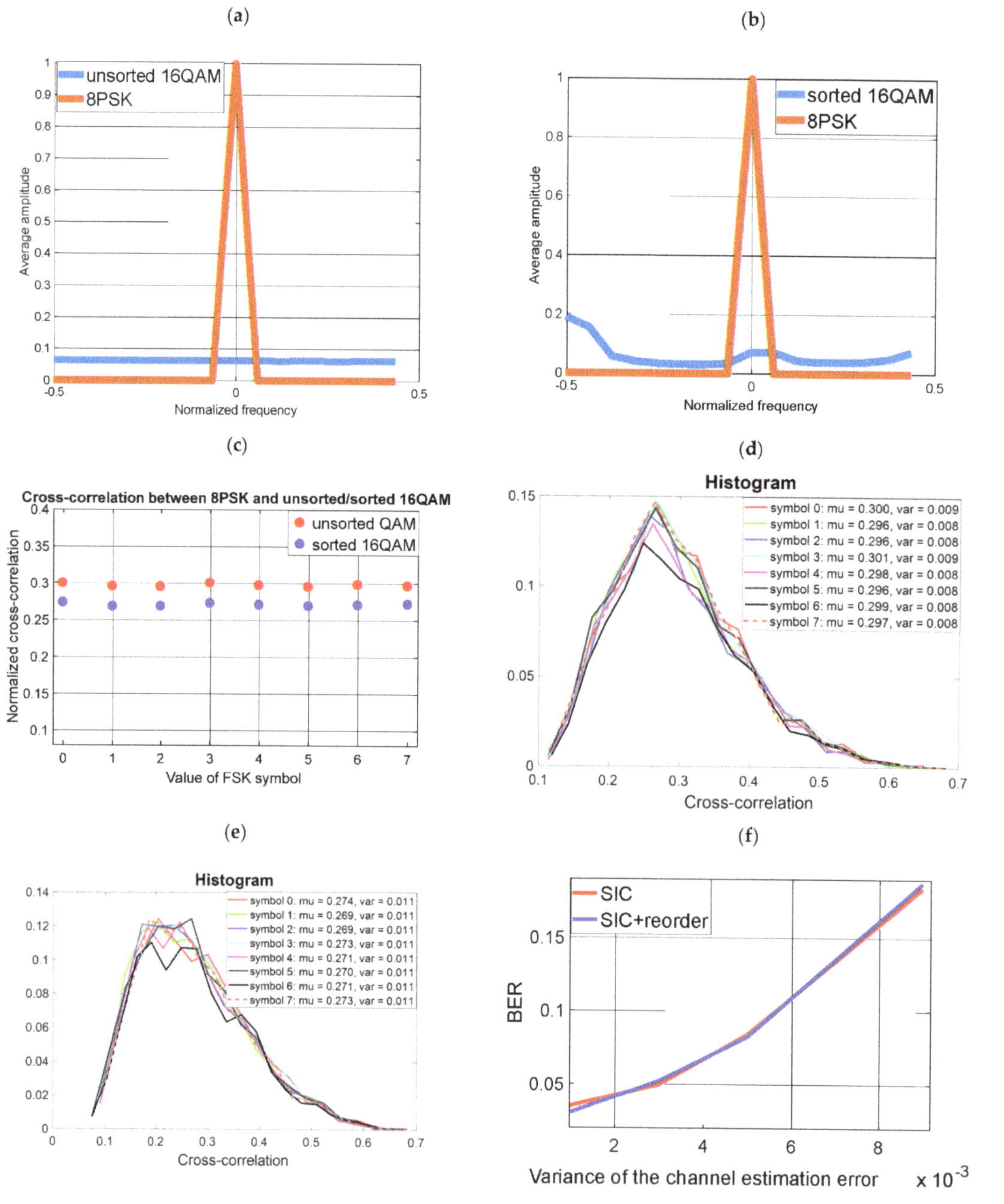

Figure 10. Use of 8-PSK modulation for covert transmission (**a**) Averaged FFT spectrum of unsorted cover and 8PSK (**b**) Averaged FFT spectrum of sorted cover and 8PSK (**c**) Averaged cross-correlation value of unsorted cover and 8PSK (**d**) Histogram of the cross-correlation value of the unsorted cover and 8PSK (**e**) Histogram of the cross-correlation value of the sorted cover and 8PSK (**f**) Reception of covert information (8PSK) for the unsorted and sorted cover (non-ideal SIC).

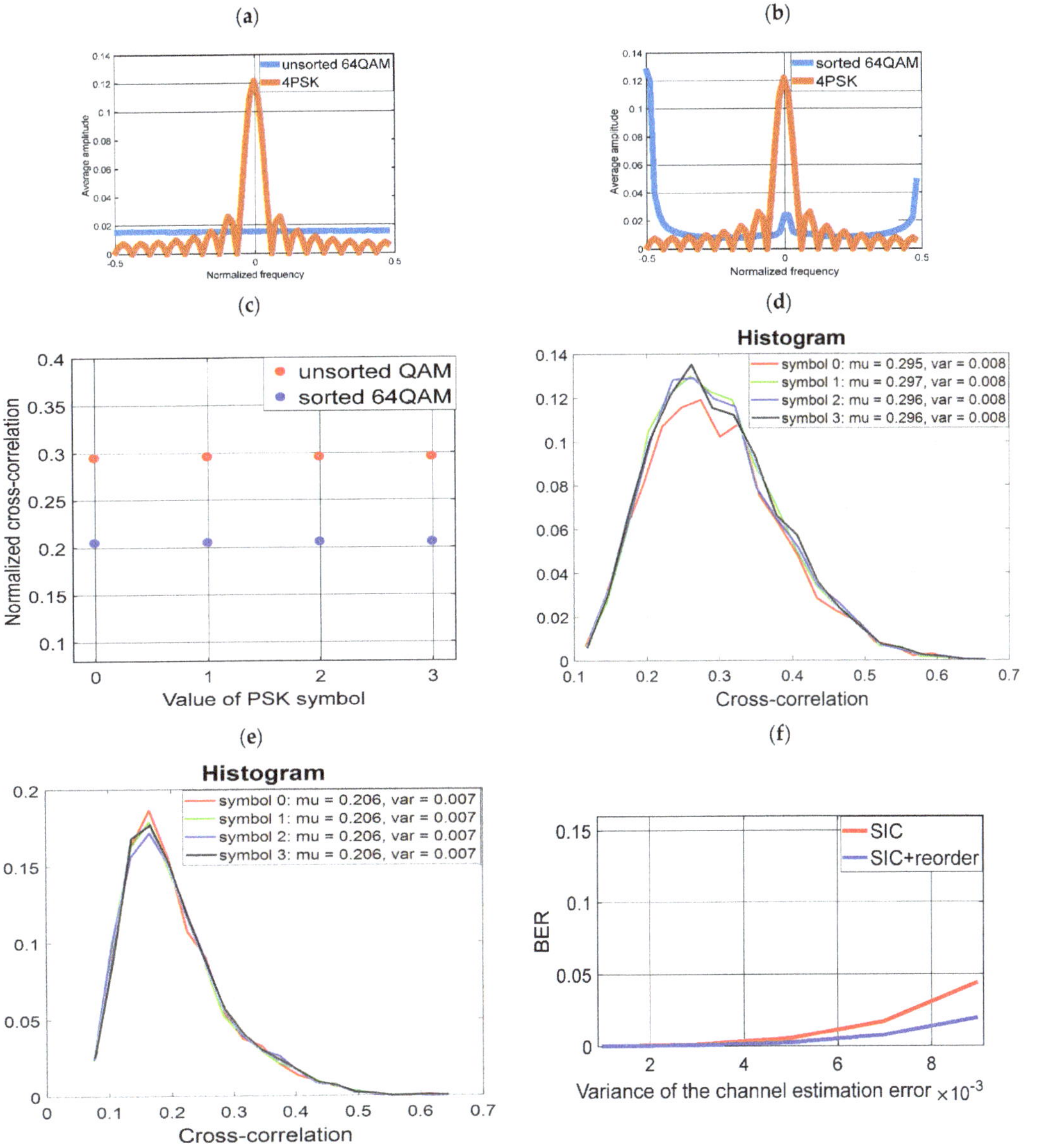

Figure 11. Use of 4-PSK modulation for covert transmission (**a**) Averaged FFT spectrum of unsorted cover and 8PSK (**b**) Averaged FFT spectrum of sorted cover and 4PSK (**c**) Averaged cross-correlation value of unsorted cover and 4PSK (**d**) Histogram of the cross-correlation value of the unsorted cover and 4PSK (**e**) Histogram of the cross-correlation value of the sorted cover and 4PSK (**f**) Reception of covert information (4PSK) for the unsorted and sorted cover (non-ideal SIC).

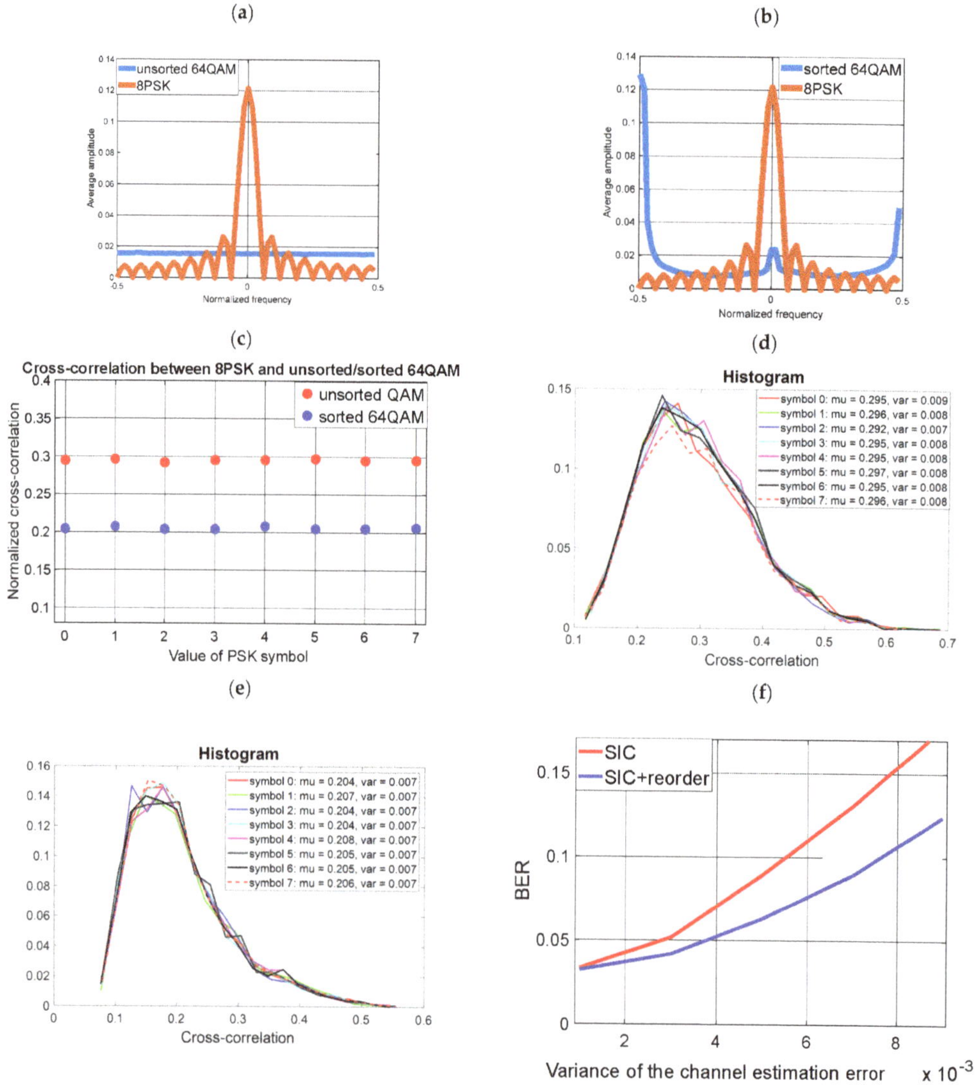

Figure 12. Use of 8-PSK modulation for covert transmission (**a**) Averaged FFT spectrum of unsorted cover and 8PSK (**b**) Averaged FFT spectrum of sorted cover and 8PSK (**c**) Averaged cross-correlation value of unsorted cover and 8PSK (**d**) Histogram of the cross-correlation value of the unsorted cover and 8PSK (**e**) Histogram of the cross-correlation value of the sorted cover and 8PSK (**f**) Reception of covert information (8PSK) for the unsorted and sorted cover (non-ideal SIC).

5. Steganographic Analysis

The choice of FSK modulation as the modulation for covert information is not only due to its good transmission properties and easy orthogonalization (quasi-orthogonalization) process with respect to the cover. The use of FSK modulation provides better properties in terms of low probability of detection (LPD), which is due to increased immunity to steganographic analysis compared to amplitude–phase modulations with constant constellation points.

By steganographic analysis we mean testing of probability density distributions and cumulative distribution function estimated by means of a histogram and cumulative histogram. Quantitatively, a measure of the difference in distributions can be calculated using the Kolmogorov–Smirnov test. To do this, the receiver must have noise information in the radio channel [30–32] and statistics formed from the signal from the SIC system output (we assume that the receiver is able to demodulate the cover information). If we denote the cumulative histogram distribution of the noise and signal after performing the SIC operation by F_w and $F_{\hat{x}_2}$, respectively, the Kolmogorov–Smirnov distance $KSTEST$ is expressed by the formula [10,33]:

$$KSTEST = max\left|F_w - F_{\hat{x}_2}\right| \quad (15)$$

Results of $KSTEST$ calculated on the basis of 200,000 IQ samples for $SNR = 45$ dB and 50 dB conditions relative to the cover in the form of 64QAM, $\sqrt{P_1} = 1$, $\sqrt{P_2} = 0.005$ and zero channel estimation error are shown in the Table 1.

Table 1. KSTEST calculation.

Covert Modulation	*KSTEST*	
	***SNR* = 45 dB**	***SNR* = 50 dB**
2PSK	0.153	0.321
4PSK	0.087	0.214
8PSK	0.082	0.188
2FSK	0.081	0.186
4FSK	0.081	0.186
8FSK	0.080	0.185

Example histograms and cumulative histograms for 2PSK and 2FSK modulations are presented in Figure 13. It was assumed that $SNR = 45$ dB.

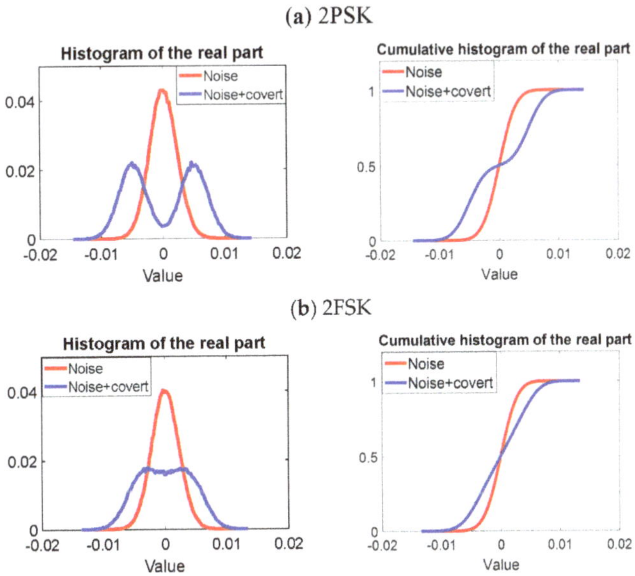

Figure 13. Histograms and cumulative histograms for (**a**) 2PSK and (**b**) 2FSK signals.

The analysis indicates that the 8FSK signal for the additive white Gaussian noise (AWGN) channel and ideal SIC provides the highest bit rate for the covert signal, while providing the best (calculated in K-S distance) steganographic protection. In addition, simulation tests have proven (Section 4) that sorting can be successfully used, which effectively reduces the impact of channel estimation error.

6. Practical Implementation

The concept of a covert channel based on quasi-orthogonal coding was implemented by using the USRP-2920 [34] hardware platform manufactured by National Instruments. USRP is the essential hardware part for generating a radio signal, while the software part is provided by the LabView software (with Matlab scripts) installed on a personal computer (PC). An Ethernet network adapter with a bit rate of 1 Gb/s is used to provide communication between USRP and the PC. Two USRP-2920 were used to implement a test stand (Figure 14) for detectors (in the transmitter–receiver system) connected with the computer by an unmanaged switch. The system was placed in an office room, and the distance between the transmitter and receiver was 5 m. The line-of-sight (LOS) propagation conditions were ensured disturbance only by office equipment such as PCs and monitors.

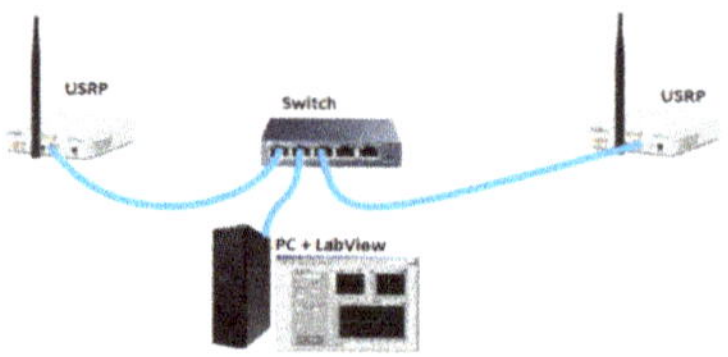

Figure 14. Test system used for examinations.

The data were preceded by a short and long training sequence (Figure 15) as defined in [35]. On this basis, transmission channel parameters were estimated and synchronization and frequency and phase correction were made. For performance analysis, there was no channel encoding during the signal transmission. In order to compare the results obtained, tests were performed for the case of transmission with and without sorting. Sorting was done as in simulation #2 in Section 4. The results obtained are shown in Figure 16. Estimated SNR value refers to cover signal. Cover detection are intended to show that a certain minimum SNR for the cover channel is required to receive the covert channel. The cover signal has to be detected correctly first and then the covert signal can be received. During the test, a low power covert signal was selected deliberately. First, the authors intended to make the signal as difficult to detect as possible, and second, to obtain conditions under which it is more sensitive to channel estimation errors. The test verified the previously assumed and simulation-validated thesis that sorting aimed at mutual quasi-orthogonality of signals can improve the bit error rate. The difference for a signal with sorting versus without sorting for parameters defined in Table 2 is about 3 dB. The proposed algorithm effectively reduces channel estimation error and improves SIC operation. Significant gain was achieved, although the channel parameters were estimated every 64 blocks, which should give small channel estimation error. Improved CSI (lower error) could be achieved by, for example, additional pilot signals and training sequences, however, this would come at a cost of system resources and maximal bit rate.

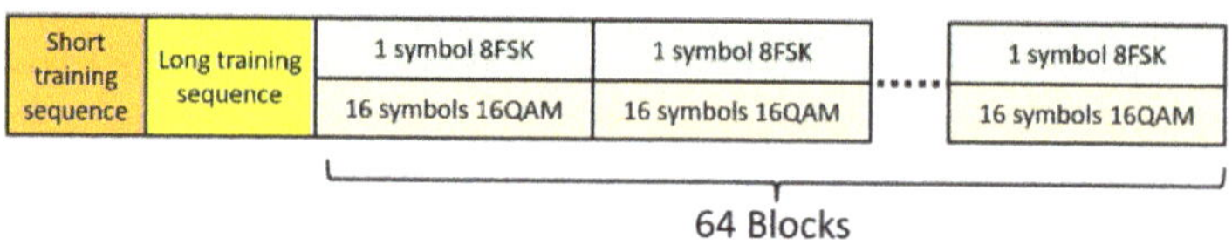

Figure 15. Structure of the transmitted signal.

Table 2. Parameters of radio signal.

Carrier Frequency	850 MHz	
Cover (carrier)	Modulation	16QAM
	Bandwidth	8 MHz
	Transmission rate	32 Mb/s
	Block length	16
	Power	P_0
Covert information	Modulation	8FSK
	Number samples per symbol	16
	Number symbols in block	1
	Transmission rate	1.5 Mb/s
	Power	$0.01 \cdot P_0$

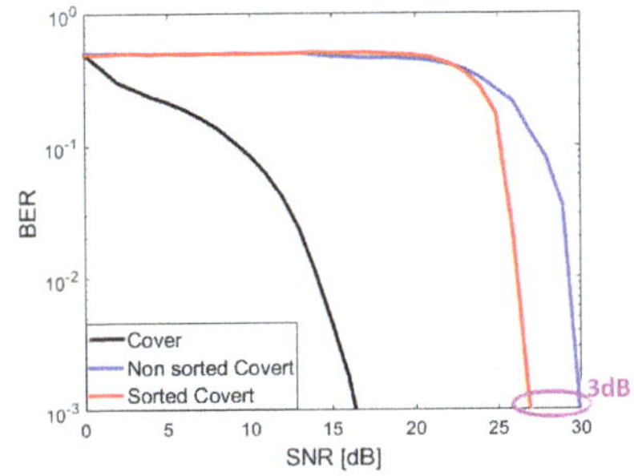

Figure 16. BER versus SNR for covert signal transmission with and without sorting.

7. Summary

Creating a covert channel in the physical layer of wireless communications is an issue that is difficult to implement in practice. The low probability of detecting such a channel and the need to affect the cover's signal as little as possible entails the low power that can be allocated to covert transmission. The natural solution in such a situation is to increase energy per bit by increasing its duration while accepting a lower transmission speed. Such a solution encounters a serious problem, arising from the estimation of channel parameters, which becomes apparent in the inability to extract the covert signal. The purpose of the article was to identify solutions to this type of problem. First, it was noted that higher transmission speed can be achieved by using FSK modulation, which does not require an increase in signal power if covert modulation order is increased, since the energy per bit remains constant, and this is done at the expense of signal bandwidth. However, as noted, FSK modulation is more difficult to receive in non-perfect SIC compared to low-value PSK modulation. The solution in such a case may be the use of sorting, which aims to more easily extract the signal through greater separability of signals in the frequency domain. Although, at the transmitter, the primary FSK signal is converted to a pseudo-noise sequence, the final reception is performed by a traditional FSK demodulator. The proposed solution for creating quasi-orthogonal signals can also be applied to other modulations, which was also simulated in this paper. Importantly, the sorting method is determined by the cover signal, hence there is no need to send additional information between the transmitter and receiver. Although a correct decoding of a block of cover data is required to receive a single or several covert symbols, this is not a major limitation, since a cover signal is a strong signal by its very definition.

Author Contributions: Conceptualization, K.G.; Methodology, Z.P. and J.M.K.; Validation, K.G., Z.P. and J.M.K.; Formal analysis, K.G.; Investigation, Z.P.; Writing-original draft, K.G. and J.M.K.; Writing-review & editing, Z.P. and J.M.K.; Visualization, K.G.; Funding acquisition, J.M.K. All authors have read and agreed to the published version of the manuscript.

Funding: This research received no external funding.

Institutional Review Board Statement: Not applicable.

Informed Consent Statement: Not applicable.

Data Availability Statement: The data presented in this study are available on request from the corresponding author.

Acknowledgments: The authors would like to express their great appreciation to the sensors journal editors and anonymous reviewers for their valuable suggestions, which have improved the manuscript quality.

Conflicts of Interest: The authors declare no conflict of interest.

References

1. Chen, O.; Meadows, C.; Trivedi, G. Stealthy Protocols: Metrics and Open Problems. In *Concurrency, Security, and Puzzles. Lecture Notes in Computer Science*; Gibson-Robinson, T., Hopcroft, P., Lazić, R., Eds.; Springer: Cham, Switzerland, 2017; Volume 10160. [CrossRef]
2. Classen, J.; Schulz, M.; Hollick, M. Practical covert channels for wifi systems. In Proceedings of the 2015 IEEE Conference on Communications and Network Security (CNS), Florence, Italy, 28–30 September 2015; pp. 209–217.
3. Dutta, A.; Saha, D.; Grunwald, D.; Sicker, D. Secret Agent Radio: Covert Communication through Dirty Constellations. In *Information Hiding. IH 2012. Lecture Notes in Computer Science*; Kirchner, M., Ghosal, D., Eds.; Springer: Berlin/Heidelberg, Germany, 2013; Volume 7692. [CrossRef]
4. Grzesiak, K.; Piotrowski, Z.; Kelner, J.M. A Wireless Covert Channel Based on Dirty Constellation with Phase Drift. *Electronics* **2021**, *10*, 647. [CrossRef]
5. Piotrowski, Z. Drift Correction Modulation Scheme for Digital Signal Processing. *Math. Comput. Model.* **2013**, *57*, 2660–2670. [CrossRef]
6. Grzesiak, K.; Piotrowski, Z. NN-Based 8FSK Demodulator for the Covert Channel. *Sensors* **2022**, *22*, 7181. [CrossRef] [PubMed]
7. D'Oro, S.; Restuccia, F.; Melodia, T. Hiding Data in Plain Sight: Undetectable Wireless Communications Through Pseudo-Noise Asymmetric Shift Keying. In Proceedings of the IEEE INFOCOM 2019–IEEE Conference on Computer Communications, Paris, France, 29 April–2 May 2019; pp. 1585–1593.
8. Bonati, L.; D'Oro, S.; Restuccia, F.; Basagni, S.; Melodia, T. SteaLTE: Private 5G cellular connectivity as a service with full- stack wireless steganography. In Proceedings of the IEEE INFOCOM, Vancouver, BC, Canada, 10–13 May 2021.
9. Qiao, S.; Liu, G.; Shi, J.; Ji, X.; Liu, W. Wireless Covert Channel with Polarized Dirty Constellation in Backscatter Communication. *Res. Square* **2021**. [CrossRef]
10. Cao, P.; Liu, W.; Liu, G.; Ji, X.; Zhai, J.; Dai, Y. A Wireless Covert Channel Based on Constellation Shaping Modulation. *Secur. Commun. Netw.* **2018**, *2018*, 1214681. [CrossRef]
11. Oyerinde, O.O.; Mneney, S.H. Review of channel estimation for wireless communication systems. *IETE Tech. Rev.* **2012**, *29*, 282–298. [CrossRef]
12. Yue, X.; Liu, Y.; Kang, S.; Nallanathan, A.; Chen, Y. Modeling and Analysis of Two-Way Relay Non-Orthogonal Multiple Access Systems. *IEEE Trans. Commun.* **2018**, *66*, 3784–3796. [CrossRef]
13. Do, D.T.; Nguyen, T.T.T. Impacts of imperfect SIC and imperfect hardware in performance analysis on AF non-orthogonal multiple access network. *Telecommun. Syst.* **2019**, *72*, 579–593. [CrossRef]
14. Yue, X.; Qin, Z.; Liu, Y.; Dai, X.; Chen, Y. Outage Performance of a Unified Non-Orthogonal Multiple Access Framework. In Proceedings of the 2018 IEEE International Conference on Communications (ICC), Kansas City, MO, USA, 20–24 May 2018; pp. 1–6. [CrossRef]
15. Kara, F.; Kaya, H. BER performances of downlink and uplink NOMA in the presence of SIC errors over fading channels. *IET Commun.* **2018**, *12*, 1834–1844. [CrossRef]
16. Grzesiak, K.; Piotrowski, Z. From Constellation Dithering to NOMA Multiple Access: Security in Wireless Systems. *Sensors* **2021**, *21*, 2752. [CrossRef] [PubMed]
17. Ikki, S.; Aissa, S. Two-way amplify-and-forward relaying with Gaussian imperfect channel estimations. *IEEE Commun. Lett.* **2012**, *16*, 956–959. [CrossRef]
18. Wang, C.; Liu, T.-K.; Dong, X. Impact of channel estimation error on the performance of amplify-and-forward two-way relaying. *IEEE Trans. Veh. Technol.* **2012**, *61*, 1197–1207. [CrossRef]
19. Ma, Y.; Jin, J. Effect of channel estimation errors on M-QAM with MRC and EGC in Nakagami fading channels. *IEEE Trans. Veh. Technol.* **2007**, *56*, 1239–1250. [CrossRef]
20. Yang, Z.; Ding, Z.; Fan, P.; Karagiannidis, G.K. On the Performance of Non orthogonal Multiple Access Systems with Partial Channel Information. *IEEE Trans. Commun.* **2016**, *64*, 654–667. [CrossRef]
21. Keel, B.M.; Baden, J.M.; Heath, T.H. A Comprehensive Review of Quasi-Orthogonal Waveforms. In Proceedings of the 2007 IEEE Radar Conference, Waltham, MA, USA, 17–20 April 2007; pp. 122–127. [CrossRef]

22. Strinati, E.C.; Simoens, S.; Boutros, J. New error prediction techniques for turbo-coded OFDM systems and impact on adaptive modulation and coding. In Proceedings of the 2005 IEEE 16th International Symposium on Personal, Indoor and Mobile Radio Communications, Berlin, Germany, 11–14 September 2005; Volume 2, pp. 1116–1119. [CrossRef]
23. Desset, C.; Ahmed, N.; Dejonghe, A. Energy Savings for Wireless Terminals through Smart Vertical Handover. In Proceedings of the 2009 IEEE International Conference on Communications, Dresden, Germany, 14–18 June 2009; pp. 1–5. [CrossRef]
24. Hamed, E. Practical distributed MIMO for WiFi and LTE. Doctoral Dissertation, Massachusetts Institute of Technology, Cambridge, MA, USA, 2018.
25. Bob, W. *FSK: Signals and Demodulation*; Watkins-Johnson Company Technotes 7.5: Palo Alto, CA, USA, 1980.
26. Boonrungruedee, T.; Khumsat, P. 27-MHz FSK Wireless System Resilient to In-band Interference for IoT Applications. In Proceedings of the 2021 18th International Conference on Electrical Engineering/Electronics, Computer, Telecommunications and Information Technology (ECTI-CON), Phuket, Thailand, 24–27 June 2020; pp. 692–695. [CrossRef]
27. Saadeh, W.; Altaf, M.A.B.; Alsuradi, H.; Yoo, J. A Pseudo OFDM With Miniaturized FSK Demodulation Body-Coupled Communication Transceiver for Binaural Hearing Aids in 65 nm CMOS. *IEEE J. Solid-State Circuits* **2017**, *52*, 757–768. [CrossRef]
28. Shang, Z.; Zhao, Y.; Lian, Y. A Low Power Frequency Tunable FSK Receiver Based on the N-Path Filter. *IEEE Trans. Circuits Syst. II Express Briefs* **2019**, *66*, 1708–1712. [CrossRef]
29. Chiu, C.Y.; Zhang, Z.C.; Lin, T.H. Design of a 0.6-V 429-MHz FSK Transceiver Using Q-Enhanced and Direct Power Transfer Techniques in 90-nm CMOS. *IEEE J. Solid-State Circuits* **2020**, *55*, 3024–3035. [CrossRef]
30. Park, D.; Ahn, J.; Choe, C.; Woo, S.; Ahn, S.; Choi, J. A Noise-Shaped Signaling Method for Vehicle-to-Everything Security. *IEEE Access* **2021**, *9*, 75385–75397. [CrossRef]
31. Choi, J.; Park, D.; Kim, S.; Ahn, S. Implementation of a Noise-Shaped Signaling System through Software-Defined Radio. *Appl. Sci.* **2022**, *12*, 641. [CrossRef]
32. Xu, Z.; Jin, W.; Zhou, K.; Hua, J. A Covert Digital Communication System Using Skewed α-Stable Distributions for Internet of Things. *IEEE Access* **2020**, *8*, 113131–113141. [CrossRef]
33. Ahmaderaghi, B.; Kurugollu, F.; Rincon, J.M.D.; Bouridane, A. Blind Image Watermark Detection Algorithm Based on Discrete Shearlet Transform Using Statistical Decision Theory. *IEEE Trans. Comput. Imaging* **2018**, *4*, 46–59. [CrossRef]
34. 20 MHz Bandwidth, 50 MHz to 2.2 GHz USRP Software Defined Radio Device. Available online: https://www.ni.com/pl-pl/support/model.usrp-2920.html (accessed on 5 January 2023).
35. IEEE. 802.11a-1999. *IEEE Standard for Telecommunications and Information Exchange Between Systems—LAN/MAN Specific Requirements—Part 11: Wireless Medium Access Control (MAC) and physical layer (PHY) specifications: High Speed Physical Layer in the 5 GHz band.* 1999. Available online: https://pdos.csail.mit.edu/archive/decouto/papers/802.11a.pdf (accessed on 10 May 2023).

electronics

MDPI

Article

Grouping Sensors for the Key Distribution of Implicit Certificates in Wireless Sensor Networks

Ray-I Chang [1,*], Chien-Wen Chiang [1] and Yu-Hsin Hung [2,*]

1 Department of Engineering Science and Ocean Engineering, National Taiwan University, No. 1, Sec. 4, Roosevelt Road, Taipei 10617, Taiwan
2 Department of Industrial Engineering and Management, National Yunlin University of Science and Technology, Yunlin 64002, Taiwan
* Correspondence: rayichang@ntu.edu.tw (R.-I.C.); hungyh@yuntech.edu.tw (Y.-H.H.)

Abstract: As sensor nodes communicate via wireless channels, information security is essential for wireless sensor networks. Efficient protection mechanisms to ensure data security among nodes are critical. This study developed the bi-directed grouping (top-down grouping (TDG) and bottom-up grouping (BUG)) methods. In this study, we propose a group-based key distribution method, "aggregator-based grouping" (ABG), which combines the advantages of TDG and BUG to address the security issues of nodes. It employs horizontal and vertical searches, which are based on breadth-first and aggregator searches, respectively. A node performs encryption and decryption only when it requires either data aggregation or inter-group communication. The secure aggregation method can be applied to key-grouping management. We compared the proposed method with TDG and BUG using the same number of groups and network structure. For a network with maximum group members of 50 (total sensor nodes = 1000), compared with TDG and BUG, ABG reduced the number of encryption and decryption operations by ~36%. ABG avoids unnecessary encryption and decryption in the network.

Keywords: aggregator-based grouping; wireless sensor networks; encryption and decryption operations

Citation: Chang, R.-I.; Chiang, C.-W.; Hung, Y.-H. Grouping Sensors for the Key Distribution of Implicit Certificates in Wireless Sensor Networks. *Electronics* **2023**, *12*, 2815. https://doi.org/10.3390/electronics12132815

Academic Editors: Djuradj Budimir and Francisco Falcone

Received: 26 February 2023
Revised: 15 June 2023
Accepted: 23 June 2023
Published: 26 June 2023

1. Introduction

According to a report on the Internet of Things (IoT) analytics, the number of connected devices will increase to ~27 billion by 2025 [1]. The increasing number of devices needs to be integrated and communicated; thus, network technologies are essential for the connectivity of several devices. Various network technologies are used to connect devices. The wireless sensor network (WSN) is a major network technology. It comprises a group of sensor nodes with tiny shapes and limited power resources. It is used to measure the physical conditions of an environment, and the collected data are forwarded to data storage. Since WSNs can easily deploy a local network at a low cost, they can be employed in many domains for different novel applications [2–5]. WSN is essential to the IoT evolution, which will become a mainstream part of the Internet in the future [6]. In WSNs, each sensor node collects data from the environment and then delivers them to the base station (BS) by multi-hops via other nodes. As wireless communication in WSN is power consuming, some specific nodes on the multi-hops path are "data aggregators," aggregating data received from children nodes to reduce the data volume and the number of transmissions to the parent node [7]. Notably, communication in WSN is conducted through wireless channels. It is more vulnerable to malicious attacks, such as eavesdropping, camouflage, and modification. Thus, there is a need for a protection mechanism to ensure data security between nodes [8,9].

As the keys used for encryption and decryption are different in the asymmetric encryption approach, heavy computation is required, which is unsuitable for WSN [10,11].

The symmetric encryption approach, where encryption and decryption operations use the same key, is more favorable to WSN. However, an efficient key management mechanism to distribute keys to nodes is vital in such an approach. Herein, we propose an implicit security scheme [12] for key distribution as it is easy to implement and requires only linear complexity. This scheme employs a group-based key management framework [13], where, based on the grouping algorithm applied, a WSN is divided into several groups. Then, it generates and delivers correspondent encryption parameters for each group to calculate its key for later encryption and decryption operations. Each group uses a unique key to encrypt data. Designing a good grouping algorithm is vital because the group topology determines the number of encryption and decryption operations needed. Unfortunately, existing key management techniques demand high computation, a large amount of memory, and complex communication rounds. However, the secure aggregation method has good management efficiency in WSNs. In our previous study [14], we proposed an online query scheme (OBEQ) with a tree topology of WSN, which uses aggregators between BS and sensors for communication processes. Further, we developed two intuitive grouping strategies: top-down grouping (TDG) and bottom-up grouping (BUG). In this study, we propose aggregator-based grouping (ABG), a generic design for efficient private key management, which protects sensor node data using a group aggregation scheme. It combines TDG and BUG and executes horizontal and vertical searches separately based on the breadth-first and aggregator-based searches from TDG and BUG, respectively. It is adaptable to the efficiency and security requirements of various sensor node distributions.

Furthermore, we developed a group-based key distribution strategy to address the security issues of nodes. Using the proposed ABG, a node executes encryption and decryption operations if and only if it must do either data aggregation or inter-group communication. Thus, the proposed algorithm can minimize unnecessary encryption and decryption operations. The remainder of this study is organized as follows: Section 2 presents typical key distribution approaches and protection schemes for data aggregators in WSN; Section 3 introduces the details of the proposed method; an experiment and a performance evaluation are presented in Section 4; Section 5 presents the conclusion and significance of the study.

2. Related Work

Data security for wireless communication between nodes is a crucial issue in WSNs. Considering data security, data have been protected using keys while they are communicated among different sensor nodes [13–20]. However, efficient and lightweight security mechanisms are needed because each sensor node has limited power computation and storage capacities. The matrix-based structure and group key set-up protocols are used in key management to secure multicast communications in heterogeneous WSNs, but it has high energy consumption [21]. Key distribution and management approaches are categorized into four types: single master key (SMK), all pair wise key (APWK), random pair wise key (RPWK), and group-based key (GBK) [13,14,16]. SMK is the simplest for distributing keys [22–24], where all sensor nodes use the same encryption key to protect their communications. Despite its simplicity, the entire protection mechanism is broken when a hacker breaks and captures one of the sensor nodes. In APWK, a unique encryption key protects communication between any two nodes [17]. A break in any communication does not proliferate to other communications, but this approach is not energy efficient, especially during calculations [25]. In RPWK, a key pool is created, from which each node randomly chooses some keys to make its key ring [18–26]. Nodes broadcast their key ring identities (ID) to other nodes. A secure channel is established if either of the two nodes within the wireless communication coverage has the same ID. For GBK [19,20], an entire WSN is divided into several groups. Sensor nodes' connections are of two types: in group and inter-group. All sensor nodes in the same group use the same key to protect their communications. One node (or some nodes) has the key of its neighboring group to ensure secure communication between the two groups. Hereafter, such a node is called a "boundary node." Among these four approaches, GBK has good security and resilience since cracking one node will not endanger the entire network. Moreover, GBK is scalable since each node needs to store only one or two keys. In addition, the entire network is more difficult to crack because the damage caused by any single attack will only be confined to its group nodes. An algorithm based on the implicit certificate has been proposed to solve the security problem among the access points in a dynamic access point group and between the users' equipment [27]. The hybrid-session key management scheme for WSN was proposed to reduce power consumption by minimizing public key cryptography [28]. The uneven clustering approach improves the energy efficiency load balance in WSN [29]. In summary, group- and matrix-based management approaches have quick reaction times and high connectivity with networks, respectively. However, they have high computation overhead time and memory consumption [30]. To address these challenges, herein, we use the aggregating method to solve the large memory consumption and constrained computation performance of the existing key management methods.

In WSNs, secure aggregation methods are widely employed in inference attack protection [31] and smart grids [32]. Secure aggregation has been employed in federated learning systems, and the results show that the method needs fewer training iterations and is flexible [31]. Regarding data aggregation, Secure Information Aggregation (SIA) [33,34] targets a flatter WSN hierarchy with only one data aggregator to which all sensor nodes send data. In addition to aggregating the received data, the aggregator uses the hash tree [35–37] to convert individual data to an authentication code. As SIA considers only one data aggregator, it is unsuitable for large-scale network deployment. Some studies have been conducted to improve the computation efficiency of aggregation methods [12,37–41]. The turbo-aggregate method achieves $O(nLogn)$ for a secure aggregation in a network with n users, and it can speed up the network's dispatching efficiency [37].

Energy consumption is another issue in data aggregation. The facility derived from data aggregation is not fully utilized, and much power is consumed during communication. For energy efficiency in WSN, a fuzzy-based node arrangement is used to select the parent node and format the tree topology. Secure Reference-based Data Aggregation (SRDA), in which each sensor node compares its sensed data with the averaged value of previous sensed data ("reference value"), has been proposed [39]. Each node transmits and encrypts only the differential value between the sensed data and the reference value to reduce bandwidth and power consumption. The main disadvantage of this scheme is that only the cluster root can aggregate sensed data; thus, the aggregation effectiveness is reduced. In [40], the authors proposed "Concealed Data Aggregation" to build upon privacy homomorphism, which can directly perform calculations on encrypted data. Thus, all encrypted data are directly fed into the aggregation function, and the aggregated result is delivered to the BS. However, this scheme cannot render high-level security. The logical key hierarchy is used to speed up the encryption and decryption for implementing an effective key-numbering approach [41,42]. In Ref. [43], the logic operations are used to develop the lightweight authenticated group-key distribution scheme to speed up the encryption and decryption operation. In Ref. [12], a promising scheme called implicit security was proposed. The term "implicit security" implies that data protection comes from the partitioning of data d into pieces using mathematical polynomial operations (instead of relying on cryptography). The time complexity for this implicit security scheme is only $O(n)$ for data partitioning or recovery. In this study, we used this scheme to partition and reconstruct our encryption key.

Compared with existing key management approaches, the algorithm proposed herein employs the aggregating method to group keys, and the bi-directed search improves the computation performance. In contrast to previous studies, we account for data aggregation and develop clustering algorithms with the minimum number of encryption/decryption operations to generate a group key. This enables secure communication among nodes within a group and increases scalability.

3. The Proposed Methods

The sensing nodes are safe before deployment, and their positions cannot be changed afterward. Moreover, after deployment, the position of the nodes cannot be predicted in advance. We divided the sensor nodes into three roles: the BS, aggregation node (Aggregator), and general sensor node. Therefore, it can store information for each node. It also controls the selection of aggregation nodes. Unless a replacement instruction is issued, the aggregation node that performs the aggregation function is fixed every time a user makes a request. Generally, the sensing node senses parameters in the environment or serves as an intermediate node that transmits data received from its child nodes to the aggregation node. Each node generates a group key, which is shared between a single node and all nodes in the group to which it belongs for secure data communication.

The proposed network approach undergoes six stages to ensure data security between nodes, namely, "network deployment," "key partitioning and distribution," "network grouping," "group key generation," "key and node management," and "data encryption, transmission, and aggregation," in that order. The aggregation nodes are generated by random selection. In network deployment, the algorithm proposed in [44] is used to construct the tree topology for the WSN. First, BS broadcasts a request to build up tree topology for all sensor nodes. The request contains a BS ID and a parameter depth indicating how many levels the tree topology should have. When node i receives this request from node j, it adds one to the parameter depth, sets node j as a parent node, and broadcasts the request to other nodes. This step constructs a tree topology for "all" network nodes in a normal case. Thus, the process continues until all nodes have received the request. This constraint can be refined to ensure the process does not terminate. Figure 1 shows a schematic of the network deployment.

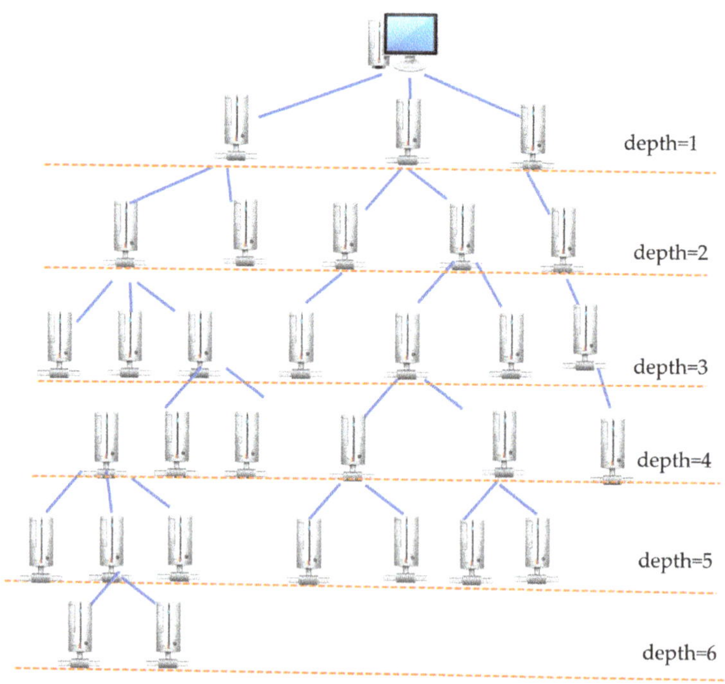

Figure 1. Architecture of the network.

After the network accomplishes deployment, keys need to be added to the network to secure the data. The key partition is a critical element in creating a group key in the next stage. When WSN deployment is completed, the BS produces a specific key partition for each sensor node following the framework in Figure 2 [12]. As expressed in Equations (1) and (2), we assume there are k roots, $r_1, r_2, \ldots$, and r_k, and p is a large prime.

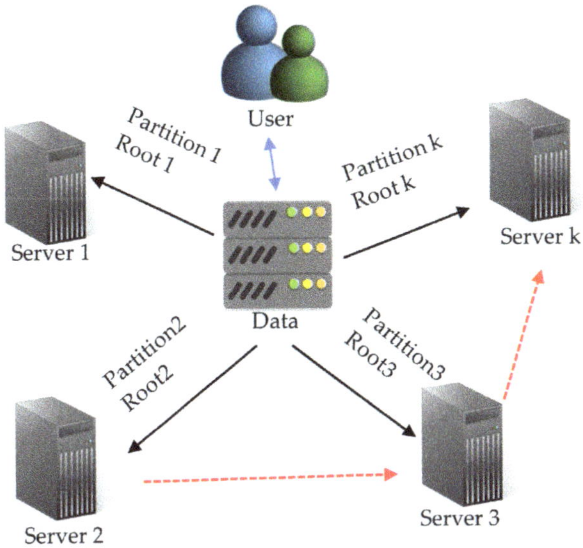

Figure 2. Framework of the key partitioning and distribution.

In this study, we adopt implicit security techniques to handle keys and to construct a tree-based network that dynamically adjusts its topology. With our mechanism, the network is divided into several subtrees, with each subtree being a group of multiple nodes. Each node within a subtree uses key partitioning to generate a group key, which is used for data transmission between group nodes. Additionally, we use symmetric encryption algorithms and hash functions to achieve secure data aggregation with reduced computational overhead. Notably, the leaking of any piece or partition of information will not expose the original data since its reconstruction requires access to each server and the knowledge about the servers that store data partitions. Accordingly, every partition is implicitly secure without the need for encryption.

Each root in Equation (2) represents a partition. Based on [12], all partitions must be acquired to perform the calculation in Equation (3) to recover the original data (key). Parakh [12] found the roots in polynomials over a finite field (which was a large prime number), as presented in Equation (1), where there are a total of n roots. To utilize it in encryption algorithms, transformation to a finite field is carried out based on Equation (2). Subsequently, the data is substituted in Equation (1), and k roots ($r_1 \ldots r_k$) are obtained. Each root is referred to as a partition, and the partitions are randomly distributed among various network servers. Only the owner is aware of the storage locations of the partitions, and to access them, one must know the server's password (the password for accessing the server and not the data password). Notably, all partitions are required to recover the original data. Figure 2 illustrates the concept of data partitions. To recover the original data, the partitions are multiplied as presented in Equation (3). Thus, these pieces can be stored randomly on different network servers.

$$x^k + a_{k-1}x^{k-1} + a_{k-2}x^{k-2} + \ldots + a_1 x + d = 0, \quad (1)$$

$$(x - r_1)(x - r_2)(x - r_3)\ldots(x - r_k) \equiv 0 \bmod p \quad (2)$$

$$d = r_1 \bullet r_2 \bullet\bullet\bullet r_k \bmod p. \quad (3)$$

The BS delineates the production of key partitions. It randomly generates a key d and a large prime p. For N sensor nodes, BS randomly generates N-1 integers (i.e., $r_1, r_2, \ldots,$ and $r_{N\text{-}1}$) smaller than p and uses them to perform an inverse multiplication operation. Then, the kth key partition can be obtained from Equation (4). Finally, the BS distributes the N key partitions to the corresponding N nodes. Each node receives one key partition.

$$r_k = d\bullet(r_1\bullet r_2\bullet\ldots\bullet r_{k-1})^{-1} \bmod p \tag{4}$$

The key is arranged in the network. We employed the group key management approach as described in the previous section to improve the network scalability and reduce the memory space. In the network grouping phase, our tree network is further divided into several subtrees. Each subtree represents a group, and all nodes in a group use a shared group key to protect their in group communications. We propose two intuitive strategies, TDG and BUG, to divide a network into groups, with MN_G being the maximum group size.

In TDG, a breadth-first search starts from BS (Figure 3). If the number of nodes joining the group equals the maximum number of group members, the search stops, and the group is completed. The node with the lowest depth in a group is delegated as the root of the group, and the search restarts from the deepest node to its child node. The above process continues until all sensor nodes are assigned to a group, and all groups can connect with other groups via a shared node. Considering the tree topology in Figure 4 as an example, we assume a maximum number of group members of six. With TDG, the entire network can be divided into eight groups.

```
Top-Down Grouping (TDG)
Input: nodes S, C, MN_G, the number of groups i, depth
1:    If (no group has not been created)
                Set BS as node S, BS's child node as C
                The default value of i is set as 1
2:    If the group size is smaller than MN_G
                Join node S to Gi
        Else    i = i + 1, join node S to Gi
3:    If node S has a child node
                Set the child node of node S as node S
                Go to step 2
4:    If there exists one node that has not been assigned to any group
            Set the parent node of this node as node S
                Go to step 2
            i = i + 1, join node S to Gi
                Go to step 2
5:      If there exists one group wherein one node having the lowest depth does
        not connect with its parent node
                i = i + 1,
                join the node having the lowest depth and its parent node to Gi
6:    End: Output: The node grouping results in the network
```

Figure 3. Pseudocode of the top-down grouping (TDG) algorithm.

In BUG, the search starts from the deepest node to its parent node (Figure 5). If the number of nodes joining the group is smaller than MN_G, other child nodes of parent node y are searched for and added to the group. If the parent node y has no child nodes, it will conduct the breadth-first search from its parent node to find other child nodes. When the number of group sizes equals MN_G, the search stops, and the group is completed. The node with the lowest depth is delegated as the root of the group, and the search restarts from its parent node to create another group. This process continues until all sensor nodes are assigned to a group, and all groups can connect with other groups via a shared node. Similarly to the previous example, with BUG, the whole network can be divided into nine groups, and the grouping results are shown in Figure 6.

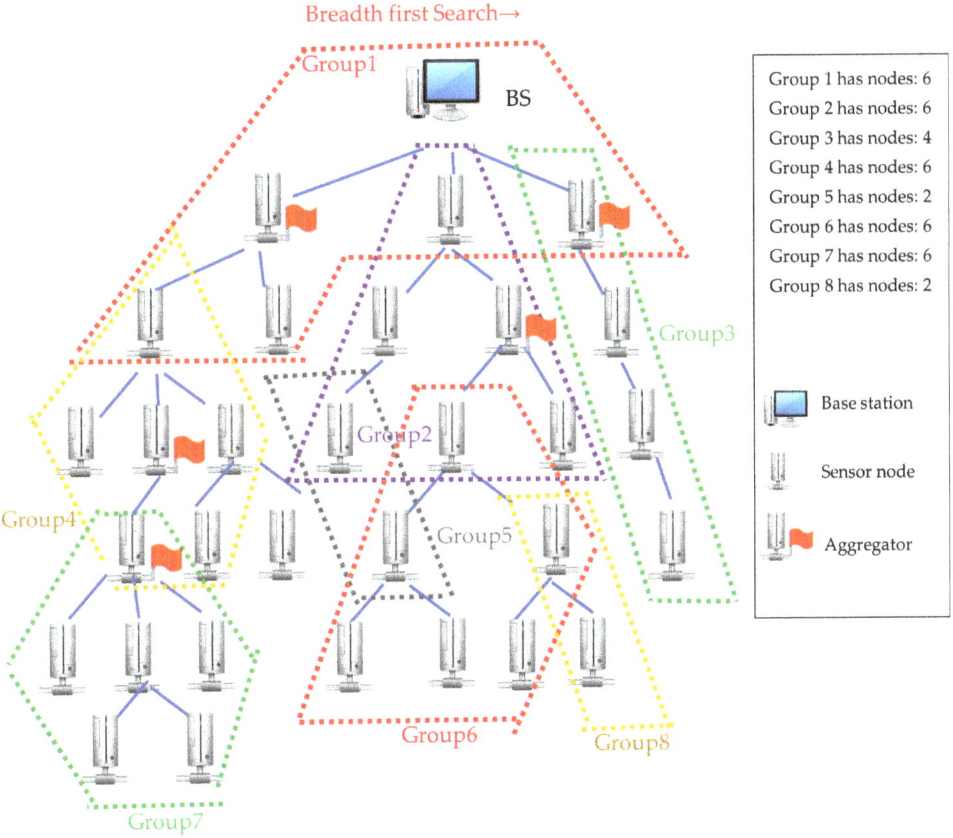

Figure 4. Grouping results of the TDG algorithm.

To integrate the aggregation and encryption functions more effectively, we propose an alternative grouping algorithm, ABG (Figure 7). The breadth- and left-first searches start from BS to its child node x. If the number of nodes added to the group is smaller than MN_G, it continues searching for other child nodes of node x. When the aggregator is met, the search is targeted at other child nodes, which are non-aggregators. If the number of nodes added to the group is smaller than MN_G, the search stops, and the group is completed. The node with the lowest depth is delegated as the root of the group, and the search restarts from the deepest node to its child node. This process continues until all sensor nodes are assigned to a group, and all groups can connect with other groups via a shared node. For the previous example, ABG can divide the entire network into nine groups. The grouping results are shown in Figure 8.

According to the grouping results obtained in the earlier phase, each node in each group is assigned to a corresponding group key parameter by BS. The group key parameter of each node is composed of the key partitions of other nodes in the same group. Specifically, the group key parameter g_{pi} for node i is calculated from Equation (5), where k_{pj} represents the key partition of the node, and m and N are the total number of sensor nodes in the group and the entire network, respectively. Then, the group key GK is calculated from Equation (6).

$$g_{pi} = \prod_{j=1}^{m} k_{pj} \text{mod} p, i \neq j, 0 < m \leq N \tag{5}$$

$$\text{GK} = k_{pi} \bullet g_{pi} \bullet r \bmod p. \tag{6}$$

Since the group key parameters and partitions may not be altered frequently, the last random number r is generated and used in each user request to change the group key, thus enhancing the security level for the entire system. Therefore, nodes residing in the same group will obtain the same GK by referencing the corresponding key partitions, group key parameters, and a random number. To sustain the WSN operation, the actions of removing and inserting some sensor nodes in the network are unavoidable. For example, a node must be replaced when it runs out of power. This alters the network topology; thus, the network grouping and group key parameters must be adjusted. This phase is out of the scope of this study; thus, no further details are provided.

```
Bottom-Up Grouping (BUG)
Input: nodes S, C, MN_G, the number of groups i, depth
1:    If there exists one node having the highest depth that has not been
          assigned to any group
                  Set this node as node S,
            Set the node's parent node as node P
                  i = i + 1
2:    If the group size is smaller than MN_G
                  Join nodes S, P to Gi
        Else      i = i + 1, join node S to Gi
3:    If node P has a child node
                  Go to step 4
        Else      Go to step 1
4:    If the group size is smaller than MN_G
                  Join P's child node to Gi
        Else      i = i + 1, join P's child to Gi
5:    If there exists one node that has not been assigned to any group
            Set the parent node of this node as node P
                  Go to step 2
            i = i + 1, join node S to Gi
6:      If there exists one group wherein one node having the lowest depth does
        not connect with its parent node
                  i = i + 1,
                  join the node having the lowest depth and its parent node to Gi
7:    End: Output: The node grouping results in the network
```

Figure 5. Pseudocode of the bottom-up grouping (BUG) algorithm.

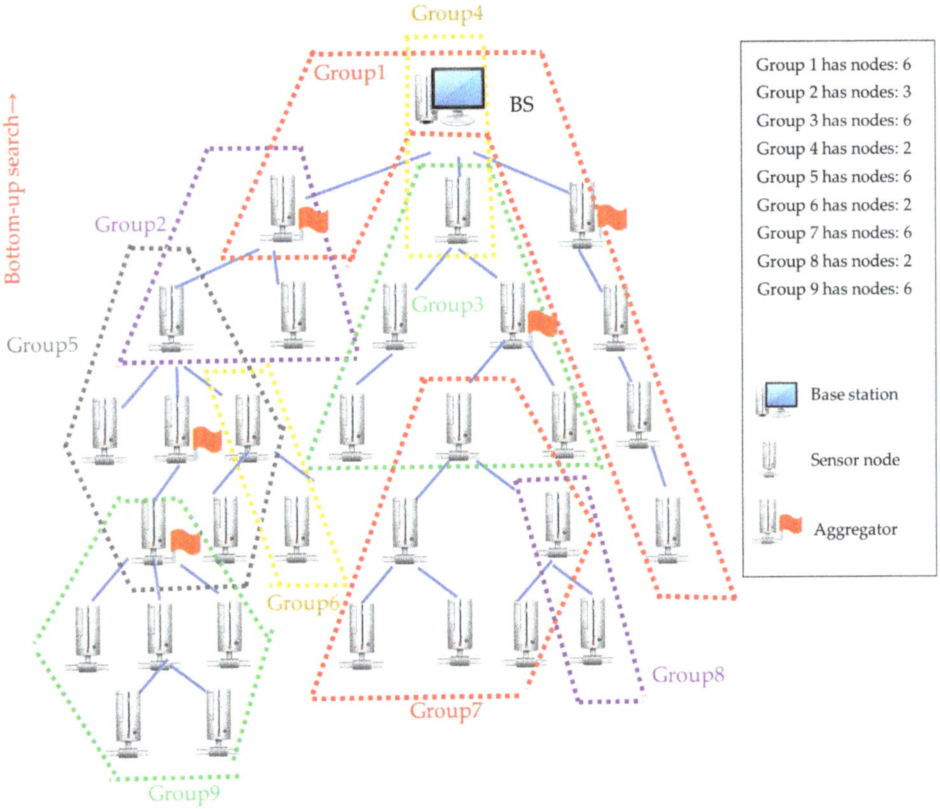

Figure 6. Grouping results of the BUG algorithm.

When completing the above grouping processes, secure communication channels are established for the entire WSN. Each sensor node senses, encrypts, and transmits data to its parent nodes. The data aggregator decrypts all received data, performs the data aggregation function, re-encrypts the aggregation result, and delivers it to the parent nodes. The boundary node in charge of inter-group communications decrypts the received data using the same group key and encrypts them using the key of the group to which the data are delivered. We can change only the random number and leave the rest unchanged to relieve the work on the generation of key partitions and group key parameters for BS to update or alter group keys. If higher security is required, changing the random number during a shorter period is a convenient way to update group keys frequently.

We propose a group-based key distribution mechanism, symmetric encryption systems, and hash functions for encryption and the verification of transmitted data. We summarize the common security requirements and explain how these requirements are met by our mechanisms:

(1) Confidentiality: Provide a secure and confidential channel for wireless network communication to prevent eavesdropping. We use secret-sharing techniques to divide the group key into "n" shares, which also corresponds to the number of nodes. These shares are used to generate a group key required for symmetric data encryption.
(2) Integrity: Verify that the message has not been altered during transmission. Each node and base station stores a one-way hash function. Before data transmission, the node hashes the data, and the receiving end verifies data integrity using the same

hash function. If the node loses a message, it can send a request to the base station to retransmit the message.

(3) Availability and authentication: Ensure that the entire wireless sensor network, or even individual sensor nodes, can provide service. In addition, authenticate the primary nodes in a cluster or the base station. The base station periodically broadcasts a random number.

```
Aggregator-based grouping (ABG)
Input: nodes S, C, MN_G, the number of groups i, depth
1:  If (no group has not been created)
            Set BS as node S, BS's child node as C
            i = 1
2:  If the group size is smaller than MN_G
            Join node S to Gi
      Else    i = i + 1, join node S to Gi
3:  If node S is an aggregator and has a child node
        i = i + 1, join nodes S and C to Gi
        Set node C as node S
        Go to step 2
4:  If node S has a child node
            Set the child node of node S as node S
            Go to step 2
5:  If there exists one node that has not been assigned to any group
        Set the parent node of this node as node S
            Go to step 2
        i = i + 1, join node S to Gi
            Go to step 2
6:    If there exists one group wherein one node having the lowest depth does
      not connect with its parent node
            i = i + 1,
            join the node having the lowest depth and its parent node to Gi
7:  End: Output: The node grouping results in the network
```

Figure 7. Pseudocode of the aggregator-based grouping (ABG) algorithm.

When a node receives a random number broadcast from the base station, it combines the number with its key share and encrypts the number using the group key before sending it back to the base station. Since the base station has the key shares of all nodes, it can compare the received value with its calculation to verify the legitimacy of a node and to detect malicious nodes in the network. A non-authenticated node must be confirmed by the user before being allowed to join the network. The base station allocates a share to the added node and redistributes the group key parameters to the nodes within the group.

(4) Scalability: The system must be able to support large-scale network architectures, particularly considering the key distribution mechanism for large networks.

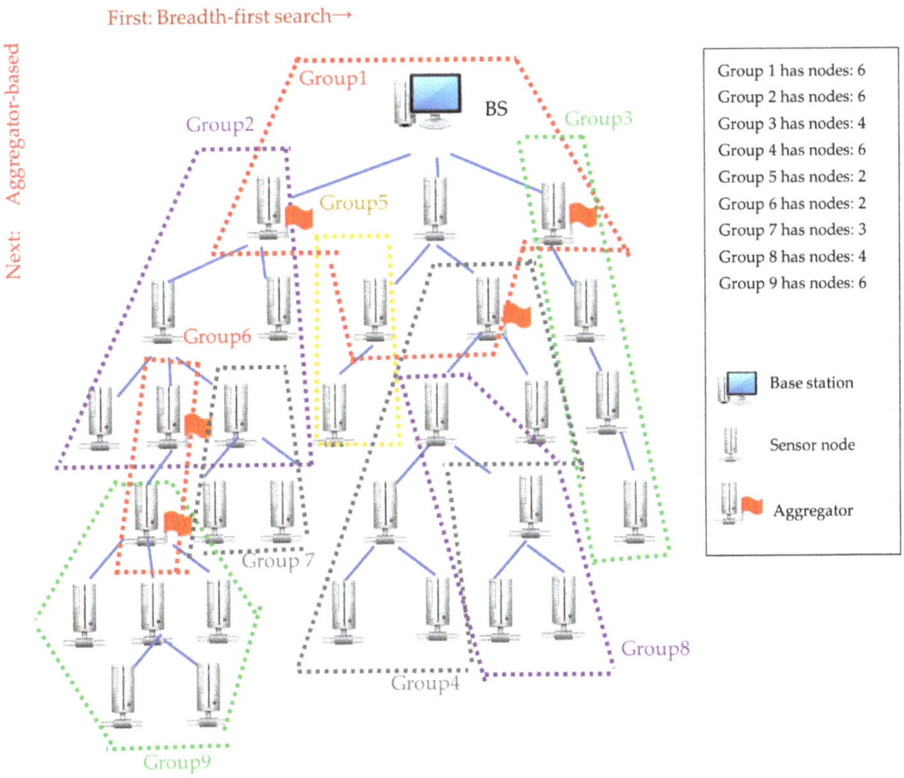

Figure 8. Grouping results of the ABG algorithm.

We adopt a group key distribution mechanism that uses less storage. As the number of nodes increases, it does not directly lead to an increase in the number of keys in the network.

(5) Efficiency: Wireless sensor networks must consider storage capacity, processing power, and communicational capability. The critical distribution mechanism should manage the number of keys, overly complex key computation, and excessive key exchanges between nodes. We use a group key distribution model and implement computationally efficient symmetric encryption to protect data and minimize computational overhead. WSN may encounter issues after use, such as node power depletion and the addition of new nodes. During re-initialization, group changes, and other unexpected errors may cause the number of nodes to change. When nodes are added or removed, the network membership changes, and new keys need to be generated and distributed. The capability of adding and removing nodes is therefore necessary. This paper proposes a mechanism for dynamic node management and key distribution, which consists of three parts: node addition, node and key deletion, and key update.

■ Node addition: After a specific period of network usage, some nodes may have depleted their power, and the system may deploy new nodes to replace the old ones to maintain the operation of the wireless sensor network. When a new node joins the network, the base station assigns a key partition block to that node and notifies all nodes in the network, including the new node. The node is then added to a designated group. If the group has reached its maximum size, other groups that can admit the node are identified or the grouping process is re-executed. Otherwise, the new node and its neighbors generate the group key, establish a data transmission channel within the group, and complete the node addition process.

- Node and key deletion: When a node depletes its power or experiences a failure, the base station removes that node from the network and broadcasts the node's ID to all nodes in the network for deletion. Additionally, the base station deletes all related key information of the node, including the key partition block and the group key, to ensure the security of future data transmissions. After node deletion, the child nodes point to the parent node, and a new group key is generated.
- Key update: This mechanism applies to the subsequent stages of node addition and node and key deletion. Therefore, when a node joins or leaves a group, new keys can be generated using the previously described group key parameter allocation method.

Furthermore, aggregate nodes, which perform aggregation function calculations, have higher power consumption than regular nodes. Accordingly, the base station may designate nodes with sufficient power as new aggregate nodes to maintain the network lifetime. Since our grouping method is based on aggregate nodes, grouping must be performed for the new aggregate nodes, and the group keys for each group are generated according to our key update mechanism.

Additionally, if the network experiences poor transmission path efficiency, high packet loss, or nodes not returning data to the aggregate node or base station, the base station can designate a path for data transmission through other nodes. As changing paths may result in different group traversals, the base station can execute the key update mechanism to generate new group keys.

(6) Data freshness: Ensure that each data record is up to date and prevent replay attacks. Users include a random number in each request packet, combining it with key shares or group key parameters. This ensures that each generated group key is unique for every request.

4. Results

In this study, we used a tree topology to simulate a WSN environment. The simulated network has a tree structure in which simulated sensor nodes are connected like the branches of a tree. The tree structure network has three branches, and 1000 nodes were implanted randomly in the network with an increment of 100. Each sensor node was assumed to transmit one data. The sensor nodes were grouped into 10, 15, 20, 25, and 50 using TDG, BUG, and ABG. The aggregators were the reference of node grouping because power volume and memory space are critical resources for sensor nodes. The aggregator proportion was set from 10% to 50%. Space consumption can be evaluated from the number of group key parameters needed to be stored in the node. Table 1 lists the network parameter setting.

Table 1. Network parameter setting.

Parameters	Values
Topology	Tree topology
Deployment way	Random deployment
The number of sensor nodes	100–1000 nodes (100-node increases)
The number of aggregators	10–50% (10% increases each time)
MN_G	10, 15, 20, 25, and 50

Encryption and decryption follow the same estimation rules. Figure 9 shows the encryption and decryption procedures and how to estimate the number of encryption and decryption operations. Data are transmitted from one sensor node ($Node_1$) to the target node (Aggregator). The data may be transmitted between the nodes from different groups. The data transmitted from $Node_1$ to the aggregator passes through the node from three different groups (Figure 10), and they need three encryptions and two decryptions. We infer that one node transmits data to the aggregator through nodes from different t

groups, and $t + 1$ encryptions and t decryption are needed. The node from the same bottom level ($Node_{6\text{–}7}$) must be aggregated as one aggregator ($Aggregator_2$) using the aggregate function, and the data are encrypted as one data. The aggregator ($Aggregator_2$) transmits the data to the previous level (depth = 2), repeating the aggregation procedure until the data are transmitted to the top level (depth = 1). Thus, one encryption is needed. The aggregator needs to decrypt each child non-aggregator node (green node). If one aggregator has y child non-aggregator nodes, it needs y decryption.

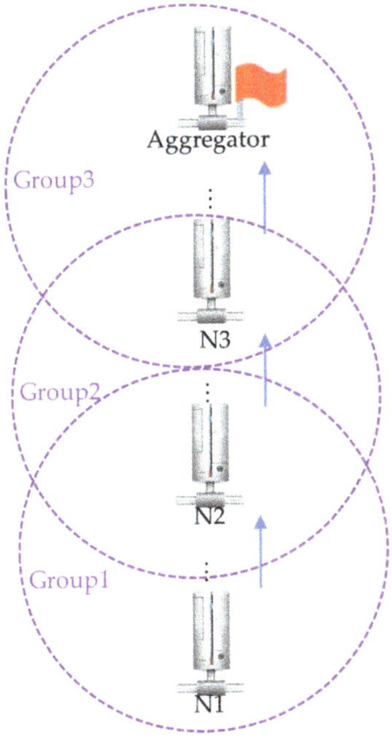

Figure 9. Data transmission to the aggregator.

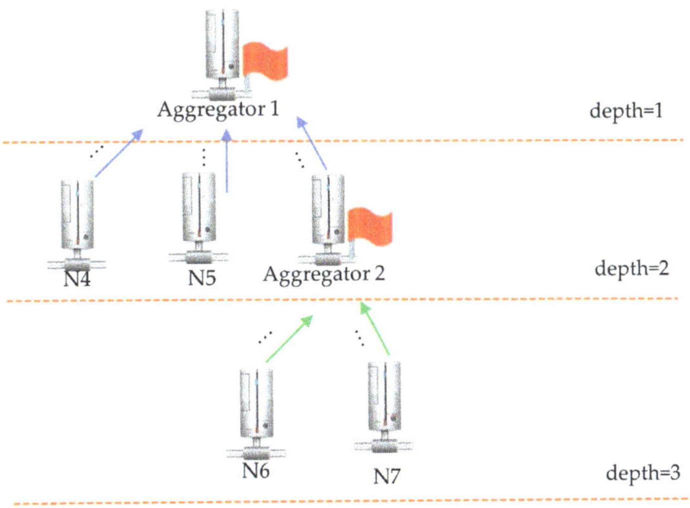

Figure 10. Data receival by the aggregator.

In this experiment, we simulated wireless sensor networks using C programming language by randomly deploying 100–1000 nodes in a tree structure with 100-node increments in each simulation, and the degree of the tree was set to three. The aggregator nodes accounted for 10–50% of all network nodes. Based on the algorithm used for calculating the number of encryption and decryption operations in the simulations, we inferred that as the number of groups between transmitting and aggregator nodes increases, the number of encryption and decryption operations required for data transmission by a given node increases. We assumed that each node would transmit a piece of data, and the maximum group size was set to 10, 15, 20, 25, and 50 individuals. The experimental parameters included the number of encryption and decryption operations required for the entire network to transmit data to a base station and the average number of group key parameters that the nodes need to store.

After reading data, nodes perform encryption before transmitting the data to the upper layer. Notably, there exist two cases that require additional decryption and encryption operations during the transmission. The first case pertains to data transmission from one group to another: when the connecting node receives the data, it is decrypted using the current group key and then encrypted using the next group key until all data are transmitted to the base station. The second case pertains to data aggregation: upon receiving the data, the aggregator node decrypts the data, performs aggregation with other data to reduce packet size, and encrypts the aggregated result for further transmission. In this case, if the data aggregator and connecting nodes are the same, the number of encryption and decryption operations required to transmit the data to the base station will be reduced. This is the primary concept behind the proposed algorithm that performs grouping based on the aggregator nodes. In this study, we used a three-depth tree structure network as an instance (Figure 10). The maximum number of groups was nine. The sensor nodes were grouped using TDG, BUG, and ABG. The most significant results determined by different grouping algorithms are the number of encryption and decryption. Generally, most nodes perform encryption for sensed data only once and send them to BS or relay the encrypted data to BS without decryption, except for the following boundary and aggregator nodes. These two nodes decrypt and then encrypt received data using the group key(s) and neighboring group key, respectively. The boundary node is responsible for inter-group connections. If the sensed data go through b boundary nodes before reaching the aggregator, there would be b encryption and $b-1$ decryption executed in the boundary nodes. Before executing the aggregation function, the aggregator decrypts all received data from its child nodes and executes encryption once. Table 2 lists the grouping results from the experiment under the same conditions.

Table 2. Grouping results.

Group Id	TDG		BUG		ABG	
	Encryption	Decryption	Encryption	Decryption	Encryption	Decryption
1	5	7	2	9	5	17
2	8	11	2	0	5	1
3	3	0	8	10	3	0
4	9	10	1	0	5	0
5	3	2	9	10	2	1
6	10	5	3	2	1	5
7	5	0	10	5	4	2
8	3	2	3	2	6	3
9			5	0	5	0
Total	46	37	43	38	*36*	*26*

We used the saved percentage, which compares control and experimental methods (Equation (7)), as criteria to evaluate the performance of the algorithms.

$$\text{Saved percentage} = \frac{N_{\text{Experimental}} - N_{\text{Control}}}{N_{\text{Control}}} \tag{7}$$

where $N_{\text{Experimental}}$ indicates the number of encryption and decryption operations using ABG algorithm. N_{Control} represents the number of encryption and decryption operations using TDG algorithm or BUG algorithm.

We evaluated the three network grouping algorithms under different network scales, including the number of sensor nodes and group size. The number of sensor nodes was auto-increased in increments of 100 until 1000 nodes were reached. The maximum number of group members was set to 10, 15, 20, and 25 (Figure 11). When the node grouping method was constrained by a maximum of 10 nodes of a group, ABG could reduce the number of encryptions and decryptions by 16% and 18% compared with those of TDG and BUG, respectively. The maximum number of upward group adjustments was 25. ABG could also reduce the number of encryptions and decryptions required for TDG and BUG by 22% and 28%, respectively. When more aggregation nodes also served as boundary nodes simultaneously, the number of encryption and decryption decreased further. Thus, if a network allows severe nodes of group members, ABG would have better computation performance than TDG and BUG. When a maximum of 50 group members were considered, ABG could reduce the encryption and decryption by 35% and 36% compared with those of TDG and BUG, respectively.

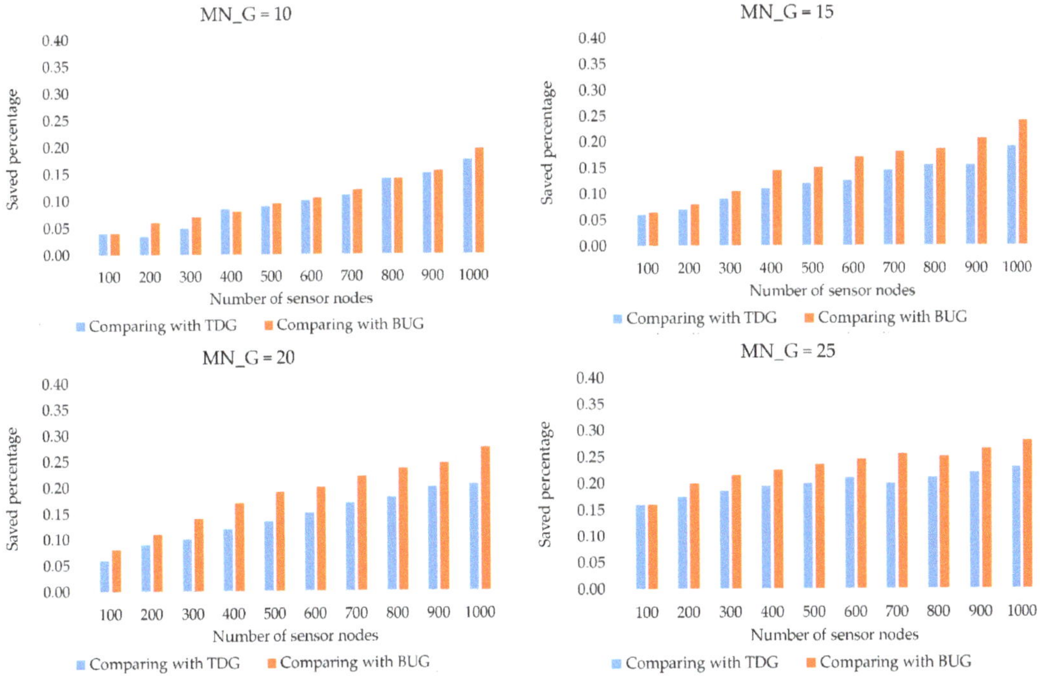

Figure 11. *Cont.*

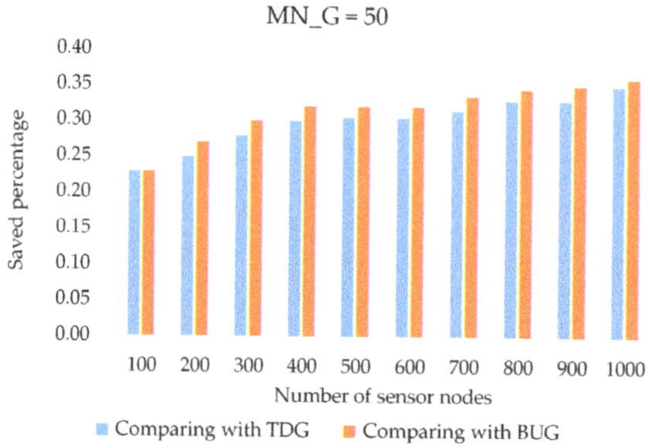

Figure 11. Comparison between the percentage reduction in the number of encryption and decryption for different network scales.

We investigated the influence of encryption and decryption using the group algorithms between different proportions of aggregator nodes, and the results are shown in Figure 12. TDG, BUG, and ABG were used to group 1000 nodes in the network. When a network allows fewer group members, more groups are created. Also, only a few nodes can share the same key in the group; thus, the security level of the entire network can be increased. Contrarily, when a large number of nodes exist in each group, all nodes can share the same key in the group. Thus, the number of encryptions and decryptions can be reduced, but the security level of the network reduces. Therefore, choosing an appropriate MN_G is an important issue when launching WSN applications. Herein, the proposed ABG could further reduce the number of encryptions and decryptions when the percentage of aggregators occupied in the network approached 30% with MN_G = 10, 15, 20, and 25. When the percentage was over 30%, the grouping results for all MN_G values were similar. Thus, the number of encryption and decryption operations was also similar.

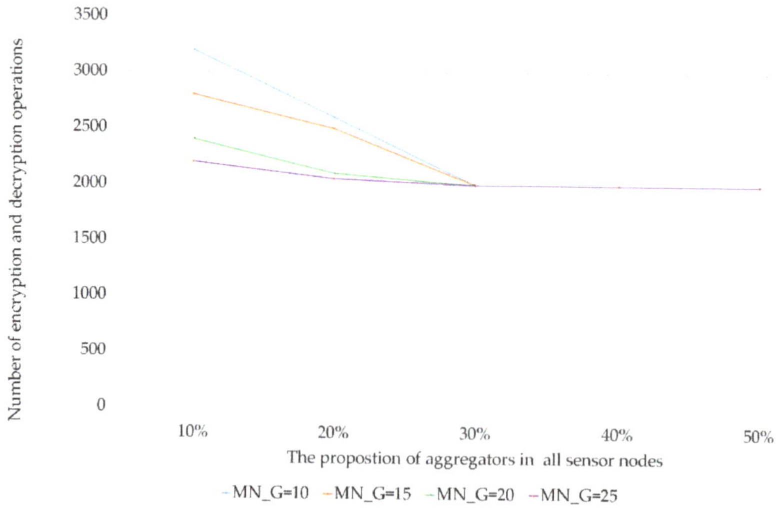

Figure 12. Influence of the number of aggregators on the number of encryptions and decryptions.

In the proposed scheme, the memory consumption for each node is attributed mainly to the storing of the group key parameter. Unlike normal nodes in WSNs, the boundary node has more than one group key parameter to create correspondent group keys and render secure communication among different groups. When the maximum number of group members is small, more groups will be created, resulting in more boundary nodes. The aggregators have two types of arrangements: neighboring and non-neighboring. Theoretically, to reduce storage requirements, the group sizes should be as close to the maximum number of group members as possible so that the group number can be lowered to reduce the requirement for deploying boundary nodes. For ABG, since the addition of sensor nodes to one group stops when the aggregation node is met, the group size in each group is likely lower than the maximum number of group members. Thus, more groups are generated, and more boundary nodes would be created accordingly, resulting in more storage requirements for group key parameters. This problem can be alleviated by evenly distributing the network aggregators and preventing them from being close to one another. For MN_G = 6 (Figure 13a), with neighboring aggregators in the network, only three nodes were in Group 1. With non-neighboring aggregators, there were six nodes in Group 1 (Figure 13b). The results show that the average storage of group key parameters for ABG with the neighboring aggregator arrangement is smaller than that of TDG and BUG.

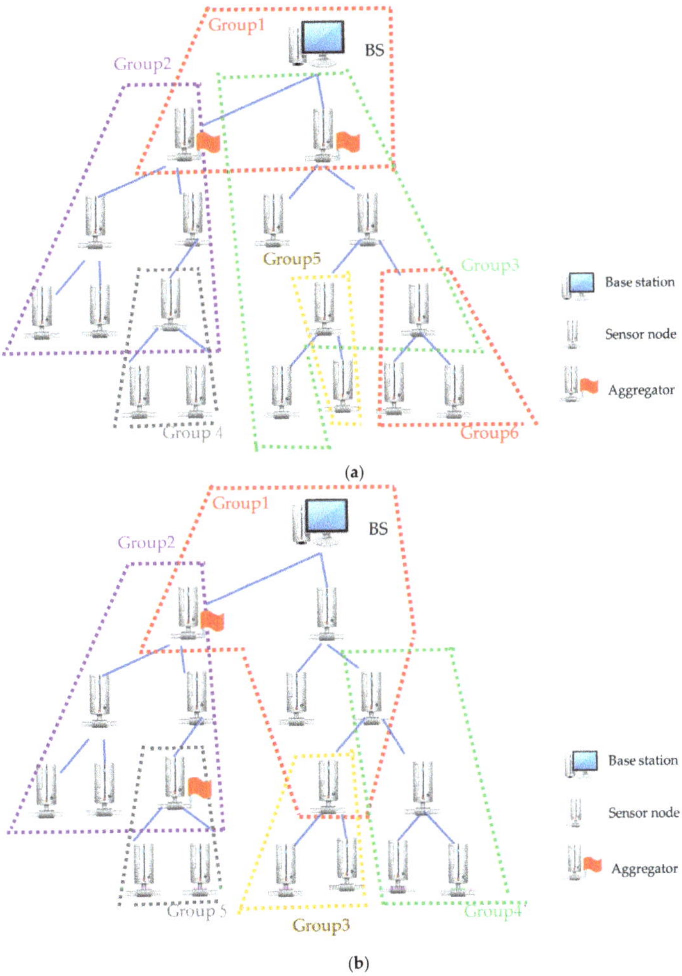

Figure 13. Aggregator arrangements: (**a**) neighboring and (**b**) non-neighboring.

5. Conclusions

There is an increasing number of sensor nodes in WSNs. The sensor node grouping directly affects the communication efficiency of WSNs. Herein, we developed a grouping method for the key distribution of implicit certificates in WSNs to enhance computing performance. Further, we propose a novel key distribution mechanism for WSN applications. Following the management mechanism along with its above-average performance on storage requirement, the resilience of node capture, key connectivity, and scalability, we developed an efficient network grouping strategy, ABG, which can minimize encryptions and decryptions by combining data aggregation and inter-group communication (both of which require encryptions and decryptions for the sensed data) in the same node. We investigated the influence of the number of aggregators in the network on the number of encryptions and decryptions. We found that ABG can reduce EN compared to other grouping algorithms. Moreover, the average storage of group key parameters for ABG with a neighboring aggregator arrangement is also smaller than that of TDG and BUG. The performance evaluation results suggest that the proposed scheme can be employed in ubiquitous WSN application domains. Experimental design conditions include same power, memory capacity, CPU processing power, and communicational capability, and the results depend on these conditions. Thus, if the conditions are changed, the results may differ: there are the experiment's limitations.

In our future studies, in addition to investigating the encryption and decryption operations issues, we will investigate other issues, such as the average storage of group key parameters and the effect of non-neighboring and neighboring aggregators on encryption and decryption. This will provide a more efficient way to communicate and secure data in WSN. This study proposed an experimental environment that may not adapt to the actual network environment. In the future, a system simulation method, such as steady-state simulation, is expected to be employed in the experiment to make the investigation fit the realistic network environment.

Author Contributions: Conceptualization, R.-I.C.; data curation, C.-W.C.; formal analysis, C.-W.C.; funding acquisition, R.-I.C. and Y.-H.H.; investigation, R.-I.C., C.-W.C. and Y.-H.H.; methodology, R.-I.C. and C.-W.C.; software, R.-I.C. and C.-W.C.; supervision, R.-I.C. and Y.-H.H.; validation, Y.-H.H.; writing—original draft, R.-I.C. and Y.-H.H.; writing—review and editing, R.-I.C. and Y.-H.H. All authors have read and agreed to the published version of the manuscript.

Funding: This work was supported in part by Taiwan National Science and Technology Council (NSTC), under Grant no. 110-2410-H-002-094-MY2, 111-2221-E-224-033-MY2 and MOE "Teaching Practice Research" Subsidies Program grant number PBM1110139.

Institutional Review Board Statement: Not applicable.

Informed Consent Statement: Not applicable.

Data Availability Statement: Not applicable.

Conflicts of Interest: The authors declare no conflict of interest.

References

1. Hasan, M. State of IoT 2022: Number of Connected IoT Devices Growing 18% to 14.4 billion Globally. IoT Analytics. Available online: https://iot-analytics.com/number-connected-iot-devices/ (accessed on 18 May 2022).
2. Akyildiz, I.; Su, W.; Sankarasubramaniam, Y.; Cayirci, E. Wireless sensor networks: A survey. *J. Comput. Netw.* **2022**, *38*, 393–422. [CrossRef]
3. Li, W.; Kara, S. Methodology for monitoring manufacturing environment by using wireless sensor networks (WSN) and the internet of things (IoT). *Procedia CIRP* **2017**, *61*, 323–328. [CrossRef]
4. García-Hernández, C.F.; Ibargüengoytia-González, P.H.; García-Hernández, J.; Pérez-Díaz, J.A. Wireless Sensor Networks and Applications: A Survey. *Int. J. Comput. Sci. Netw. Secur.* **2007**, *7*, 264–273.
5. Culler, D.; Estrin, D.; Srivastava, M. Guest Editors' Introduction: Overview of Sensor Network. *IEEE Comput. Soc.* **2004**, *37*, 41–49. [CrossRef]

6. Zahariadis, T.; Trakadas, P.; Leligou, H.; Papadopoylos, K.; Ladis, E.; Tselikis, C.; Vangelatos, C.; Besson, L.; Manner, J.; Loupis, M.; et al. Securing wireless sensor networks towards a trusted Internet of Things. In *Towards the Future Internet—A European Research Perspective*; IOS Press: Amsterdam, The Netherlands, 2009; pp. 47–56.
7. Cheng, S.; Lai, Y.; Wang, C.; Lin, R. Hierarchical data aggregation model and group management for wireless sensor network. In Proceedings of the 2009 the Fourth International Conference on Communications and Networking, Xi'an, China, 26–28 August 2009.
8. Ren, X.; Yu, H. Security Mechanisms for Wireless Sensor Networks. *Int. J. Comput. Sci. Netw. Secur.* **2006**, *6*, 155–161.
9. Belinda, M.; Dhas, C. A Study of Security in Wireless Sensor Networks. *Masaum J. Rev. Surv.* **2009**, *1*, 91–95.
10. Wander, A.S.; Gura, N.; Eberle, H.; Gupta, V.; Shantz, S.C. Energy Analysis of Public-Key Cryptography for Wireless Sensor Networks. In Proceedings of the IEEE Pervasive Computing and Communication, Kauai, Hawaii, 8–12 March 2005; pp. 324–328.
11. Amin, F.; Jahangir, A.H.; Rasifard, H. Analysis of Public-Key Cryptography for Wireless Sensor Networks Security. *Int. J. Comput. Syst. Sci. Eng.* **2009**, *5*, 107–112.
12. Parakh, A.; Kak, S. Online Data Storage using Implicit Security. *Inf. Sci.* **2009**, *179*, 3323–3331. [CrossRef]
13. Chang, C.C.; Arafa, S.; Muftic, S. Key Establishment Protocol for Wireless Sensor Networks. In Proceedings of the IEEE International Conference on Mobile Ad-Hoc and Sensor Systems, Pisa, Italy, 8–11 October 2007.
14. Chang, R.-I.; Tsai, J.-H.; Wang, C.-H. Edge Computing of Online Bounded-Error Query for Energy-Efficient IoT Sensors. *Sensors* **2022**, *22*, 4799. [CrossRef]
15. Jones, J.; Haas, Z.J. Predeployed Secure Key Distribution Mechanisms in Sensor Networks: Current State-of-the-Art and a New Approach using Time Information. *IEEE Wirel. Commun.* **2008**, *15*, 42–51.
16. Dustin, M.; Shankarappa, J.; Petrowski, M.; Weerasingha, H.; Fu, H. Analysis of Key Management in Wireless Sensor Networks. In Proceedings of the IEEE Electro/Information Technology Conference, Chicago, IL, USA, 17–20 May 2007.
17. Cheng, Y.; Agrawal, D.P. Efficient Pairwise Key Establishment and Management in Static Wireless Sensor Networks. In Proceedings of the Mobile Ad Hoc and Sensor System Conference, Washington, DC, USA, 25–27 May 2005.
18. Eschenauer, L.; Gligor, V. A Key-Management Scheme for Distributed Sensor Networks. In Proceedings of the 9th ACM Conference on Computer and Communications Security, Washington, DC, USA, 18–22 November 2002; pp. 41–47.
19. Du, W.; Deng, J.; Han, Y.S.; Chen, S.; Varshney, P.K. *A Key Management Scheme for Wireless Sensor Networks Using Deployment Knowledge*; IEEE Infocom: Hong Kong, China, 2004.
20. Liu, D.; Ning, P.; Du, W. Group-Based Key Pre-distribution in Wireless Sensor Networks. In Proceedings of the 4th ACM Workshop on Wireless Security (WiSe'05), Cologne, Germany, 2 September 2005; pp. 11–20.
21. Alshammari, M.R.; Elleithy, K.M. Efficient and secure key distribution protocol for wireless sensor networks. *Sensors* **2018**, *18*, 3569. [CrossRef] [PubMed]
22. Iqbal, U.; Shafi, S. A provable and secure key exchange protocol based on the elliptical curve diffe–hellman for wsn. In *Advances in Big Data and Cloud Computing: Proceedings of ICBDCC18*; Springer: Singapore, 2019; pp. 363–372.
23. Lin, H.Y. Integrate the hierarchical cluster elliptic curve key agreement with multiple secure data transfer modes into wireless sensor networks. *Connect. Sci.* **2022**, *34*, 274–300. [CrossRef]
24. Tehseen, M.; Javed, H.; Shah, I.H.; Ahmed, S. A lightweight key negotiation and authentication scheme for large scale wsns. In *Recent Trends and Advances in Wireless and IoT-Enabled Networks*; Springer: Berlin/Heidelberg, Germany, 2019; pp. 225–235.
25. Elhoseny, M.; Hassanien, A.E.; Elhoseny, M.; Hassanien, A.E. Secure data transmission in WSN: An overview. In *Dynamic Wireless Sensor Networks: New Directions for Smart Technologies*; Springer: Berlin/Heidelberg, Germany, 2019; pp. 115–143.
26. Robinchandra Singh, U.; Roy, S. Survey on key management schemes and cluster based routing protocols in wireless sensor network. *Int. J. Comput. Intell. IoT* **2019**, 576–594.
27. Chen, Z.; Chen, S.; Xu, H.; Hu, B. A security scheme of 5G ultradense network based on the implicit certificate. *Wirel. Commun. Mob. Comput.* **2018**, *2018*, 1–11. [CrossRef]
28. Mehmood, G.; Khan, M.S.; Waheed, A.; Zareei, M.; Fayaz, M.; Sadad, T.; Kama, N.; Azmi, A. An Efficient and Secure Session Key Management Scheme in Wireless Sensor Network. *Complexity* **2021**, *2021*, 6577642. [CrossRef]
29. Vijayan, K.; Raaza, A. A novel cluster arrangement energy efficient routing protocol for wireless sensor networks. *Indian J. Sci. Technol.* **2016**, *9*, 1–9. [CrossRef]
30. Srinivas, K.; Sagar, K. Secure-Key Management Protocols for Wireless Sensor Networks in IoT. *Ind. Eng. J.* **2022**, *15*.
31. Fereidooni, H.; Marchal, S.; Miettinen, M.; Mirhoseini, A.; Möllering, H.; Nguyen, T.D.; Rieger, P.; Sadeghi, A.-R.; Schneider, T.; Yalame, H.; et al. SAFELearn: Secure aggregation for private federated learning. In Proceedings of the 2021 IEEE Security and Privacy Workshops (SPW), San Francisco, CA, USA, 27 May 2021; IEEE: Piscataway, NJ, USA, 2021; pp. 56–62.
32. Guan, Z.; Zhang, Y.; Zhu, L.; Wu, L.; Yu, S. EFFECT: An efficient flexible privacy-preserving data aggregation scheme with authentication in smart grid. *Sci. China Inf. Sci.* **2019**, *62*, 32103. [CrossRef]
33. Przydatek, B.; Song, D.; Perrig, A. SIA: Secure information aggregation in sensor networks. In Proceedings of the 1st ACM Conference on Embedded Networked Sensor Systems, Los Angeles, CA, USA, 5–7 November 2003.
34. Przydatek, B.; Chan, H.; Perrig, A.; Song, D. SIA: Secure information aggregation in sensor networks. *J. Comput. Secur.* **2007**, *15*, 69–102.
35. Li, H.; Lu, R.; Zhou, L.; Yang, B.; Shen, X. An Efficient Merkle-Tree-Based Authentication Scheme for Smart Grid. *IEEE Syst. J.* **2013**, *8*, 655–663. [CrossRef]
36. Nesa, N.; Banerjee, I. A lightweight security protocol for IoT using Merkle hash tree and chaotic cryptography. *Adv. Comput. Syst. Secur.* **2020**, *10*, 3–16.

37. So, J.; Güler, B.; Avestimehr, A.S. Turbo-aggregate: Breaking the quadratic aggregation barrier in secure federated learning. *IEEE J. Sel. Areas Inf. Theory* **2021**, *2*, 479–489. [CrossRef]
38. Bhushan, S.; Kumar, M.; Kumar, P.; Stephan, T.; Shankar, A.; Liu, P. FAJIT: A fuzzy-based data aggregation technique for energy efficiency in wireless sensor network. *Complex Intell. Syst.* **2021**, *7*, 997–1007. [CrossRef]
39. Sanli, H.O.; Ozdemir, S.; Cam, H. SRDA: Secure Reference-Based Data Aggregation Protocol for Wireless Sensor Networks. In Proceedings of the IEEE Vehicular Technology Conference, Los Angeles, CA, USA, 26–29 September 2004; pp. 4650–4654.
40. Girao, J.; Westhoff, D. CDA: Concealed Data Aggregation for Reverse Multicast Traffic in Wireless Sensor Networks. In Proceedings of the IEEE International Conference on Communications, Seoul, Republic of Korea, 16–20 May 2005.
41. Ozdemir, S.; Xiao, Y. Secure Data Aggregation in Wireless Sensor Networks: A Comprehensive Overview. *Comput. Netw.* **2009**, *53*, 2022–2037. [CrossRef]
42. Jiao, R.; Ouyang, H.; Lin, Y.; Luo, Y.; Li, G.; Jiang, Z.; Zheng, Q. A computation-efficient group key distribution protocol based on a new secret sharing scheme. *Information* **2019**, *10*, 175. [CrossRef]
43. Harn, L.; Hsu, C.; Xia, Z. General logic-operation-based lightweight group-key distribution schemes for Internet of Vehicles. *Veh. Commun.* **2022**, *34*, 100457. [CrossRef]
44. Madden, S.; Franklin, M.J.; Hellerstein, J.M.; Hong, W. TAG: A Tiny Aggregation Service for Ad-Hoc Sensor Networks. In Proceedings of the 5th Symposium on Operating Systems Design and Implementation, Boston, MA, USA, 9–11 December 2002; pp. 131–146.

MDPI
St. Alban-Anlage 66
4052 Basel
Switzerland
Tel. +41 61 683 77 34
Fax +41 61 302 89 18
www.mdpi.com

Electronics Editorial Office
E-mail: electronics@mdpi.com
www.mdpi.com/journal/electronics

www.ingramcontent.com/pod-product-compliance
Lightning Source LLC
Chambersburg PA
CBHW061807210926
51850CB00006B/632
9783036582283